DATE DUE

DEMCO 38-296

Weather Wisdom

Weather Wisdom

Proverbs, Superstitions, and Signs

Compiled by
Stewart A. Kingsbury,
Mildred E. Kingsbury,
and
Wolfgang Mieder

PETER LANG
New York • Washington, D.C./Baltimore
Bern • Frankfurt am Main • Berlin • Vienna • Paris

...loging-in-Publication Data

...erstitions, and signs / compiled by
... Kingsbury, and Wolfgang Mieder.
   p. cm.
Includes bibliographical references and index.
1. Weather—Folklore. I. Kingsbury, Stewart A. II. Kingsbury, Mildred E.
III. Mieder, Wolfgang.
QC998.W43   551.6'31—dc20   96-6366
ISBN 0-8204-3326-8

Die Deutsche Bibliothek-CIP-Einheitsaufnahme

Weather wisdom: proverbs, superstitions, and signs / comp. by Stewart A.
Kingsbury… –New York; Washington, D.C./Baltimore; Bern; Frankfurt am
Main;
Berlin; Vienna; Paris: Lang.
ISBN 0-8204-3326-8
NE: Kingsbury, Stewart A. [Hrsg.]

Cover design by James F. Brisson.

The woodcut on the cover dates from 1854. It is entitled "Unwetter"
(stormy weather) and is by Franz Graf von Pocci (1807–1876).
The illustration is included in Franz Graf von Pocci, *Die gesamte Druckgraphik,*
ed. by Marianne Bernhard. Herrsching: Manfred Pawlak, 1974, p.432.

The paper in this book meets the guidelines for permanence and durability
of the Committee on Production Guidelines for Book Longevity
of the Council of Library Resources.

© 1996 Peter Lang Publishing, Inc., New York

Printed in the United States of America.

Contents

Preface

The idea of putting together a major collection of weather wisdom expressed in the form of proverbs, superstitions, and signs was conceived about two years ago. All three of us had worked together on *A Dictionary of American Proverbs* (1992) and *A Dictionary of Wellerisms* (1994), and we decided that a book of weather wisdom expressed in the English language was definitely a desideratum not only for professional meteorologists but also for people of all walks of life interested in the weather.

This collection of 4,435 texts is based on oral and written sources, with each text being annotated by references to major collections of weather lore from the 19th and 20th centuries. While some of the weather observations have become current as proverbs expressing folk wisdom in a metaphorical fashion, others quite clearly are superstitions without any rational or scientific basis. But there are also many so-called weather signs which state upcoming weather patterns based on the observation of natural phenomena such as cloud formations, wind directions, temperatures, animal behavior, etc. All the texts have been handed down over generations to help people deal with the various problems and challenges that the weather may bring on any given day, week, month, season, or year.

While our combined work on this book was progressing at a solid pace, Stewart A. Kingsbury was diagnosed with cancer. He fought this terrible disease valiantly, but he passed away on October 23, 1994, leaving his widow Mildred E. Kingsbury and me to finish this project with the computer and word-processing assistance of Ms. Billie L. Schuller and Prof. George B. Bryan. Stewart Kingsbury would so very much have liked to have seen this book completed before his untimely death. Now that his and our labors are finished, we would like to dedicate this collection of weather wisdom to Stewart A. Kingsbury (1923-1994) in memory of his untiring commitment to the study of folk language and folk wisdom. We hope that we succeeded in finalizing this book the way he wished to have it done. Doubtlessly our book would have been more perfect if Stewart A. Kingsbury could have been a member of our scholarly team to the very end.

September 1995   M. E. K., W. M.

Introduction

People throughout the world have summarized their weather observations and experiences into easily memorable proverbs for centuries. To this day the acute farmer and sailor rely on such formulaic bits of weather wisdom to plan various agricultural tasks or to navigate a ship. The ordinary citizen, no matter whether residing in a rural or urban setting, also recalls these traditional proverbs to forecast the weather. While they lack the scientific precision of modern meteorology, these proverbs contain the collective wisdom of generations of people who have depended on knowing at least to some degree of certainty what the weather might bring. There have been those who have relegated weather proverbs to mere beliefs or superstitions lacking any kind of scientific reliability. Yet in recent years numerous scholarly treatises by meteorologists have appeared in defense of the truth element in many weather proverbs.

Proverbs in general have been defined as concise statements of apparent truths with currency among the folk, or more elaborately stated, proverbs are short, generally known sentences of the folk which contain wisdom, truths, morals, and traditional views in a metaphorical, fixed, and memorizable form and which are handed down orally from generation to generation. Weather proverbs are a subgenre of this larger class of proverbial wisdom. Their major function is to predict the weather so that people might be able to plan the daily affairs of life without too many climatic uncertainties or surprises. They are based on keen observation and scrutiny of natural phenomena by experienced people who formulated and promulgated their wisdom in proverbial form. Since weather proverbs usually contain prognostic statements, they have also been called predictive sayings, weather rules, farmers' rules, and weather signs. Their intent is to establish a causal or logical relationship between two natural events which will lead to a reasonable statement concerning the weather of the next hour, day, week, month or even year.

Structurally, weather proverbs follow a rather distinct pattern which can be summarized with the formula "If A then B," each proverbial text consisting of a clear antecedent in the first part and a consequent in the second part. The well-known proverb "April showers bring May flowers" clearly illustrates this common structure. "April showers" is "A," and the ensuing "May flowers" is element "B." By looking at this structure in a bit more detail, it can also be stated that the binary form of weather proverbs includes "A" in the form of a weather sign and a date as well as "B" stating a prognostication and its particular date. Thus the proverb "A green Christmas, a white Easter" basically states that if there is a thaw in the middle of the winter, there will be some more severe snow fall in spring. It should be noted, however, that many weather proverbs do not necessarily express a precise time element in the second part, especially if the texts refer to agricultural cycles. The proverb "A wet March makes a sad harvest" is a case in point, but it does, of course, indirectly project from the month of March to harvest time in the summer.

While weather proverbs contain plenty of colorful images or metaphors, they usually do not take on a figurative meaning as is customarily the case with other proverbs. In fact, some scholars prefer not to use the genre description of "proverb" for weather signs at all, arguing that *bona fide* proverbs take on various meanings depending on the contexts in which they appear. A good example for such a figurative proverb with multiple semantic possibilities is "A rolling stone gathers no moss," which might imply that mobility avoids stagnancy (gathering moss) or that mobility never establishes roots and wealth (moss = money). Predictive sayings, on the other hand, are usually interpreted literally, i.e., they have but one distinct meaning. Thus "A snow year, a rich year" means nothing but that a lot of snow in winter will act as a fertilizer (wetness) later on when the plants need moisture to grow. There is here no metaphorical shift of meaning, and the weather proverb means what it says and no more.

Yet it would be wrong to deny the thousands of predictive sayings any proverbial status merely because they are more literal than other proverbs. It has even been argued that they are nothing but superstitions in proverbial form, but then there are also plenty of very common proverbs which are not at all figurative, as for example "Honesty is the best policy." In addition, some of the most popular weather proverbs are in fact figurative as well. Such texts as "One swallow does not make a summer" and "Lightning never strikes twice in the same place" can take on different semantic shades in various situations. This is also the case with the most frequently used English weather proverb "Make hay while the sun shines," which, in addition to its solid advice to farmers, is also used to encourage entrepreneurs of whatever kind to take advantage of a good situation. There appears to be no sound reason to reclassify weather proverbs as proverbial superstitions, i.e., as superstitions expressed in proverbial language using rhyme ("A sunshiny shower never lasts half an hour"), parallelism ("Frost year, fruit year"), personification ("Winter spends what summer lends"), etc. Such scholarly genre differentiation ignores the close connections and overlap of proverbs *per se* and weather proverbs. In addition, it is a proven fact that native informants both in oral societies of Africa and urban environments of the United States, for example, consider such texts as "Rain before seven, clear before eleven" as traditional folk proverbs and not as superstitions.

Nevertheless, this collection does include dozens of superstitions which can not be regarded as either figurative or literal proverbs. Yet texts as "Kill a beetle and it will rain," "If Christmas Day on a Monday fall,a troublous winter we shall have all," "If all the food at the table is consumed or none of it is wasted, weather next day will be afair," "The first Friday in July is always wet," "If you cover a mirror when a storm is raging, the lightning will not strike you," "Never look at the new moon over your left shoulder, it is bad luck," "The burial of an old man causes rain," "It never rains if you wear your rubbers," "If you kill a spider on Friday, it will rain on Sunday," "If the first thunder is in the north, the bear has stretched his left leg in his winter bed," "When it thunders the angels are bowling," "If you leave your umbrella at home, it is sure to rain," and "If you see many women pushing baby buggies, a rain is predicted" are indeed traditional statements of which some continue to be repeated in oral and written communication today, and they have commonly been included in collections of weather lore. The same is true for such weather signs as "A mackerel sky, not twenty-four hours dry," "Much dew after a fair day indicates another fair day," "A winter's fog will freeze a dog," "Ring around the moon, rain before noon," and "If the sun in red should set, the next day surely will be wet; if the sun should set in gray, the next will be a rainy day." While quite a few of these texts contain proverbial markers as rhyme, parallel structure, personification, etc., they are limited to precise weather conditions and have no figurative meaning. They are thus not proverbs in the pure sense of that folkloric genre, but they nevertheless express folk beliefs or wisdom about the weather in more or less proverbial language.

It should also be noted that just as there are contradictory proverb pairs as "Absence makes the heart grow fonder" and "out of sight, out of mind," there exist some weather sayings which express opposite views. The following two weather observations are clearly contradictory: "When the fog falls fair weather follows; when it rises rain follows" and "When fog goes up, the rain is o'er; when fog comes down, 'twill rain some more." Certain weather signs are simply interpreted in diametrically opposed fashion, as in "The blooming of flowers late in the autumn presages a bitter winter say some; a mild winter say others" and "A large number of caterpillars in autumn signifies either a cold or a warm winter." Yet such ambiguity is actually rather the exception than the rule. For all general purposes, the folk have in fact observed the weather quite well for centuries, and their wisdom in the form of proverbs and predictive signs appears to make considerable sense when compared with the insights of modern scientific meteorology.

Weather proverbs as well as superstitions and signs can be classified into various groups according to their content or referents. Collections range from limited regional listings to major national or international compendia of several thousand texts, the three best comparative reference works being Otto von Reinsberg-Düringsfeld, *Das Wetter im Sprichwort* (1864; rpt. 1976), Rev. C. Swainson, *A Handbook of Weather Folk-Lore* (1873; rpt. 1974), and Alexis Yermoloff, *Die landwirtschaftliche Volksweisheit* (1905). Most collectors have grouped their massive materials as they relate to the year, seasons, individual months, days of the week, sun, moon, stars, rainbow, fog, clouds, frost, snow, rain, thunder and lighting, wind, animals, plants, and saints' days. The following examples from different countries illustrate at least a few representative categories, indicating at times the broad geographical and linguistic distribution of some weather proverbs by citing identical texts in various languages:

> *year*: ·"A snow year, a rich year" (English), "Année neigeuse, année fructueuse" (French), "Schnee Jahr—reich Jahr" (German), "An nervòs, an frutòs" (Italian).
> *seasons*: "A cold spring kills the roses" (Arabic), "Mild winter—dry spring" (French), "When the butterfly comes,

the summer comes also" (Native American), "Warm fall—long winter" (German).

months: "March windy and April rainy, makes May the pleasantest month of any" (English), "Mars venteux et avril pluvieux font le may gay et gracieux" (French), "Märzenwind, Aprilregen verheißen im Mai großen Segen" (German).

clouds: "Mackerel scales and mare's tails make lofty ships carry low sails" (English), "Black clouds bring rain" (Persian), "Every cloud has a silver lining" (English), "Clouds are the sign of rain" (African).

wind: "When there is wind, it is always cold" (Italian), "Every wind has its weather" (American), "Much wind, little rain" (Dutch), "Let the sail come down if the wind becomes strong" (Philippine).

animals: "When dogs bark at the moon, a strong frost will follow" (German), "When ducks are driving through the burn, that night the weather takes a turn" (Scottish), "When swallows fly low or when geese fly, expect storm or cold" (English).

plants: "Many acorns bring a strong winter" (German), "Onion's skin very thin, mild winter's coming in; onion's skin thick and rough, coming winter cold and rough" (English), "Ill weeds are not hurt by frost" (Portuguese).

The origin, history and dissemination of weather proverbs are quite complex, each individual text deserving a special historical study. Many of them, having been transmitted orally from generation to generation, go back to non-literate times. They have been recorded in classical antiquity by Aristotle in his *On Meteors* and in the *Inquiry into Plants and Minor Works in Odours and Weather Signs* by his student Theophrastus. The wisdom literature of the Bible also contains early references of weather proverbs, notably "When it is evening, ye say, It will be fair weather: for the sky is red. And in the morning, It will be foul weather today: for the sky is red and lowering" (Matthew 16:2-3). Through translations of the Bible, this proverb has become current in many languages, albeit in many variants predicting the weather by these signs of the sky for the sailor, shepherd, traveler, etc. A few examples in English are "Evening red and morning gray will send the sailor [or traveler] on his way, but evening gray and morning red will bring rain down upon his head," "When the evening is gray and morning is red, the sailor is sure to have a wet head," and "Red sky at night, sailor's [or shepherd's] delight; red sky in the morning, sailors take warning." The last example is also known as a variant with "red sky" being replaced by "rainbow." Modern meteorologists have gone to some lengths proving scientifically that there is a kernel of truth to this weather sign. In simple terms, a red or pink hue in the evening sky is a result of dry dust particles, indicating fair weather ahead. On the other hand, a gray evening sky means that it is heavy with water droplets that will probably fall the next day from the west. Other proverbs were coined as people observed certain weather signs locally and regionally. Many older English weather proverbs have been registered in John Claridge's many editions of *The Shepherd of Banbury's Rules to Judge of the Changes of the Weather* (London 1748) which itself goes back to *The Shepherd's Legacy* of 1670. Various farmers' almanacs, notably Benjamin Franklin's *Poor Richard's Almanac* (1733-1758), helped to spread traditional weather wisdom in the New World, and immigrants brought their native weather proverbs and superstitions along as well.

A whole sub-class of weather proverbs or better weather signs are those which are attached to saints' days, a phenomenon which spread throughout the Christian world. The following examples should be taken *cum grano salis*, i.e., not even the folk originators of this proverbial wisdom meant it to be taken as sacrosanct. All that these country folk did is to summarize a general weather or agricultural time sequence by attaching its beginning to a memorable saint's day and adding a reasonable prognosis: "He who shears his sheep before St. Servatius [May 13] loves more his wool than his sheep" makes sense in an environment in which sudden cold weather might still set in during the middle of May. However, the observation that "If St. Martin's Day [November 11] is dry and cold, the winter will not be long lasting" is rather questionable. Chance might have it that in a certain region a dry and cold day on November 11 was in fact followed by a relatively short winter. People, however, took these proverbs with them when they moved about, and clearly these sayings lost even more of their slight validity. Other such proverbs are attached to church festivals like Candlemas Day (February 2, feast of the purification of Mary). Due to the date involved, most of these proverbs contain spring predictions, as for example this somewhat elaborate "If Candlemas Day be mild and gay, go saddle your horses and buy them hay; but if Candlemas Day be stormy and black, it carries the winter away on its back."

Even more unscientific or based on mere folkloric superstition are the numerous proverbial interpretations of the meteorological phenomenon of the sun shining while it is raining at the same time. Several scholars have studied the folk beliefs attached to this particular weather situation, and Matti Kuusi from Finland published a 420-page book on *Regen bei Sonnenschein* (1957) in which he lists and discusses the weather proverb "When it rains and the sun shines the devil is beating his grandmother" and its many variants from all around the world. Clearly these texts are folk beliefs couched in proverbial language, but this study also shows the complexity of the origin, history, dissemination, and meaning of weather lore. A few international variants in English translation must suffice, but similar studies can and should certainly be undertaken for other weather proverbs as well. Most of the variants begin with a statement that says in general "When it rains and the sun shines ..." followed by explanatory comments like "... foxes are on a marriage parade" (Japanese), "... the devil is getting married" (Bulgarian), "... the devil is beating his wife" (Hungarian), "... witches are doing their wash" (Polish), "... the gypsies are washing their children" (Finnish), "... a tailor is going to hell" (Danish), "... mushrooms are growing" (Russian), "... good weather is coming" (German), "... husband and wife are quarreling" (Vietnamese), "... the sheep get scared" (Spanish), etc. Many of these variants are current throughout Europe and other parts of the world, while others are known only locally. They all show a fascination with this peculiar weather sign which has led to the most astounding folkloric explanations. Belief in superstitions is decreasing today, but that does not mean that these proverbial statements are not uttered anymore, even if only to comment in jest on a bizarre natural phenomenon.

Such proverbial superstitions are the exact opposite to the modern science of meteorology. They lack any validity and appear nonsensical to the modern educated person who has a certain meteorological sophistication due to the modern mass media and their involved weather reports. Nevertheless, the general proverb "Some are weather-wise, and some are otherwise" continues to have truth value today. Some people do indeed know more about natural weather signs than others, and they certainly continue to employ those weather proverbs that make sense even in the modern world. After all, many of them are based on ancient empiricism, and some of these basic weather truths have not changed and will not change in the foreseeable future. Even the folk are well aware of the obvious shortcomings of some weather proverbs, claiming with considerable proverbial irony and insight that "All signs fail in dry weather." There are even such parodies as the German "Wenn der Hahn kräht auf dem Mist, ändert sich das Wetter oder bleibt wie's ist" (When the rooster crows on the dunghill, the weather will change or stay as it is). Other humorous reactions to the human preoccupation with the weather are such gems as "Change of weather is the discourse of fools" from the early seventeenth century and the American standard of "Everybody talks about the weather, but nobody does anything about it" from the turn of this century. In the long-run, those who delight in citing weather proverbs, superstitions, and signs as well as those who pride themselves on being meteorological scientists all will have to agree with the medieval English proverb that simply recommends "Take the weather as it comes." Weather prediction will likely always be a proverbial mixture of truth and superstition, of sense and nonsense, and of science and folklore.

Bibliography

Arora, Shirley L. "Weather Proverbs: Some 'Folk' Views." *Proverbium* 8 (1991): 1-17.

Bonser, Wilfrid. *Proverb Literature: A Bibliography of Works Relating to Proverbs*. London: Glaisher, 1930; rpt. Nendeln, Liechtenstein: Kraus, 1967. 422-429 (nos. 3675-3746).

Brunt, D. "Meteorology and Weather Lore." *Folklore* (London) 57 (1946): 66-74.

Dallet, Gabriel. *La prévision du temps et les prédictions météorologiques*. Paris: Baillière, 1887.

Delsol, Paula. *La météorologie populaire*. Montréal: Presses Select, 1973.

Doctor, Raymond D. "Predictive Sayings in Gujerati [India]." *Folklore* (London) 97 (1986): 41-55.

Dundes, Alan. "On Whether Weather 'Proverbs' are Proverbs." *Proverbium* 1 (1984): 39-46; also in Alan Dundes. *Folklore Matters*. Knoxville, Tennessee: University of Tennessee Press, 1989. 92-97.

Dunwoody, H.H.C. *Weather Proverbs*. Washington, D.C.: Government Printing Office, 1883.

Freier, George D. *Weather Proverbs*. Tucson, Arizona: Fisher Books, 1992.

Frick, R.-O. "Le peuple et la prévision du temps." *Schweizerisches Archiv für Volkskunde* 26 (1926): 1-21, 89-100, 171-188 and 254-279.

Garriott, Edward B. *Weather Folk-Lore and Local Weather Signs.* Washington, D.C.: Govenrment Printing Office, 1903; rpt. Detroit, Michigan: Grand River Books, 1971.

Hall, Hermann Christian van. *Spreekwoorden en voorschriften in spreuken, betreffende Landbouw en Weêrkennis.* Haarlem: Kruseman, 1872.

Hauser, Albert. *Bauernregeln. Eine schweizerische Sammlung mit Erläuterungen.* Zürich: Artemis, 1973.

Hellmann, Gustav. "Über den Ursprung der volkstümlichen Wetterregeln (Bauernregeln)." *Sitzungeberichte der Preussischen Akademie der Wissenschaften*, no vol. (1923): 148-170.

Helm, Karl. "Bauernregeln." *Hessische Blätter für Volkskunde* 38 (1939): 114-132.

Heyd, Werner P. *Bauernweistümer.* 2 vols. Memmingen: Maximilian Dietrich, 1971 and 1973.

Hoyos Sáinz, Luis de. "Bases metódicas y técnicas para un refranero agricola." *Revista de Dialectologia y Tradiciones Populares* 7 (1951): 242-253.

Humphreys, William Jackson. *Weather Proverbs and Paradoxes.* Baltimore, Maryland: William & Wilkins, 1923.

Inwards, Richard. *Weather Lore. A Collection of Proverbs, Sayings, and Rules Concerning the Weather.* London: Tweedie, 1869; 3rd ed. London: Elliot Stock, 1898.

Knapp, Elisabeth. *Volkskundliches in den romanischen Wetterregeln.* Tübingen: Karl Bölzle, 1939.

Kuusi, Matti. *Regen bei Sonnenschein. Zur Weltgeschichte einer Redensart.* Helsinki: Suomalainen Tiedeakatemia, 1957.

Lee, Albert. *Weather Proverbs.* Garden City, New York: Doubleday, 1976; Garden City, New York: Dolphin Books, 1977.

Legros, Elisée. "A propos de dictons météorologiques." *Enquêtes du Musée de la Vie Wallone*, no vol. (1972): 17-46.

Liebl, Elsbeth. "Wetterregeln." *Atlas der schweizerischen Volkskunde*, edited by Walter Escher, Elsbeth Liebl, and Arnold Niederer. Basel: Krebs, 1979. II, 842-916 and maps nos. 266-270.

Malberg, Horst. *Bauernregeln: Ihre Deutung aus meteorologischer Sicht.* Berlin: Springer, 1989.

Matzen, Raymond. "Die elsässischen Wetter- und Bauernregeln als lebendiger Ausdruck von Land, Sprache und Kultur." *Europhras 88. Phraséologie Contrastive*, edited by Gertrud Gréciano. Strasbourg: Université des Sciences Humaines, 1989. 301-311.

Moll, Otto. *Sprichwörterbibliographie.* Frankfurt am Main: Vittorio Klostermann, 1958. 540-551 (nos. 8454-8671e).

Munzar, Jan. *Medardova kápe: Aneb pranostiky očima meteorologa.* Praha: Horizont, 1986.

Page, Robin. *Weather Forecasting: The Country Way.* London: Davis-Poynter, 1977; London: Penguin Books, 1981.

Profantová, Zuzana. *Dúha vodu pije. Slovenské l'udové pranostiky.* Bratislava: Tatran, 1986.

Reinsberg-Düringsfeld, Otto von. *Das Wetter im Sprichwort.* Leipzig: Hermann Fries, 1864; rpt. Leipzig: Zentralantiquariat, 1976.

Rodríguez Marín, Francisco. *Los refranes del almanaque.* Sevilla: Dias, 1896.

Röhrich, Lutz, and Wolfgang Mieder. *Sprichwort*. Stuttgart: Metzler, 1977. 7-10.

Roucy, François de. *Dictons popularies sur le temps ou Recueil de proverbes météorologiques de la France*. Paris: Plon, 1877.

Sinha, R.N. "Climatic Cycle, Agricultural Rhythm and Local Proverbs in the Canal Irrigated Areas of the Patna District [India]." *Patna University Journal* 18 (1963): 10-27.

Swainson, Rev. C. *A Handbook of Weather Folk-Lore*. Edinburgh: William Blackwood, 1873; rpt. Detroit, Michigan: Gale Research Company, 1974.

Taylor, Archer, *The Proverb*. Cambridge, Massachusetts: Harvard University Press, 1931; rpt. Hatboro, Pennsylvania: Folklore Associates, 1962; rpt. again with an introduction and bibliography by Wolfgang Mieder. Bern: Peter Lang, 1985. 109-121.

Tibau, Georges. "Zestig Vlaamse weerspreuken onder de loep van de statistik." *Volkskunde* 78 (1977): 33-59.

Ting, Nai-tung. "Chinese Weather Proverbs." *Proverbium*, no. 18 (1972): 649-655.

Wurtele, M.G. "Some Thoughts on Weather Lore." *Folklore* 82 (1971): 292-303.

Yermoloff, Alexis. *Die landwirtschaftliche Volksweisheit in Sprichwörtern, Redensarten und Wetterregeln*. Leipzig: F.A. Brockhaus, 1905.

How to Use This Book

The work consists of an introductory essay on the nature, language, origin, history, dissemination, and meaning of proverbs, superstitions, and signs concerning the weather. This is followed by a bibliography of the most important scholarship on traditional weather wisdom. The heart of the book is comprised of 4,435 texts from the English speaking world. A list of sources as well as a comprehensive index are included as well.

The proverbs, superstitions, and signs are arranged in alphabetical order according to the most significant key-word, usually a noun, in the text. For some of the most popular key-words, such as "moon," "rain," or "sky," sub-categories as for example "moon (new)," "moon (old)," "moon (pale)," "moon (red)," etc. are used to facilitate the location of specific texts. When available, variants are registered following the main entry. Each text has been verified by consulting forty-one earlier collections from the 19th and 20th centuries which are listed in the "List of Sources." These references are cited by the authors' last name and the appropriate page number. In the majority cases, where more than one reference could be listed, they are listed in chronological order reflecting the recorded history of the particular text. In addition to these citations from published sources many texts also indicate the fact that they are recorded in the large archival "Collection of the Committee on Proverbial Sayings of the American Dialect Society" which was assembled under the direction of Margaret M. Bryant between circa 1945 and 1980. These unpublished materials are now housed in the archives of the University of Missouri at Columbia, Missouri.

This collection of weather wisdom also contains an extensive "Index of Indications" to make each text as accessible as possible. The key-words used for arranging this collection usually refer to the indicators of a certain weather condition, as is the case for example for the proverb "When apples have thick skins, the winter will be severe" which is listed under the key-word "apple." Yet in the index under "apple" one is also referred to such texts as "March dust on an apple leaf, brings all kinds of fruit to grief" or "If the sun smiles on St.

Eulalie's Day [February 12th], it is good for apples and cider," where the word "apple" appears only in the indication part of the text without being the key-word. Readers wishing to find weather wisdom on any subject should definitely check both under the specific key-word in the collection itself and in this index at the end of the book.

acorn

1. A hard shelled acorn is a sign of a hard winter.
   Recorded in Bryant (Wis.).
2. A lot of acorns on the trees means a severe winter. Var.: If the oak bear much mast, it foreshows a long hard winter. Comment: Mast means acorns.
   Cited in Inwards, p. 210; Smith, p. 12.
3. Many acorns means a good spring.
   Cited in Smith, p. 11.
4. When beech acorns thrive well and oak trees hang full, a hard winter will follow with much snow.
   Cited in Dunwoody, p. 92.

air

5. Early in the morning a clear atmosphere (air) with little or no humidity promises a fine day.
   Cited in Hyatt, p. 12.
6. Much undulation in the air on a hot day in May or June foretells cold.
   Cited in Inwards, p. 150.

algae

7. A lot of algae on a pond in the spring means a dry summer.
   Cited in Smith, p. 11.

All Saints' Day

8. All Saints' Day will bring out the winter, St. Martin's Day will bring out Indian Summer. Comment: All Saints' Day is November 1st, and St. Martin's Day is November 11th.
   Cited in Dunwoody, p. 104; Garriott, p. 43; Inwards, p. 66; Hand, p. 43.
9. If on All Saints' Day the beech acorn is dry, we will stick behind the stove in winter, but if it is wet and not light, the winter will not be dry, but wet. Vars.: (a) If on All Saints' Day the beechnut be found dry, we shall have a hard winter; but if the nut be wet and not light, we may expect a wet winter. (b) If on All Saints' Day the beechnut is dry we shall have a hard winter; but if the nut be wet and not light, we may expect a wet winter.

Cited in Dunwoody, p. 92; Garriott, p. 43; Inwards, p. 66; Hand, p. 43.

10. On All Saints' Day, cut off some of the bark from a beech tree, and after that, a little piece of wood; cut it, if it be dry, then the ensuing winter will be dry, but pretty warm and temperate; if moist, a wet winter.
Cited in Swainson, p. 142.

11. On All Saints' Day there is snow on the ground.
Cited in Swainson, p. 142.

12. On the first of November, All Saints' Day, if the weather holds clear, an ending of wheat sowing do make for this year. Var.: On All Saints' Day, if the weather hold clear, an end of wheat sowing do make for the year.
Cited in Denham, p. 61; Garriott, p. 43; Whitman, p. 40; Hand, p. 43.

13. Raw weather on All Saints' Day and All Souls' Day warns you of the approach of winter, but fair weather on these days will last for six weeks. Comment: All Souls' Day is November 2nd.
Cited in Hyatt, p. 13.

All Saints' Rest

14. Some weather that is warm is called "All Saints' Rest" (Indian Summer).
Cited in Sloane 2, p. 96.

almanac

15. The almanac maker makes the almanac, but God makes the weather.
Cited in Whitman, p. 51; Inwards, p. 23; Mieder, p. 15. Recorded in Bryant (Nebr.).

anemone

16. The yellow wood anemone and the wind flower close their petals and droop before rain.
Cited in Inwards, p. 214.

anemone (sea)

17. The sea anemone closes before rain and opens for fine, clear

weather.
Cited in Inwards, p. 199.

angel
18. Rain is caused by the angels crying.
Cited in Smith, p. 9.

animal
19. Animals making unusual noise indicates change of weather.
Cited in Dunwoody, p. 32.
20. Animals moving are a sign of bad weather.
Cited in Smith, p. 1.
21. Domestic animals stand with their heads from the coming storm.
Cited in Dunwoody, p. 31.
22. Heavy coats of fur on animals portend a cold winter; light coats indicate a warm winter. Var.: Heavy hair on animals means a long winter.
Recorded in Bryant (Nebr., Wis.).
23. If animals crowd together, rain will follow.
Cited in Inwards, p. 178.
24. If animals, domestic or wild, have a thick coat, you can look for a strong hard winter.
Cited in Smith, p. 12.
25. When animals seek sheltered places instead of spreading over their usual range, an unfavorable change of weather is probable, if not certain, rain or stormy weather.
Cited in Steinmetz 1, p. 111; Dunwoody, p. 14; Inwards, p. 178.

ant
26. Ants are very busy, gnats bite, crickets are lively, spiders come out of their nests, and flies gather in houses just before rain. Var.: Ants are very busy before a rain.
Cited in Dunwoody, p. 55; Garriott, p. 24; Hand, p. 24; Freier, p. 28.
27. Ants building sand cones around holes, expect rain.
Cited in Freier, p. 61.
28. Ants conveying their eggs from their cells indicates rain. Vars.: (a) Ants that move their eggs and climb, rain is coming anytime. (b) If ants carry eggs to high ground, expect rain.

Cited in Mitchell, p. 233; Lee, p. 56; Freier, p. 61.

29. Ants herding together and running about in circles warn you of rain.
    Cited in Hyatt, p. 17.

30. Ants sometimes get down fifteen inches from the surface before very hot weather.
    Cited in Inwards, p. 204.

31. Expect stormy weather when ants travel in lines, and fair weather when they scatter. Vars.: (a) Expect stormy weather when ants travel in a straight line; when they scatter all over, the weather is fine. (b) Stormy weather when ants travel in lines; and when they scatter weather's fine.
    Cited in Dunwoody, p. 55; Garriott, p. 24; Inwards, p. 205; Hand, p. 24; Lee, p. 56; Freier, p. 61. Recorded in Bryant (Oreg.).

32. If ants are more than ordinarily active, or if they remove their eggs from small hills, it will surely rain. Vars.: (a) An unusual activity among ants will be succeeded by rain. (b) If the ants are extremely busy, it will rain soon.
    Cited in Inwards, p. 205; Hyatt, p. 17; Reeder, no pp.

33. If ants put dirt on one side of the ant hill, rain is coming.
    Cited in Smith, p. 5.

34. If ants their walls do frequent build, rain will from the clouds be spilled. Vars.: (a) If ants raise the sides of their mounds higher, rain is in the air. (b) If the small ants build mounds around their holes, it is to keep out the flood water.
    Cited in Dunwoody, p. 55; Inwards, p. 204; Sloane 2, p. 124; Hyatt, p. 17; Lee, p. 51; Reeder, no pp.

35. If in the beginning of July the ants are enlarging and building up their piles, an early and cold winter is at hand. Vars.: (a) If ants increase the size of their mounds at the beginning of July, they are enlarging the tunnels of their nests in expectation of an early and severe winter. (b) If in the beginning of July the ants are enlarging and building up their piles, an early and cold winter will follow.
    Cited in Dunwoody, p. 55; Inwards, p. 205; Hyatt, p. 17.

36. It is said that ants run faster or slower as the atmosphere grows warmer or colder, and that an approximation to the air temperature may be obtained by timing the speed of these insects.

Cited in Inwards, p. 205.

37. Look for rain after you see ants traveling in a straight column.
Cited in Hyatt, p. 17.

38. Step on an ant and it will rain. Vars.: (a) If you step on an ant,
it will rain. (b) To kill an ant or to tread on an ant-hill, intention-
ally or unintentionally, causes a rain.
Cited in Sloane 2, p. 42; Hyatt, p. 17; Smith, p. 5.

39. Summer arrives while the first ants are throwing up their
mounds.
Cited in Hyatt, p. 17.

40. The appearance of red ants announces the arrival of spring.
Cited in Hyatt, p. 17.

41. There is never any frost after red ants have appeared.
Cited in Hyatt, p. 17.

42. When ants are situated in low ground, their migration may be
taken as an indication of approaching heavy rains. Var.: When
ants migrate away from low ground, it forbodes rain.
Cited in Dunwoody, p. 55; Garriott, p. 24; Inwards, p. 205;
Hand, p. 24; Sloane 2, p. 124.

ant hole

43. An open ant hole indicates clear weather; a closed one an approaching storm. Var.: Ant holes are open and surrounded with
much sand in clear weather and closed before an approaching
storm.
Cited in Dunwoody, p. 55; Inwards, p. 205; Alstad, p. 99; Lee,
p. 56.

anthill

44. If you find anthills, it's a sign of fair weather.
Cited in Boughton, p. 125.

apple

45. If apples bloom in April, you'll have your fill; but if in May,
they'll all go away. Comment: If apple trees bloom in April they
will produce a good crop.
Cited in Wilson, p. 70.

46. If the fall apples are one sided, with thick, rough skins, a severe winter may be expected. Var.: When apples have thick skins, the winter will be severe.
    Cited in Dunwoody, p. 66; Smith, p. 14.
47. Till St. Swithin's Day be past, the apples be not fit to taste.
    Comment: St. Swithin's Day is July 15th.
    Cited in Northall, p. 453.
48. When the apple blooms in March, you need not for barrels search; but when the apple blooms in May, search for barrels every day. Vars.: (a) If apples bloom in March, in vain for 'um you'll sarch; if apples bloom in April, why then they'll be plentiful; if apples bloom in May, you may eat 'um night and day. (b) If apples bloom in May, you may eat 'um night and day. (c) If the apple tree blossoms in March, for barrels of cider you need not search; if the apple tree blossoms in May, you can eat apple dumplings every day.
    Cited in Northall, p. 484; Wright, p. 35.

apple blossom
49. When apple blossoms bloom at night, for fifteen days no rain in sight.
    Cited in Sloane 2, p. 30.

April
50. A cold and moist April fills the cellar and fattens the cow. Var.: Cellars are filled and cattle fattened by a cold wet April.
    Cited in Dunwoody, p. 96; Hyatt, p. 34.
51. A cold April the barn will fill. Vars.: (a) A cold April and a full barn. (b) A cold April and a wet May fill the barn with grain and hay. (c) The colder the April, the better the farm crops.
    Cited in Denham, p. 40; Swainson, p. 77; Cheales, p. 21; Dunwoody, p. 95; Northall, p. 437; Garriott, p. 44; Wright, p. 43; Humphreys 2, p. 9; Whitman, p. 43; Inwards, p. 51; Hand, p. 44; Sloane 2, p. 30; Wilson, p. 132; Page, p. 46; Wilshere, p. 10. Freier, p. 81; Mieder, p. 25.
52. A dry April not the farmer's will; rain in April is what he wills. Var.: A dry April not the farmer's will, April wet is what he would get.
    Cited in Dunwoody, p. 95; Inwards, p. 51.

53. A green April brings a flourishing May.
    Cited in Inwards, p. 52.
54. A wet April, a heavy wheat crop.
    Cited in Hyatt, p. 34.
55. A wet cold April will fill the wine kegs in the cellar.
    Cited in Hyatt, p. 34.
56. After a wet April follows a dry June. Vars.: (a) After a wet
    April, a dry June. (b) Moist April, clear June.
    Cited in Dunwoody, p. 95; Garriott, p. 44; Whitman, p. 42;
    Inwards, p. 52; Hand, p. 44; Freier, p. 84.
57. After warm April and October, a warm year next.
    Cited in Inwards, p. 52.
58. An April flood carries away the frog and her brood.
    Cited in Denham, p. 43; Swainson, p. 80; Northall, p. 431;
    Northall, p. 437; Inwards, p. 51; Wilson, p. 17.
59. An April shower and May sun, will make cloth white, and fair
    maids dun. Var.: An April shower and a May sun will make
    cloth white and maids dun. Comment: Dun means dingy.
    Cited in Denham, p. 41; Swainson, p. 82.
60. April and May are the keys of the whole year. Var.: April and
    May are the keys to the year.
    Cited in Denham, p. 40; Dunwoody, p. 95; Inwards, p. 52.
61. April borrows three days from March, and they are ill.
    Cited in Dunwoody, p. 96; Mieder, p. 25.
62. April cold and wet fills barn and barrel.
    Cited in Dunwoody, p. 96; Inwards, p. 51.
63. April cold with dropping rain willows and lilacs brings again, the
    whistle of returning birds, and trumpet-lowing of the herds.
    Cited in Wright, p. 43.
64. April moist and warm makes the farmer sing like a nightingale.
    Cited in Inwards, p. 52.
65. April showers bring May flowers. Vars.: (a) April showers bring
    forth May flowers. (b) April showers bring May flowers, but
    who wants to wade through mud to see a bud. (c) April showers
    bring summer flowers. (d) April showers impregnate the flowers.
    Cited in Denham, p. 42; Mitchell, p. 230; Steinmetz 1, p. 19;
    Dunwoody, p. 96; Northall, p. 437; Taylor 1, p. 114; Inwards,
    p. 52; Smith, p. 8; Wilson, p. 17; Whiting 1, p. 11; Simpson,

p. 6; Dundes, p. 94; Whiting 2, p. 16; Mieder, p. 25. Recorded in Bryant (Idaho).

66. April snow breeds grass.
    Cited in Dunwoody, p. 96; Inwards, p. 51.

67. April snow stays no longer than water on a trout's back.
    Cited in Inwards, p. 51.

68. April wears a white hat. Var.: No April is so good that it won't snow on a farmer's hat.
    Cited in Inwards, p. 51; Mieder, p. 25.

69. April weather, rain and sunshine, both together.
    Cited in Swainson, p. 77; Northall, p. 437; Wilson, p. 18; Wilshere, p. 10.

70. April with his hack and his bill, plants a flower on every hill.
    Var.: April with his hack and his bill, sets a flower on every hill.
    Cited in Denham, p. 41; Northall, p. 437.

71. As long as the cherries bloom in April, the grapevine will be in bloom.
    Cited in Dunwoody, p. 65.

72. Betwixt April and May if there be rain, 'tis worth more than oxen or wain. Comment: Wain means wagon or cart.
    Cited in Swainson, p. 82; Northall, p. 438; Inwards, p. 52.

73. Fogs in April foretell a failure of the wheat crop next year.
    Cited in Inwards, p. 51.

74. If in April a northeast wind shifts to the northwest and returns to the northeast, you may look for rain with hail.
    Cited in Hyatt, p. 9.

75. If the first three days of April be foggy, there will be a flood in June. Var.: If the first three days of April be foggy, rain in June will make the lanes boggy.
    Cited in Swainson, p. 84; Marvin, p. 209; Inwards, p. 52; Wilshere, p. 10.

76. Sharp April kills the pig.
    Cited in Mieder, p. 25.

77. Sweet April smiling through her tears, shakes raindrops from her hair and disappears.
    Cited in Wright, p. 51.

78. Till April's dead, change not a thread.
    Cited in Cheales, p. 23; Northall, p. 438; Garriott, p. 44; Whitman, p. 43; Inwards, p. 51; Hand, p. 44; Mieder, p. 25.

79. Wet April brings milk to cows and sheep.
    Cited in Mieder, p. 25.

80. When April blows his horn, it's good for both hay and corn.
    Vars.: (a) April thunder indicates a good hay and corn crop. (b)
    When April blows his horn then 'tis good for hay and corn. (c)
    When April blows its horn then it stands good with hay, rye, and
    corn. (d) When April makes much noise we will have plenty of
    rye and hay.
    Cited in Denham, p. 1; Steinmetz 1, p. 137; Swainson, p. 80;
    Cheales, p. 21; Dunwoody, p. 96; Northall, p. 430; Marvin, p.
    216; Inwards, p. 52; Sloane 2, p. 30; Wilson, p. 17; Page, p.
    46; Wilshere, p. 10; Freier, p. 77; Mieder, p. 25.

April 1

81. First of April fair and mild, the nill may be so much more wild.
    Comment: Nill is an archaic spelling of needle and refers to the
    needle of a barometer.
    Cited in Dunwoody, p. 95.

April 15

82. April fifteenth, first swallow day.
    Cited in Wright, p. 42.

Ascension Day

83. As the weather on Ascension Day, so may be the entire autumn.
    Cited in Inwards, p. 72.

Ascension Thursday

84. If it rains on Ascension Thursday, which comes forty days after
    Easter, it will rain for forty days.
    Cited in Boughton, p. 124.

ash

85. Avoid an ash, it courts a flash. Comment: A flash is lightning.
    Cited in Page, p. 27.

86. When the ash is out before the oak, then we may expect a choke;
    when the oak is out before the ash, then we may expect a splash.
    Vars.: (a) Ash out before oak, there'll be a smoke; oak before
    ash there'll be a splash. (b) If the ash before the oak comes out,

there has been, or there will be drought. (c) If the ash before the oak, we shall have a summer of dust and smoke. (d) The ash before the oak; choke, choke, choke; the oak before the ash; splash, splash, splash. Comment: Choke means drought.
Cited in Northall, p. 475.

87.   When the ash leaves come out before the oak, expect a wet season. Vars.: (a) If buds the ash before the oak, you'll surely have a summer soak; but if behind the oak the ash is, you'll only have a few light splashes. (b) If the ash is out before the oak; you may expect a thorough soak; if the oak is out before the ash, you'll hardly get a single splash.
Cited in Dunwoody, p. 64; Inwards, p. 210.

Ash Wednesday

88.   As Ash Wednesday, so the fasting time. Comment: Ash Wednesday is the first day of Lent.
Cited in Dunwoody, p. 99; Inwards, p. 70.

89.   Wherever the wind lies on Ash Wednesday, it continues all Lent. Vars.: (a) The wind that prevails on Ash Wednesday will remain throughout Lent. (b) Wherever the wind lies on Ash Wednesday, it continues during all Lent.
Cited in Whitman, p. 37; Inwards, p. 70; Lee, p. 151.

aspen

90.   Trembling of aspen leaves in calm weather indicates an approaching storm.
Cited in Dunwoody, p. 64; Inwards, p. 212.

ass

91.   Hark! I hear the asses bray, we shall have some rain today. Vars.: (a) When the ass begins to bray, be sure we shall have rain that day. (b) When the ass begins to bray, surely rain will come that day.
Cited in Swainson, p. 229; Dunwoody, p. 14; Northall, p. 472; Marvin, p. 208; Whitman, p. 47; Inwards, p. 180; Lee, p. 44; Wilshere, p. 22; Freier, p. 49.

92.   If asses bray violently, it will rain. Var.: If asses bray more frequently than ususal, it foreshows rain.
Cited in Steinmetz 1, p. 111; Swainson, p. 229.

93. If asses hang their ears downward and forward, and rub against walls, rain is approaching.
    Cited in Swainson, p. 229; Inwards, p. 180; Freier, p. 67.
94. Young asses rolling and rubbing their backs on the ground indicates heavy showers.
    Cited in Freier, p. 67.

August

95. A wet August never brings dearth.
    Cited in Swainson, p. 119.
96. As August, so next February. Var.: A chilly August, a cold February; a sultry August, a mild February.
    Cited in Dunwoody, p. 98; Garriott, p. 45; Inwards, p. 61; Hand, p. 45; Hyatt, p. 13; Freier, p. 79.
97. Cool August nights reveal hot weather for September.
    Cited in Hyatt, p. 13.
98. Dry August and warm, doth harvest no harm. Var.: Dry August, arid, warm, doth harvest no harm.
    Cited in Swainson, p. 118; Cheales, p. 22; Dunwoody, p. 98; Northall, p. 442; Garriott, p. 45; Wright, p. 97; Whitman, p. 43; Inwards, p. 60; Hand, p. 45; Wilson, p. 206; Wilshere, p. 12; Freier, p. 83. Recorded in Bryant (Wis.).
99. If the first week in August is unusually warm, the winter will be white and long. Var.: Hot weather during the first week of August means a white winter, but cool weather on these days means an open winter.
    Cited in Dunwoody, p. 98; Inwards, p. 61; Hyatt, p. 13.
100. When in August the sun shines warm and the moon and stars are bright, it is good for grapes because they then ripen well. Var.: August sunshine and bright nights ripen the grapes.
    Cited in Dunwoody, p. 98; Inwards, p. 60.
101. When in beginning of August thunderstorms are passing, they will generally last to the end of the month. Var.: Thunderstorms in the beginning of August will generally be followed by others all the month.
    Cited in Dunwoody, p. 98; Inwards, p. 61.
102. When it rains in August it rains honey and wine.
    Cited in Dunwoody, p. 98; Whitman, p. 43; Hand, p. 45.

103. When the dew is heavy in August the weather generally remains fair.
Cited in Dunwoody, p. 98; Inwards, p. 61.

August 24

104. If the twenty-fourth of August be fair and clear, then hope for a prosperous autumn that year.
Cited in Cheales, p. 24; Dunwoody, p. 98; Northall, p. 442; Garriott, p. 43; Whitman, p. 39.

aurora

105. Aurorae are almost invariably followed by stormy weather in from ten to fourteen days.
Cited in Inwards, p. 101; Sloane 2, p. 127.

106. If an aurora appears during warm weather, cold and cloudy weather is to follow.
Cited in Inwards, p. 101.

107. The aurora, when very bright, indicates approaching storm.
Cited in Dunwoody, p. 106; Inwards, p. 101; Sloane 2, p. 127.

aurora borealis

108. Aurora borealis denotes cold. Vars.: (a) Northern lights bring cold weather with them. (b) Northern lights foretell cold weather. (c) Northern lights mean cold weather or change in the weather.
Cited in Dunwoody, p. 76; Sloane 2, p. 49. Recorded in Bryant (Mich., Wis.).

109. The aurora borealis indicates change, especially if lurid and fiery.
Cited in Mitchell, p. 223; Inwards, p. 101.

autumn

110. A long autumn, a long winter.
Cited in Hyatt, p. 14.

111. A moist autumn, with a mild winter, is followed by a cold and dry spring, retarding vegetation. Var.: A pleasant autumn and a mild winter is followed by a cold and dry spring, retarding vegetation.
Cited in Dunwoody, p. 88; Garriott, p. 45; Whitman, p. 45; Inwards, p. 33; Hand, p. 45; Freier, p. 77.

112. A pleasant autumn and a mild winter will cause the leaves to fall next September.
   Cited in Dunwoody, p. 88; Garriott, p. 45; Whitman, p. 45.

113. A severe autumn denotes a windy summer, a windy winter a rainy spring, a rainy spring a severe summer, a severe summer a windy autumn; so the air balance is seldom debtor unto itself.
   Cited in Whitman, p. 45.

114. A wet fall (autumn) indicates a cold and early winter.
   Cited in Dunwoody, p. 89; Inwards, p. 33.

115. After a warm autumn, a long winter. Var.: A warm autumn; a cold winter.
   Cited in Dunwoody, p. 92; Hyatt, p. 14.

116. Clear autumn, windy winter.
   Cited in Dunwoody, p. 92.

117. Flowers in bloom late in autumn indicate a bad winter.
   Cited in Dunwoody, p. 99.

118. If the autumn warm, bright and clear, we may expect a fertile year.
   Cited in Dunwoody, p. 98.

119. In autumn a spoonful of rain makes a tub of mud.
   Cited in Inwards, p. 30.

120. Long fall (autumn), late spring.
   Recorded in Bryant (Miss. Can. [Ont.]).

121. No autumn fruit without spring blossoms.
   Cited in Wilson, p. 23.

122. Warm clear days in autumn breed bad weather.
   Recorded in Bryant (Mich.).

autumn equinox

123. A quiet week before the autumn equinox and after, the temperature will continue higher than usual into the winter. Comment: The autumn equinox occurs on or about September 21st.
   Cited in Inwards, p. 63.

azalea

124. When wild azalea shuts its doors, that's when winter's tempest roars.
   Cited in Sloane 1, p. 154.

bake

125. Never bake on a rainy day, it won't turn out right.
Cited in Smith, p. 9.

bark

126. Thick bark on trees means there is a hard winter ahead.
Cited in Smith, p. 13.
127. When a lot of bark falls off trees, a bad winter is ahead.
Cited in Smith, p. 13.
128. When bark on trees is thin it means a mild winter is coming.
Cited in Smith, p. 13.
129. When the bark is thick on the north side of tree trunks, the winter will be long, cold and severe.
Cited in Smith, p. 14.

barley

130. Dry your barley land in October, or you'll always be sober.
Cited in Denham, p. 60; Northall, p. 443.
131. It is always windy in barley harvests; it blows off the heads of the poor.
Cited in Wright, p. 96.

barn

132. Set your barn upon the ground, winter's heaves will move it round.
Cited in Sloane 1, p. 21.

barometer

133. A fall of mercury (barometer) with a south wind is invariably followed by rain in greater or less quantities.
Cited in Inwards, p. 172.
134. A high and steady barometer is indicative of settled weather.
Cited in Inwards, p. 175.
135. A rapid rise (of the barometer) in winter, after bad weather, is usually followed by clear skies and hard white frosts.
Cited in Steinmetz 2, p. 261.
136. A steady and considerable fall in the mercury (barometer) during an east wind denotes that the wind will soon go round to the

south unless a heavy fall of snow or rain immediately follow: in this case, the upper clouds usually come from the south.
Cited in Inwards, p. 172.

137. A sudden and considerable rise of the barometer after several hours of heavy rain, accompanied by a drying westerly wind, indicates more rain within thirty hours, and a considerable fall of the barometer.
Cited in Inwards, p. 173.

138. A sudden rise in the barometer is very nearly as dangerous as a sudden fall, because it shows that the atmosphere is unsteady. In an ordinary gale the wind often blows hardest when the barometer is just beginning to rise, directly after having been very low.
Cited in Garriott, p. 15; Inwards, p. 172; Hand, p. 15.

139. A very low barometer is usually attendant upon stormy weather, with wind and rain at intervals, but the latter not necessarily in great quantity. If the weather, notwithstanding a very low barometer, is fine and calm, it is not to be depended upon; a change may come on very suddenly.
Cited in Inwards, p. 173; Hand, p. 15.

140. After heavy rains from southwest, if the barometer rises steadily upon the wind shifting to the northwest, expect three or four fine days.
Cited in Inwards, p. 174.

141. Falling fast (barometer), fine at last.
Cited in Inwards, p. 175.

142. First rise after very low indicates a stronger blow. Vars.: (a) First rise after low foretells a sharp blow. (b) First rise after low, foretells stronger blow. (c) First rise after very low indicates a strong blow; first rise after a low precedes a stormy blow.
Comment: Refers to the rising and falling of the barometer.
Cited in Steinmetz 1, p. 155; Cheales, p. 26; Dunwoody, p. 86; Garriott, p. 16; Whitman, p. 33; Inwards, p. 175; Lee, p. 23; Wilshere, p. 19.

143. If after a storm of wind and rain the mercury (barometer) remain steady at the point to which it had fallen, serene weather may follow without a change of wind; but on the rising of the mercury rain and a change of wind may be expected.
Cited in Inwards, p. 173.

144. If the barometer and thermometer both rise together, it is a very sure sign of coming fine weather. Var.: During summer, if pressure (barometer) and temperture increase together, expect several fine days.
Cited in Garriott, p. 15; Whitman, p. 33; Inwards, p. 174; Hand, p. 15.

145. If the barometer fall gradually for several days during the continuance of fine weather, much wet will probably ensue in the end. In like manner, if it keep rising while the wet continues, the weather, after a day or two, is likely to set in fair for some time.
Cited in Inwards, p. 173; Hand, p. 15.

146. If the barometer falls two or three tenths of an inch in four hours, expect a gale of wind.
Cited in Garriott, p. 15; Inwards, p. 174; Hand, p. 15.

147. If the mercury (barometer) fall during a high wind from the southwest, south-southwest, or west-southwest, an increasing storm is probable; if the fall be rapid, the wind will be less violent, but of longer continuance.
Cited in Inwards, p. 172.

148. If the mercury (barometer) fall with the wind at the west, northwest, or north, a great reduction of temperature will follow: in winter severe frosts: in the summer cold rains.
Cited in Inwards, p. 172.

149. If the weather gets warmer while the barometer is high and the wind north-easterly, we may look for a sudden shift of wind to the south. On the other hand, if the weather becomes colder while the wind is south-westerly and the barometer low, we may look for a sudden squall or a severe storm from the northwest, with a fall of snow if it be winter.
Cited in Inwards, p. 174.

150. If you observe that the surface of the mercury in the barometer vibrates upon the approach of a storm, you may expect the gale to be severe.
Cited in Garriott, p. 16; Inwards, p. 174.

151. In fair weather, when the barometer falls much and low, and thus continues for two or three days before the rain comes, you may expect much rain, and probably high winds.
Cited in Inwards, p. 173.

152. In general the barometer falls before rain; and all appearances being the same, the higher the barometer, the greater the probability of fair weather.
Cited in Inwards, p. 175.

153. In summer, when the barometer falls suddenly, expect a thunderstorm; and if it does not rise again when the storm ceases, there will be several days unsettled weather.
Cited in Inwards, p. 174.

154. In wet weather, when the barometer rises much and high, and so continues for two or three days before wet weather is quite over, you may expect a continuance of fair weather for several days.
Cited in Inwards, p. 173.

155. In winter, a fall with a low barometer foretells snow.
Cited in Steinmetz 2, p. 260.

156. In winter heavy rain is indicated by a decrease of pressure (barometer) and an increase in temperature. Var.: During winter, heavy rain is indicated by a decrease of pressure, and an increase of temperature.
Cited in Garriott, p. 16; Inwards, p. 174.

157. In winter, the rising barometer indicates frost when the wind is east-north-east; and should the frost and increase of pressure continue, expect snow.
Cited in Inwards, p. 174.

158. Long foretold (falling barometer), long last; short notice, soon past.
Cited in Mitchell, p. 230; Steinmetz 1, p. 155; Cheales, p. 26; Northall, p. 464; Whitman, p. 33; Inwards, p. 175; Wilson, p. 479; Page, p. 6; Wilshere, p. 20; Simpson, p. 137; Freier, p. 43.

159. Men work better, eat more, and sleep sounder when the barometer is high.
Cited in Garriott, p. 19; Hand, p. 19.

160. Neither a sudden rise nor a sudden fall of the barometer is followed by any lasting change of weather. If the mercury rise and fall by turns, it is indicative of unsettled weather.
Cited in Inwards, p. 172.

161. Rapid changes in the barometer indicate early and marked changes in the weather.
Cited in Garriott, p. 15; Hand, p. 15.

162. Rapid fall (of the barometer) after high, sun at last, and very dry.
Cited in Inwards, p. 175.

163. Should the barometer continue low when the sky becomes clear, expect more rain within twenty-four hours. Var.: Should the barometer continue low when the sky becomes clear after heavy rain, expect more rain within twenty-four hours.
Cited in Garriott, p. 15; Inwards, p. 173; Hand, p. 15.

164. Steady high (barometer) after low, floods of rain, or hail, or snow.
Cited in Inwards, p. 175.

165. Sudden and great fluctuations of the barometer at any time of the year indicate unsettled weather for several days.
Cited in Inwards, p. 174.

166. The barometer falls lower for high winds than for heavy rain.
Cited in Garriott, p. 16; Inwards, p. 175.

167. The barometer is lowest of all during a thaw following a long frost. Vars.: (a) If it freezes, and the barometer falls two or three tenths of an inch, expect a thaw. (b) In frosty weather a falling barometer usually indicates a thaw.
Cited in Steinmetz 2, p. 260; Inwards, p. 173.

168. The barometer rises for northerly or easterly winds, and for drier, calmer, and colder weather.
Cited in Inwards, p. 173.

169. The barometer usually falls for southerly and westerly winds, and for damper, stormier, and warmer weather.
Cited in Inwards, p. 172.

170. The glass (barometer) is down, the gulls are flocked along the shore and the clouds are low'ring fast, soon the wind will roar.
Cited in Sloane 1, p. 121.

171. The variations of the barometer depend on the variations of the wind. It is highest during frost, with a northeast wind; and lowest during a thaw, with a south or southwest wind. Comment: A single barometer can never be a consistently reliable guide to coming weather. It is the relative variation of atmospheric pressure in different parts of a region that one needs to know.
Cited in Inwards, p. 172.

172. Very high and rising fast (barometer), steady rain and sure to last.
Cited in Inwards, p. 175.

173. When rise (barometer) begins after low, squalls expect and clear blow.
     Cited in Steinmetz 1, p. 155; Whitman, p. 74; Inwards, p. 175.

174. When the barometer falls considerably without any particular change of weather, you may be certain that a violent storm is raging at a distance.
     Cited in Garriott, p. 16; Inwards, p. 174.

175. When the barometer rises considerably, and the ground becomes dry, although the sky remains overcast, expect fair weather for a few days. The reverse may be expected if water is observed to stand in shallow places.
     Cited in Inwards, p. 174.

176. When the glass (barometer) falls low, prepare for a blow; when it has risen high, let all your kites fly. Vars.: (a) When the glass is low, look out for a blow; when it rises high, let your kites fly. (b) When the glass falls low, prepare for a blow. Comment: Sailors call the light sails used only in fine weather, "flying kites."
     Cited in Steinmetz 1, p. 155; Dunwoody, p. 83; Garriott, p. 16; Humphreys 2, p. 64; Whitman, p. 33; Inwards, p. 175; Lee, p. 23; Freier, p. 25; Whiting 2, p. 255. Recorded in Bryant (N.Y., Ohio).

177. When wet weather happens soon after the falling of the barometer, expect but little of it; and, on the contrary, expect but little fair weather when it proves fine shortly after the barometer has fallen.
     Cited in Inwards, p. 174.

bass

178. Lake Ontario black bass leave shoal water before a thunderstorm; this has been observed twenty-four hours before a storm. Var.: On Lake Ontario black bass leave shoal water before a thunderstorm.
     Cited in Dunwoody, p. 49; Inwards, p. 199.

bat

179. Bats flying late in the evening indicate fair weather. Vars.: (a) If bats fly abroad longer than usual after sunset, fair weather. (b)

When bats fly out after flies in the late evening, a fine day should follow.

Cited in Dunwoody, p. 34; Garriott, p. 23; Inwards, p. 185; Hand, p. 23; Page, p. 15; Freier, p. 51.

180. Bats flying low at dusk portend a fine day on the morrow.
Cited in Wilshere, p. 22.

181. Bats in flight, no rain tonight.
Recorded in Bryant (Oreg.).

182. Bats, or flying mice, coming out of their holes quickly after sunset, and sporting themselves in the open air, premonstrate fair and calm weather.
Cited in Swainson, p. 229; Dunwoody, p. 20.

183. Bats searching for a refuge are a portent of rain.
Cited in Hyatt, p. 26.

184. Bats who squeak flying tell of rain tomorrow. Vars.: (a) A loud and ceaseless squeaking among bats while they search for a refuge warns you of rain. (b) Bats who speak flying tell of rain tomorrow.
Cited in Dunwoody, p. 34; Garriott, p. 23; Hyatt, p. 26; Freier, p. 51.

185. If bats abound and are vivacious, fine weather may be expected.
Cited in Inwards, p. 185.

186. If bats are flitting high, wet weather soon follows; if low, dry weather.
Cited in Hyatt, p. 26.

187. If bats flutter, it will be fine tomorrow.
Cited in Dunwoody, p. 34.

188. It will rain if bats cry much or fly into the house. Var.: If bats cry a lot or fly into the house, it will rain.
Cited in Swainson, p. 229; Whitman, p. 49; Inwards, p. 185; Lee, p. 66.

189. When bats appear very early in the evening, expect fair weather; but when they utter plaintive cries, rain may be expected.
Cited in Inwards, p. 185.

bathe

190. He who bathes in May, will soon be laid in clay; he who bathes in June, will sing a merry tune; he who bathes in July, will dance

like a fly. Var.: They who bathe in May will soon be laid in clay; they who bathe in June will sing a another tune.
Cited in Denham, p. 45; Swainson, p. 92; Wright, p. 54; Wilson, p. 32.

bean

191.   Beans blow, before May doth grow. Vars.: (a) Be it weal or be it woe, beans must blow ere May doth go. (b) Be it weal or be it woe, beans blow before May doth go. Comment: Weal is well.
Cited in Denham, p. 48; Dunwoody, p. 64; Northall, p. 439.

192.   Go plant the bean when the moon is light, and you will find that this is right; plant the potatoes when the moon is dark, and to this line you always hark. Vars.: (a) Go plant the bean when the moon is light, and you will find that this is right; plant potatoes when the moon is dark, and to this line you always hark, but if you vary from this rule, you will find you are a fool; follow this rule to the end, and you will always have money to spend. (b) Plant your beans when the moon is light; plant potatoes when the moon is dark.
Cited in Dunwoody, p. 59; Garriott, p. 37; Whitman, p. 28; Hand, p. 37; Lee, p. 118; Freier, p. 80.

193.   If you plant garden beans when the sign is in the scales, they will hang full. Comment: The sign of the scales refers to Libra, the seventh sign of the zodiac entered by the sun at the autumnal equinox, about September 23rd.
Cited in Dunwoody, p. 59.

194.   Sow beans and peas on David and Chad, be the weather good or bad. Var.: David and Chad, sow peas good and bad. Comment: St. David's Day is March 1st and St. Chad's Day is March 2nd.
Cited in Northall, p. 450.

195.   Sow beans in the mud, and they'll grow like a wood.
Cited in Denham, p. 40; Swainson, p. 12; Dunwoody, p. 90.

196.   Sow peas and beans in the wane of the moon; who soweth them sooner, he soweth too soon. Var.: Set beans in the wane of the moon.
Cited in Denham, p. 42; Northall, p. 481; Whitman, p. 68; Wilson, p. 34; Freier, p. 83.

bear

197. Bears and coons are always restless before rain.
     Cited in Dunwoody, p. 29; Lee, p. 66.
198. If the tracks of bear are seen after the first fall of snow, an open mild winter may be expected.
     Cited in Dunwoody, p. 29; Garriott, p. 39; Hand, p. 39.
199. The bear comes out on the second of February, and if he sees his shadow, he returns for six weeks.
     Cited in Dunwoody, p. 29; Garriott, p. 39.
200. When bears lay up food in the fall, it indicates a cold winter. Vars.: (a) Bears in autumn provide against a cold winter by storing up more than the customary amount of food. (b) If bears lay up food in the fall, it indicates a cold winter.
     Cited in Dunwoody, p. 29; Garriott, p. 39; Hand, p. 39; Hyatt, p. 26; Freier, p. 78.

beast

201. If beasts eat greedily, if they lick their hooves, if they suddenly move here and there making a noise, breathing up to the air with open nostrils, rain follows.
     Cited in Inwards, p. 180.

beaver

202. Beavers in autumn build a large lodge for a cold winter or a small lodge for a mild winter.
     Cited in Hyatt, p. 26.
203. If beavers are busy building dwellings, a long, cold winter is ahead.
     Cited in Smith, p. 13.
204. In early and long winters, the beaver cuts his winter supply of wood and prepares his house one month earlier than in mild, late winters.
     Cited in Dunwoody, p. 29; Garriott, p. 39; Inwards, p. 184; Hand, p. 39; Freier, p. 81.

beaver dam

205. If a beaver dam is built high, it is supposed to be a sign of a rainy spring.
     Cited in Smith, p. 5.

206. The height above the water line of a beaver dam determines the amount of rainfall during the spring.
Cited in Boughton, p. 124.

bee

207. A bee was never caught in a shower.
Cited in Swainson, p. 254; Dunwoody, p. 56; Garriott, p. 24; Whitman, p. 48; Inwards, p. 204; Hand, p. 24; Sloane 2, p. 30; Mieder, p. 42.

208. A hive of bees in May is worth a load of hay.
Cited in Whiting 2, p. 309.

209. A swarm of bees in May is worth a load of hay; a swarm in June is worth a silver spoon; but a swarm in July is not worth a fly.
Vars.: (a) A swarm of bees in May is worth a load of hay. (b) A swarm of bees in May is worth a load of hay, but a swarm in July ain't worth a fly. (c) A swarm of bees in July is not worth a butterfly. (d) A swarm of bees in June is worth a silver spoon.
Cited in Denham, p. 45; Steinmetz 1, p. 137; Swainson, p. 254; Dunwoody, p. 97; Northall, p. 431; Sloane 2, p. 30; Wilson, p. 791; Page, p. 23; Wilshere, p. 10; Simpson, p. 218; Freier, p. 78. Recorded in Bryant (N.Y.).

210. After bees have buzzed about in March, preparations may be made for a cold spell.
Cited in Hyatt, p. 17.

211. As soon as rain impends, bees cease to leave their hives, either remaining in them all day, or else flying only to a short distance.
Vars.: (a) Bees stay in or near the hive before a rain and make journeys only when the weather will continue fair; so, more bees entering than leaving a hive betokens rain; and further, if they crowd into the house all at once, a bad storm will accompany rain. (b) If bees remain in the hive or fly but a short distance from it, expect rain. (c) If bees stay at home, rain will soon come; if they fly away, fine will be the day. (d) When bees stay close to the hive, rain is close.
Cited in Steinmetz 1, p. 112; Swainson, p. 253; Dunwoody, p. 55; Garriott, p. 19; Wright, p. 78; Inwards, p. 204; Hand, p. 19; Sloane 2, p. 30; Hyatt, p. 17; Lee, p. 52; Freier, p. 48.

212. Bees are restless before thundery weather.
Cited in Inwards, p. 204.

213. Bees never get caught in a rain.
Cited in Freier, p. 47.
214. Bees returning hastily and in large numbers are said to indicate approaching rain, although the weather may be clear.
Cited in Dunwoody, p. 56.
215. Bees will not swarm before a near storm. Vars.: (a) Bees will not swarm before a rain. (b) Bees will not swarm before a storm.
Cited in Dunwoody, p. 56; Garriott, p. 19; Inwards, p. 204; Hand, p. 19; Sloane 2, p. 30; Freier, p. 48.
216. If bees build a big hive, there will be a very long winter.
Cited in Smith, p. 12.
217. If bees build their hives high, it is a sign of a hard, cold winter.
Cited in Smith, p. 13.
218. If bees build their hives low, it is going to be a good winter.
Cited in Smith, p. 13.
219. If bees drone about as late as September, they are storing up additional honey against a long winter.
Cited in Hyatt, p. 17.
220. If you bury a bee alive it will rain within a day.
Cited in Smith, p. 5.
221. The bee doth love the sweetest flower, so doth the blossom the April shower.
Cited in Northall, p. 438.
222. The swarming of bees always occurs just before a storm.
Cited in Hyatt, p. 17.
223. When bees fly to the hive and none leave it, rain is near. Var.: When many bees enter the hive and none leave it, rain is near.
Cited in Mitchell, p. 228; Swainson, p. 253; Dunwoody, p. 14; Inwards, p. 204; Wurtele, p. 300.
224. When bees to distance wing their flight, days are warm and skies are bright; but when their flight ends near their home, stormy weather is sure to come. Var.: Bees to a distance wing their flight, when days are warm and skies are bright; but their flight ends near their home, when stormy weather is sure to come.
Cited in Dunwoody, p. 56; Garriott, p. 24; Whitman, p. 48; Inwards, p. 204; Alstad, p. 101; Hand, p. 24; Lee, p. 52; Freier, p. 48.
225. When charged with stormy matter lower the skies, the busy bee at home her labor plies.

Cited in Lee, p. 56.

226. When the bees crowd into their hive again, it is a sign of thunder and rain.
Cited in Page, p. 23; Wilshere, p. 22.

227. When the bees crowd out of their hive, the weather makes it good to be alive.
Cited in Page, p. 23; Wilshere, p. 22.

228. Whenever bees get about in February we are certain to get wind and rain next day.
Cited in Inwards, p. 204.

229. Yellow jackets (bees) building nests on top of the ground indicates an approaching dry season.
Cited in Dunwoody, p. 58.

beechnut

230. Dry beechnuts on November 1st portend a disagreeable winter.
Cited in Hyatt, p. 16.

231. When beech mast thrives well, and oak trees hang full, a hard winter will follow, with much snow. Comment: Beech mast means beechnuts.
Cited in Inwards, p. 211.

232. When the beechnuts are plenty, expect a mild winter. Var.: When beechnuts are plentiful, expect a mild winter.
Cited in Dunwoody, p. 64; Inwards, p. 210.

beetle

233. Beetles and crickets are troublesome and noisy before rain. Vars.: (a) Before rain, beetles and crickets are more troublesome than usual. (b) Before rain, beetles are more troublesome than ususal.
Cited in Steinmetz 1, p. 112; Swainson, p. 254; Inwards, p. 208.

234. If beetles fly about it means fine weather is coming.
Cited in Dunwoody, p. 34.

235. If you take a beetle and put him in a little box, he will lie on his back right before it rains.
Cited in Smith, p. 4.

236. Kill a beetle and it will rain.
Cited in Sloane 2, p. 42.

237. The clock-beetle, which flies about in the summer evenings in a circular direction, with a loud buzzing noise, is said to foretell a fine day. Var.: If the clock beetle flies circularly and buzzes, it is a sign of fine weather.
Cited in Swainson, p. 254; Inwards, p. 208; Sloane 2, p. 125.

berry
238. If there are a large number of berries on a holly tree the winter will be cold and severe. Var.: The number of berries on a holly bush foretell the winter.
Cited in Smith, p. 13.
239. When the bushes are full of berries, a hard winter is on the way Vars.: (a) Plenty of berries in autumn, mean a severe winter. (b) Plenty of berries indicates a severe winter. (c) When berries are plentiful in the hedge, on the May bush, and blackthorn, a hard winter may be expected.
Cited in Dunwoody, p. 64; Inwards, p. 209; Sloane 2, p. 131; Wilshere, p. 18.
240. When the number of red berries on a holly bush are few, the winter will be short and mild.
Cited in Smith, p. 13.

Big Dipper
241. If the Big Dipper is tilted so water may pour out, it will rain.
Cited in Smith, p. 6.
242. If the Big Dipper is upside down, there will be rain; if right-side up, no rain.
Cited in Hyatt, p. 4.
243. When the (Big) Dipper is tipped, there's going to be nice weather.
Recorded in Bryant (Wash.).

bird
244. A bird that flies back and forth in its cage is forecasting a storm.
Cited in Hyatt, p. 21.
245. A dry summer will follow when birds build their nests in exposed places.
Cited in Dunwoody, p. 34.

246. Birds and chickens flying low is a sign of rain. Vars.: (a) Birds and bats fly low before a rainy spell. (b) Birds flying low, expect rain and a blow.
Cited in Smith, p. 5; Freier, p. 21. Recorded in Bryant (S.C.).

247. Birds and fowl oiling feathers indicate rain. Vars.: (a) Birds always oil their feathers just before rain. (b) When birds oil their feathers, expect rain.
Cited in Dunwoody, p. 34; Garriott, p. 23; Hand, p. 23; Hyatt, p. 21; Freier, p. 52.

248. Birds eating a great amount of food in the morning mean rain.
Cited in Hyatt, p. 21.

249. Birds flying far to seaward foretell fair weather.
Recorded in Bryant (Mich.).

250. Birds flying in groups during rain or wind indicate hail.
Cited in Dunwoody, p. 34.

251. Birds in the lowland predict snow; birds in the highland expect fair weather.
Cited in Sloane 2, p. 130.

252. Birds on a telephone wire indicate the coming of rain. Var.: Birds sitting on a telephone line, expect rain.
Cited in Lee, p. 60; Freier, p. 53.

253. Birds return slowly to their nests before a storm.
Cited in Alstad, p. 72.

254. Birds singing during rain indicate fair weather. Var.: If birds sing during a rain, the weather will soon brighten.
Cited in Dunwoody, p. 34; Hyatt, p. 21.

255. Birds stop singing and trees are dark before a storm.
Cited in Freier, p. 57.

256. Caged birds singing in the morning before they are uncovered are a presage of a bright day.
Cited in Hyatt, p. 21.

257. Clear weather is foretold when birds venture far out over the water; stormy weather, when birds remain near the shore.
Cited in Hyatt, p. 21.

258. Generally birds and cocks pecking themselves is a sign of rain; and so when they imitate the sound of water as if it were raining.
Cited in Lee, p. 61.

259. If birds are flying together bad weather is coming.
Cited in Smith, p. 1.

260. If birds begin to whistle in the early morning in winter, it bodes frost.
Cited in Inwards, p. 186.

261. If birds caught in February are fat and sleek, it is a sign of more cold weather.
Cited in Dunwoody, p. 36.

262. If birds depart for the south in early September, the winter will be long and cold.
Cited in Hyatt, p. 21.

263. If birds in autumn grow tame, the winter will be too cold for game.
Cited in Dunwoody, p. 34.

264. If birds migrate early, we'll have a severe winter.
Cited in Freier, p. 79.

265. If birds pick their feathers, wash themselves, and fly to their nests, expect rain.
Cited in Dunwoody, p. 34.

266. If birds return slowly to their nests, rain will follow.
Cited in Inwards, p. 186.

267. If birds' wings droop, rain is coming.
Cited in Lee, p. 61.

268. If small birds seem to duck and wash in the sand, it is held to be a sign of coming rain.
Cited in Inwards, p. 186; Sloane 2, p. 129.

269. If the birds be silent, expect thunder. Var.: When birds cease to sing, rain and thunder will probably occur.
Cited in Dunwoody, p. 34; Garriott, p. 23; Inwards, p. 186; Hand, p. 23.

270. If the birds begin to sing in January, frosts are on the way. Var.: If birds begin to whistle in January, frosts to come.
Cited in Page, p. 43; Wilshere, p. 7.

271. Land birds are observed to bathe before rain. Var.: Land birds bathe before a rain.
Cited in Inwards, p. 187; Sloane 2, p. 129.

272. Mating among birds in August tells of a late winter.
Cited in Hyatt, p. 21.

273. Migrating birds fly to avoid a storm.
Cited in Freier, p. 54.

274. Migratory birds fly south from cold and north from warm weather.
Cited in Dunwoody, p. 38; Garriott, p. 23; Inwards, p. 186; Hand, p. 23.

275. The call of a spring bird late in winter is a token of colder weather.
Cited in Hyatt, p. 21.

276. The flight of birds in a southerly direction, no matter how short the distance, is a signal for falling weather.
Cited in Hyatt, p. 21.

277. When a big flock of birds are feeding in a field, there's a major weather change brewing.
Cited in Smith, p. 1.

278. When a severe cyclone is near, migratory birds become puzzled and fly in circles, dart in the air, and can be easily destroyed.
Var.: When a severe cyclone is coming, migrating birds become puzzled, fly in circles, dart, and can be easily decoyed.
Cited in Garriott, p. 23; Inwards, p. 186; Hand, p. 23; Lee, p. 61.

279. When birds and badgers are fat in October, a cold winter is expected.
Cited in Dunwoody, p. 92.

280. When birds huddle at the top of a chimney top, it is a sign of cold weather.
Cited in Freier, p. 53.

281. When birds of long flight hang about home expect a storm.
Cited in Garriott, p. 23; Hand, p. 23.

282. When birds of long flight, rooks, swallows, or others, hang around home and fly up and down, or low, rain or wind may be expected.
Cited in Dunwoody, p. 14; Inwards, p. 186.

283. When birds of passage, arrive soon from the north, it indicates the probability of an early and severe winter. Vars.: (a) If birds of passage arrive early from the north, expect frost. (b) When birds of passage arrive early in their southern passage, severe weather may be looked for soon. (c) When the feldfare, redwing, starling, swan, snowfleck, and other birds of passage, arrive soon from the north, it indicates the probability of an early and severe winter.

Cited in Mitchell, p. 227; Dunwoody, p. 34; Garriott, p. 40; Inwards, p. 187; Hand, p. 40.

284. When birds stop singing, a storm is on the way.
Cited in Freier, p. 54.

285. When numerous birds their island home forsake, and to firm land their airy voyage make, the ploughman, watching their ill-omened flight, fears for his golden fields a withering blight.
Cited in Lee, p. 61.

286. When summer birds take their flight, the summer goes with them.
Cited in Dunwoody, p. 40; Garriott, p. 40; Inwards, p. 187; Hand, p. 40.

287. When you see a multitude of small birds dusting themselves, they are preparing for a storm within three days.
Cited in Hyatt, p. 21.

blackberry

288. Blackberries that ripen late are an indication of a hard winter.
Cited in Hyatt, p. 14.

blackberry frost

289. They say blackberry frost when blackberries are first in full bloom, because there will be either no more frost or not enough frost to kill.
Cited in Hyatt, p. 34.

blackberry winter

290. The name blackberry winter is given to cold weather in May, because it makes a good blackberry crop.
Cited in Hyatt, p. 34.

blackbird

291. As soon as blackbirds gather in a cornfield, you may make ready for winter.
Cited in Hyatt, p. 21.

292. Before a snow you will always see a large flock of blackbirds on the ground.
Cited in Hyatt, p. 21.

293. Blackbird's notes are very shrill in advance of rain.
Cited in Dunwoody, p. 34.

294. Blackbirds bring healthy weather.
Cited in Dunwoody, p. 34.

295. Blackbirds flocking in the fall indicate a spell of cold weather.
Cited in Dunwoody, p. 34.

296. Blackbirds flocking together always announce a change of weather: in summer, a rain; in winter, a snow.
Cited in Hyatt, p. 21.

297. Blackbirds flying south in autumn indicate an approaching cold winter.
Cited in Dunwoody, p. 34.

298. Blackbirds singing from the treetops promise a fine day, but if they change to the bottom branches, beware. Var.: Blackbirds singing from treetops signify a fine day.
Cited in Page, p. 18; Wilshere, p. 22.

299. If a blackbird sings with its tail straight down, that is a warning, for it is waiting to shoot the water off.
Cited in Page, p. 18.

300. Spring is ushered in by the first blackbird.
Cited in Hyatt, p. 21.

301. When the blackbird sings before Christmas, she will cry before Candlemas. Comment: Candlemas Day is February 2nd.
Cited in Inwards, p. 68.

302. When the voices of blackbirds are unusually shrill, or when blackbirds sing much in the morning, rain will follow. Var.: The blackbird's call is more shrill before a rain.
Cited in Garriott, p. 24; Inwards, p. 191; Hand, p. 24; Freier, p. 52.

blackfish

303. Blackfish in schools indicate an approaching gale.
Cited in Inwards, p. 199; Hand, p. 24.

blackthorn

304. The later the blackthorn in bloom after May 1st, the better the rye and hay harvest.
Cited in Dunwoody, p. 96; Inwards, p. 55.

305. When the blackthorn's out, the cold will persist.
Cited in Wilshere, p. 17.

blossom
306. Early blossoms indicate a bad fruit year. Var.: Early blossom bad fruit.
Cited in Dunwoody, p. 65; Inwards, p. 209.
307. Late blossoms indicate a a good fruit year.
Cited in Dunwoody, p. 66.

blue jay
308. Blue jays holler for cold and snow.
Recorded in Bryant (Mich.).
309. Blue jays just before a storm become excited and cry repeatedly.
Cited in Hyatt, p. 21.
310. When blue jays call, rain is coming.
Cited in Lee, p. 64.

bluebird
311. A bluebird near your house in the morning brings a rain before night.
Cited in Hyatt, p. 21.
312. If you see a bluebird, it denotes good weather next day.
Cited in Hyatt, p. 21.
313. On hearing the first bluebird of the season, expect a rain soon.
Cited in Hyatt, p. 21.
314. One harbinger of spring is the first appearance of a bluebird.
Cited in Hyatt, p. 21.
315. When bluebirds twitter and sing, they call to each other of rain.
Var.: Bluebirds chatter when it is going to rain.
Cited in Dunwoody, p. 35; Lee, p. 65.

bog
316. Bogs draw rain as the desert draws the sun.
Cited in Alstad, p. 36.

bone
317. Broken bones ache before a rain. Var.: When a person who has recently broken a bone has pain, rain is near.

Cited in Hyatt, p. 30; Smith, p. 7.

boot

318. Boots and shoes easy to pull on and off indicate dry weather.
Cited in Dunwoody, p. 106.

bramble

319. When the bramble blossoms early in June, an early harvest is expected.
Cited in Mitchell, p. 226; Swainson, p. 257; Inwards, p. 212; Sloane 2, p. 131.

branch

320. Dead branches falling in calm weather indicate rain. Var.: Dead branches dropping from trees in fair weather are a rain warning.
Cited in Dunwoody, p. 82; Inwards, p. 209; Sloane 2, p. 112; Hyatt, p. 16; Freier, p. 41.

bread

321. A person dropping a piece of buttered bread that falls upside down on the floor is a sign of rain.
Cited in Hyatt, p. 31.
322. If in handling a loaf you break it in two parts, it will rain all the week.
Cited in Inwards, p. 223.
323. To take the last piece of bread on the plate is an omen of rain.
Cited in Hyatt, p. 31.

breeze

324. On board ship, whistle for a breeze.
Cited in Hand, p. 459.

bride

325. Happy is the bride the sun shines on. Vars.: (a) Blessed be the bride the sun shines on; cursed be the bride the rain falls on. (b) Blessed is the bride that the sun shines on. (c) Blessed is the bride that the sun shines on; blessed is the corpse that the rain rains on. (d) Blessed is the bride the sun shines on; blessed is the corpse the rain falls on. (e) Happy is the bride on a sunny day;

happy is the corpse on a rainy day. (f) Happy is the bride that the sun shines on. (g) Happy is the bride that the sun shines on; blessed are the dead that the rain falls on. (h) Happy is the bride the sun shines on; blessed is the corpse the rain falls on. (i) Happy is the bride the sun shines on; sorry is the bride the rain rains on.
Cited in Denham, p. 14; Inwards, p. 223; Wilson, p. 85; Simpson, p. 26; Mieder, p. 70.

brier

326. The sensitive brier closes up its leaves on the approach of rain.
Cited in Dunwoody, p. 67.

broom

327. If the broom be full of flower, it signifieth plenty. Var.: The broom having plenty of blossoms is a sign of a fruitful year of corn.
Cited in Inwards, p. 212.

brush

328. It will rain after you burn brush.
Cited in Hyatt, p. 32.

bubble

329. Bubbles in the water, rain tomorrow.
Recorded in Bryant (Miss.).
330. Bubbles over calm beds of water mean rain is coming.
Cited in Freier, p. 23.
331. Bubbles rising from marshy ground or from stagnant water in an old pond are a warning of rain.
Cited in Hyatt, p. 13.
332. If coffee bubbles cling to the side of the cup instead of floating on the center, rain is at hand.
Cited in Hyatt, p. 31.
333. If there's bubbles in the puddles it will rain all night.
Recorded in Bryant (Wis.).
334. If you see bubbles on top of puddles or ponds, it's going to rain the next day.
Cited in Smith, p. 8.

335. When bubbles are rising on the surface of coffee and they hold together, good weather is coming; if the bubbles break up, weather you don't need is coming.
Cited in Freier, p. 24.

336. When the bubbles of coffee collect in the center of the cup, expect fair weather. When they adhere to the cup, forming a ring, expect rain. If the bubbles separate without assuming any fixed position, expect changeable weather.
Cited in Dunwoody, p. 107; Inwards, p. 223; Lee, p. 74.

337. When water bubbles from the ground, expect rain on the following day.
Cited in Dunwoody, p. 119; Inwards, p. 219.

bucket

338. A rain is scared away or stopped by turning upside down all buckets and similar receptacles in the yard.
Cited in Hyatt, p. 32.

bud

339. Look for a heavy winter when the buds have heavy coats.
Cited in Sloane 2, p. 33.

bug

340. Fall bugs begin to chirp six weeks before a frost in the fall.
Cited in Inwards, p. 207; Sloane 2, p. 125; Freier, p. 82.

bull

341. If bulls are irritable, expect a storm.
Cited in Sloane 2, p. 121.

342. If bulls lick their hoofs or kick about, expect much rain.
Cited in Inwards, p. 181.

343. If the bull leads the cows going to pasture, expect rain.
Cited in Lee, p. 68.

344. If the bull leads the van in going to pasture, rain must be expected; but if he is careless and allows the cows to precede him, the weather will be uncertain.
Cited in Swainson, p. 229; Dunwoody, p. 29; Inwards, p. 181.

bumblebee

345. The first bumblebee humming about your door will tell you cold weather has gone and warm weather has arrived.
Cited in Hyatt, p. 17.

bunion

346. When a bunion starts to hurt in winter a blizzard is coming.
Cited in Smith, p. 10.

burnet

347. The burnet saxifrage indicates by half opening its flowers that the rain is soon to cease.
Cited in Inwards, p. 216.

butterfly

348. An early hatching among butterflies in the spring is followed by excellent weather.
Cited in Hyatt, p. 17.

349. Autumnal butterflies proclaim immediate cold weather.
Cited in Hyatt, p. 17.

350. November butterflies are an indication of an open winter.
Cited in Hyatt, p. 17.

351. The early appearance of butterflies is said to indicate fine weather. Var.: The early appearance of butterflies indicates fine weather.
Cited in Dunwoody, p. 55; Inwards, p. 207; Freier, p. 79.

352. When the white butterfly flies from the southwest, expect rain.
Cited in Dunwoody, p. 55; Inwards, p. 207.

353. Yellow butterflies during autumn presage a frost within ten days that will tint the leaves with the same color.
Cited in Hyatt, p. 17.

buzzard

354. A buzzard in flight is always a sign of rain.
Cited in Hyatt, p. 21.

355. As a herald of spring, wait for the first buzzard.
Cited in Hyatt, p. 21.

356. Buzzards flying high indicate fair weather.
Cited in Dunwoody, p. 34.

357. If a turkey buzzard is sailing through the air, the weather will turn warmer.
Cited in Hyatt, p. 21.

C

calf

358. Calves romping about in a playful mood mean a change of weather.
Cited in Hyatt, p. 27.

campfire

359. Campfires are more smoky before a rain.
Cited in Freier, p. 40.
360. Campfires burn brighter in fair weather.
Cited in Freier, p. 64.
361. Campfires sputter and spit before a rain.
Cited in Freier, p. 64.

camphor

362. Prepare for stormy weather after a camphor bottle grows cloudy or the camphor rises in a bottle.
Cited in Hyatt, p. 30.

camphor gum

363. Camphor gum will rise in alcohol before rain.
Cited in Lee, p. 75.

candle

364. Candles burn dimmer before a rain.
Cited in Freier, p. 64.
365. When the flames of candles flare and snap or burn with an unsteady or dim light, rain and frequently wind also are found to follow.
Cited in Inwards, p. 221.

Candlemas Day

366. A clear, bright Candlemas Day means a late spring. Comment: Candlemas Day is February 2nd.
Cited in Lee, p. 148.
367. After Candlemas Day the frost will be more keen, if the sun then shines bright, than before it has been.
Cited in Northall, p. 446; Inwards, p. 43.
368. As long as the bird sings before Candlemas, it will greet after it.
Cited in Denham, p. 27; Inwards, p. 43.

369. At Candlemas cold comes to us.
     Cited in Swainson, p. 46; Northall, p. 446.
370. At Candlemas Day another winter is on its way.
     Cited in Garriott, p. 42; Hand, p. 42.
371. At Candlemas Day it is time to sow beans in the clay.
     Cited in Northall, p. 447.
372. At the day of Candlemas, cold in air and snow on grass; if the
     sun then entice the bear from his den, he turns around thrice and
     goes back again.
     Cited in Freier, p. 75.
373. Candlemas shined, and the winter's behind.
     Cited in Wright, p. 18.
374. If Candlemas Day be dry and fair, the half o' winter's to come,
     and mair; if Candlemas Day be wet and foul, the half o' winter's
     gane at Yule. Var.: If Candlemas Day be wet and foul, the half
     of the winter was gone at Yule.
     Cited in Steinmetz 2, p. 303; Dunwoody, p. 102; Inwards, p. 43;
     Freier, p. 75.
375. If Candlemas Day be fair and bright, winter will have another
     flight. Vars.: (a) If Candlemas Day be fair and bright the winter
     will take another flight; but if it should be dark and drear then
     winter is gone for another year. (b) If Candlemas Day be fair and
     bright, winter will have another flight; but if Candlemas Day
     bring clouds and rain, winter is gone and won't come again. (c)
     If Candlemas Day be fair and bright, winter will have another
     flight; if on Candlemas Day, it be shower and rain, winter is
     gone, and will not come again. (d) If Candlemas Day be sunny
     and bright, winter will have another flight; if Candlemas Day be
     cloudy with rain, winter is gone and won't come again.
     Cited in Steinmetz 2, p. 211; Swainson, p. 43; Cheales, p. 20;
     Dunwoody, p. 102; Northall, p. 445; Garriott, p. 42; Wright, p.
     18; Whitman, p. 37; Inwards, p. 42; Hand, p. 42; Wilson, p.
     100; Page, p. 36; Wilshere, p. 8; Simpson, p. 30; Freier, p. 74.
376. If Candlemas Day be fair and clear, there'll be two winters in the
     year. Var.: If Candlemas Day be fair and bright, we'll have two
     winters in one year.
     Cited in Northall, p. 445; Whitman, p. 37; Inwards, p. 42;
     Sloane 2, p. 33; Smith, p. 14; Wilshere, p. 8. Recorded in
     Bryant (Pa., Vt., Wis.).

377. If Candlemas Day be fine and clear, corn and fruits will then be dear.
   Cited in Northall, p. 446; Garriott, p. 42; Whitman, p. 37; Inwards, p. 42; Hand, p. 42; Wurtele, p. 299.
378. If Candlemas Day be fine and clear, we shall have winter half the year.
   Cited in Lee, p. 139.
379. If Candlemas Day be fine, it portends a hard season to come; if Candlemas Day be cloudy and lowering, a mild and gentle season.
   Cited in Denham, p. 27.
380. If Candlemas Day be mild and gay, go saddle your horses, and buy them hay; but if Candlemas Day be stormy and black it carries the winter away on its back.
   Cited in Inwards, p. 42; Lee, p. 139; Freier, p. 75.
381. If it neither rains nor snows on Candlemas Day, you may straddle your horse and go and buy hay.
   Cited in Northall, p. 447; Brunt, p. 68; Inwards, p. 42; Freier, p. 74.
382. If on Candlemas Day it is bright and clear, the groundhog will stay in its den, thus indicating more snow and cold are to come; but if it snows or rain he will creep out, as the winter has ended.
   Cited in Dunwoody, p. 31; Hand, p. 39.
383. If the lanes are full of snow on Candlemas, so the bins will be full of corn in autumn.
   Cited in Freier, p. 76.
384. If the laverock sings afore Candlemas, she'll mourn as long after it.
   Cited in Inwards, p. 43.
385. If the sun shines bright on Candlemas Day, the half of the winter's not yet away.
   Cited in Northall, p. 445.
386. Just half your wood and half your hay should be remaining on Candlemas Day. Vars.: (a) Candlemas Day, Candlemas Day, half our fire and half our hay. (b) Half the wood and half the hay, you should have on Candlemas Day. Comment: Half through the winter only, and so half our provisions should be left.
   Cited in Dunwoody, p. 102; Northall, p. 448; Garriott, p. 41; Whitman, p. 37; Hand, p. 41; Sloane 2, p. 33; Freier, p. 74.

387. More snow and ice if the sun shines on Candlemas Day.
Cited in Freier, p. 90.

388. On Candlemas Day if the thorns hang a drop (with icicles), you are sure of a good pea crop.
Cited in Swainson, p. 48; Northall, p. 446; Whitman, p. 18.

389. On Candlemas Day just so far as the sun shines in, just so far will the snow blow in. Vars.: (a) As far as the sun shines in at the window on Candlemas Day, so deep will be the snow ere winter is gone. (b) As far as the sun shines into the cottage on Candlemas Day, so far will the snow blow in afore old May. (c) So far as the sun shines on Candlemas Day, so far the snow will blow in before the first of May.
Cited in Dunwoody, p. 102; Northall, p. 445; Inwards, p. 42.

390. On Candlemas Day the bear, badger, or woodchuck comes out to see his shadow at noon; if he does not see it he remains out; but if he does see it he goes back to his hole for six weeks, and cold weather continues six week longer.
Cited in Dunwoody, p. 1883; Garriott, p. 42; Inwards, p. 43; Hand, p. 42; Sloane 2, p. 39; Freier, p. 75.

391. On Candlemas Day, you must have half your straw and half your hay. Vars.: (a) Have on Candlemas Day one half your straw and one half your hay. (b) The farmer should have on Candlemas Day half his straw and half his hay.
Cited in Denham, p. 30; Swainson, p. 48; Dunwoody, p. 102; Northall, p. 447; Garriott, p. 41; Whitman, p. 37; Hand, p. 41; Sloane 2, p. 33; Wilson, p. 101; Freier, p. 74.

392. On the eve of Candlemas Day, the winter gets stronger or passes away.
Cited in Freier, p. 74.

393. Snow at Candlemas stops to handle us.
Cited in Northall, p. 446; Inwards, p. 41; Wilshere, p. 8.

394. Sow or set beans on Candlemas waddle. Var.: Sow beans in Candlemas waddle. Comment: Waddle means in the wane of the moon.
Cited in Denham, p. 27; Swainson, p. 48; Wilson, p. 756.

395. The badger peeps out of his hole on Candlemas Day, and when he finds it's snowing, walks abroad; but if he sees the sun shining, he draws back into the hole.
Cited in Freier, p. 75.

396. The shepherd would rather see the wolf enter his fold on Candlemas Day than the sun.
Cited in Garriott, p. 42; Hand, p. 42.

397. When Candlemas Day is come and gone, the coal lies on a red hot stove.
Cited in Dunwoody, p. 103.

398. When Candlemas Day is come and gone, the snow lies on a hot stone. Var.: Candlemas Day is come and gone when the snow lies on a hot stone.
Cited in Northall, p. 446; Inwards, p. 43; Wilson, p. 100.

399. When it rains on Candlemas Day, the cold is over.
Cited in Whitman, p. 37; Freier, p. 76.

400. When the wind's in the east on Candlemas Day, there it will stick to the second of May.
Cited in Swainson, p. 48; Northall, p. 447; Whitman, p. 18; Brunt, p. 68; Inwards, p. 43; Wilson, p. 893; Wilshere, p. 19.

401. Where the wind is on Candlemas Day, there it will stick to the end of May.
Cited in Cheales, p. 20; Dunwoody, p. 83.

car

402. As soon as you have finished washing and polishing your car a rain will arrive. Var.: Have your car washed and it will rain that same day.
Cited in Hyatt, p. 32; Smith, p. 7.

cardinal

403. A cardinal perched in a tree is a sign of rain.
Cited in Smith, p. 5.

404. Cardinals are a sign of snow.
Cited in Smith, p. 10.

405. If you see a red cardinal in the winter, look for a change in the weather.
Cited in Smith, p. 1.

carpet

406. If in walking about the house you accidentally kick up a carpet or a rug several times, a rain is foretold.
Cited in Hyatt, p. 32.

407. The carpet on the floor has a dampish feeling before a rain.
Cited in Hyatt, p. 30.

cat

408. A cat and dog getting along well together is a storm warning.
Cited in Hyatt, p. 27.
409. A cat lying on its back is rain omen.
Cited in Hyatt, p. 26.
410. A cat sleeping on its head is a sign of a storm.
Recorded in Bryant (Nebr.).
411. A cat that sleeps with its head low is presaging a rain.
Cited in Hyatt, p. 26.
412. A cat washing its face is a sign of good weather.
Recorded in Bryant (Nebr.).
413. An old cat frisking about like a kitten foretells a storm. Var.: If
an old cat runs and plays, it is a sign of bad weather.
Cited in Lee, p. 70. Recorded in Bryant (Miss.).
414. Cats are observed to scratch the wall or a post before wind, and
to wash their faces before a thaw. Vars.: (a) Cats are observed
to scratch the wall before wind, and to wash their faces before a
thaw; they sit with their backs to the fire before snow. (b) Cats
scratch a post before wind, wash their faces before rain, and sit
with backs to the fire before snow. (c) Cats scratch a wall or a
post before wind.
Cited in Mitchell, p. 228; Swainson, p. 230; Marvin, p. 218;
Inwards, p. 179; Freier, p. 30.
415. Cats clean table legs, tree trunks, etc., before storms.
Cited in Dunwoody, p. 30.
416. Cats sit with their backs to the fire before snow.
Cited in Mitchell, p. 228; Marvin, p. 218.
417. Cats wash their faces before a thaw.
Cited in Marvin, p. 218.
418. Cats with their tails up and hair apparently electrified indicate
approaching wind. Var.: As soon as you see the hair of a cat
bristling without cause, take precautions for a bad windstorm.
Cited in Dunwoody, p. 29; Inwards, p. 179; Hyatt, p. 26;
Freier, p. 30.
419. Expect bad weather when you see cats playing with their tails.
Cited in Smith, p. 1.

420. High winds are indicated, after a cat becoming frisky dashes about wildly or climbs trees.
     Cited in Hyatt, p. 26.
421. If a black cat licks her fur the wrong way, it will rain.
     Cited in Wurtele, p. 300.
422. If a cat chews grass, a rain approaches; the earlier in the day the chewing, the sooner the rain. Vars.: (a) If a cat eats grass, it is a sign of rain. (b) If the cat eats grass, it will rain that same day. (c) Cats and dogs eat grass before rain.
     Cited in Hyatt, p. 26; Smith, p. 5; Freier, p. 67. Recorded in Bryant (Nebr.).
423. If a cat is sitting in the sun and licking the bottom of a front paw, there will be rain before dark.
     Cited in Hyatt, p. 26.
424. If a cat licks against the fur instead of with the fur, it betokens bad weather. Var.: Good weather may be expected when a cat washes itself, but when it licks its coat against the grain, expect bad weather.
     Cited in Hyatt, p. 26; Lee, p. 70.
425. If a cat lies on its head it will rain.
     Cited in Smith, p. 5.
426. If a cat scratches a broom it will rain.
     Cited in Smith, p. 5.
427. If a cat scratches itself on the fence, expect rain before night.
     Cited in Lee, p. 70.
428. If a cat scrubs its bottom along the floor, it will soon storm.
     Cited in Hyatt, p. 26.
429. If a cat sits with its back or tail to the fire, colder weather is signifed. Var.: If a cat sits with its back to the fire, cold weather may be expected.
     Cited in Hyatt, p. 26. Recorded in Bryant (Nebr.).
430. If a cat sleeps on its back by the stove, bad weather is coming.
     Cited in Smith, p. 1.
431. If a cat washes its entire face, look for rain.
     Cited in Hyatt, p. 26.
432. If a cat washes its face, rain will come from the direction towards which the paw moves.
     Cited in Hyatt, p. 26.

433. If a cat washes itself, fair weather either will appear soon or continue.
Cited in Hyatt, p. 26.

434. If a cat washes itself in snow, that snow will vanish within twenty-four hours.
Cited in Hyatt, p. 26.

435. If a cat while washing itself licks upwards, clear weather is athand; if downwards, rainy weather.
Cited in Hyatt, p. 26.

436. If a sleeping cat lies in front of a fire and its nose turned upward, the weather will become colder.
Cited in Hyatt, p. 26.

437. If cats lick themselves, fair weather.
Cited in Freier, p. 30.

438. If near a river you observe a cat moving her kittens to higher ground, it denotes high water.
Cited in Hyatt, p. 27.

439. If sparks are seen when stroking a cat's back, expect a change of weather soon. Vars.: (a) If you see sparks when stroking a cat's back, the weather will change soon. (b) When sparks are seen on stroking a cat's back, expect a change of weather.
Cited in Dunwoody, p. 30; Marvin, p. 218; Lee, p. 70.

440. If the cat runs about the house and plays, storms may be expected.
Recorded in Bryant (N.Y.).

441. If the cat washes her face o'er the ear, 'tis a sign the weather 'ill be fine and clear.
Cited in Marvin, p. 218; Whitman, p. 47; Inwards, p. 179; Lee, p. 70. Recorded in Bryant (Wis.).

442. If the cat's fur looks glossy, it will be pleasant the next day.
Cited in Lee, p. 70.

443. It is a sign of rain if the cat washes her head behind her ear. Vars.: (a) If a cat washes about the ears, or back of the ears, or above the ears, rain may be predicted. (b) When a cat washes its face around its ear, expect rain.
Cited in Dunwoody, p. 29; Hyatt, p. 26; Smith, p. 4.

444. It is almost universally believed that good weather may be expected when the cat washes herself, but bad when she licks her

coat against the grain, or washes her face over her ears, or sits
with her tail to the fire.
Cited in Hand, p. 22.

445. Putting a cat under a pot brings bad weather.
Cited in Marvin, p. 218.

446. Rain always falls soon after a cat gets into the house.
Cited in Hyatt, p. 27.

447. Ship's cat unusually frisky, she has a gale of wind in her tail.
Vars.: (a) Sailors dislike to see a cat on board ship unusually
playful or quarrelsome, and they say the cat has a gale of wind
in her tail. (b) Sailor's say that "the cat has a gale of wind in her
tail."
Cited in Whitman, p. 47; Inwards, p. 179; Lee, p. 69.

448. The cardinal point to which a cat turns and washes her face afte
a rain shows the direction from which the wind will blow.
Cited in Dunwoody, p. 29; Marvin, p. 218; Inwards, p. 179.

449. The rolling of a cat outdoors in the sun is a sign of rain.
Cited in Hyatt, p. 26.

450. When a cat scratches itself, or scratches on a log or tree, it
indicates approaching rain. Var.: When a cat scratches itself, or
scratches on a log or tree, it indicates rain.
Cited in Dunwoody, p. 30; Marvin, p. 218.

451. When a cat scratches the table legs a change in the weather is
coming.
Cited in Marvin, p. 217; Inwards, p. 179.

452. When a cat sleeps on its back, there is going to be falling
weather.
Recorded in Bryant (Ohio).

453. When a cat washes her face with her back to the fire, expect a
thaw in winter.
Cited in Dunwoody, p. 30; Marvin, p. 218.

454. When cats are snoring, foul weather follows.
Cited in Dunwoody, p. 29.

455. When cats are washing themselves, fair weather follows.
Cited in Dunwoody, p. 29.

456. When cats hide under the bed, there will be a storm.
Cited in Sloane 2, p. 121.

457. When cats lie on their head with mouth turned up, expect a
storm.

Cited in Dunwoody, p. 30.

458. When cats place their paws over their ears, it is a sign of rain.
Cited in Sloane 2, p. 121.

459. When cats sit by the fire more than usual or lick their feet, it is a sign of rain.
Cited in Lee, p. 70.

460. When cats sleep on their backs, bad weather is ahead.
Cited in Smith, p. 1.

461. When cats sneeze it is a sign of rain. Vars.: (a) It is an omen of rain, when a cat sneezes while its head rests on the floor. (b) To have a cat sneeze and then wipe behind its ears means rain. (c) When a cat sneezes, it is a sign of rain.
Cited in Swainson, p. 230; Dunwoody, p. 29; Marvin, p. 217; Inwards, p. 179; Hyatt, p. 26; Lee, p. 70.

462. When cats wipe their jaws with their feet it is a sign of rain. Var.: When cats wipe their jaws with their feet, it is a sign of rain, and especially when they put their paws over their ears in wiping.
Cited in Swainson, p. 230; Marvin, p. 218; Inwards, p. 179.

463. When the cat in February lies in the sun, she will again creep behind the stove in March. Var.: A cat basking in a February sun will hug the stove in March.
Cited in Dunwoody, p. 94; Marvin, p. 218; Hyatt, p. 26.

464. When the cat lies on its brain, it is going to rain.
Cited in Marvin, p. 217; Inwards, p. 179; Sloane 2, p. 121; Lee, p. 70.

465. When the cat sits with its tail toward the fire, expect bad weather.
Cited in Lee, p. 70.

466. While rain depends, the pensive cat gives o'er her frolics, and pursues her tail no more.
Cited in Inwards, p. 179.

caterpillar

467. A large number of caterpillars in autumn signifies either a cold or a warm winter.
Cited in Hyatt, p. 17.

468. During the autumn you are warned of a severe winter by black caterpillars and of a mild winter by yellow caterpillars. Vars.: (a)

Dark-colored caterpillars in the autumn mean a harsh winter and light-colored ones a light winter. (b) More black caterpillars in the fall means a hard winter coming.
Cited in Hyatt, p. 18; Smith, p. 12.

469. If a caterpillar comes to your door in August or September and tries to enter, winter will be cold; if the caterpillar merely crawls about your door and does not try to enter, winterwill be mild.
Cited in Hyatt, p. 17.

470. If a caterpillar is dark on both ends, it will be a mild winter.
Cited in Smith, p. 12.

471. If a yellow stripe runs down the back of the autumnal caterpillar, expect cold weather for the middle of winter.
Cited in Hyatt, p. 18.

472. If caterpillars during autumn are dark-brown in the central part of the body and yellow at each end, all of the cold weather will come in the middle of winter.
Cited in Hyatt, p. 18.

473. If in autumn the front half of a caterpillar is large and the back half small, the first half of winter will be colder than the second half; conversely, if the front half of a caterpillar is small and the back half large, the second half of winter will be colder than the first half.
Cited in Hyatt, p. 17.

474. If the autumnal caterpillar is of one color, an open winter may be forecasted.
Cited in Hyatt, p. 18.

475. If the head of the autumnal caterpillar is black, the early part of winter will be cold; if the center of the body is light-colored, the middle of the winter will be light; and if the tail is black, the end of winter will be cold.
Cited in Hyatt, p. 18.

476. If there is a speck of yellow on the nose of the autumnal caterpillar, the earlier part of winter will be cold; if on the tail, the latter part of winter.
Cited in Hyatt, p. 18.

477. Late autumn caterpillers are an indication of a very mild winter.
Cited in Hyatt, p. 17.

478. The amount of brown on the woolly caterpillar foretells the severity of the coming winter.

Cited in Sloane 2, p. 63.

479. The placement of the dark spots on the caterpillar will foretell which period the heavy weather will come. Black on the front part means bad weather in the beginning of the season, in the rear part it means the last of the season.
Cited in Smith, p. 12.

480. The wider the band on the woolly caterpillar, the milder the winter.
Cited in Freier, p. 82.

catfish

481. Catfish jump out of water before rain.
Cited in Dunwoody, p. 49.

482. If the skin on the belly of the catfish is unusually thick, it indicates a cold winter; if not, a mild winter will follow.
Cited in Dunwoody, p. 49; Inwards, p. 199.

catgut

483. Strings of catgut or whipcord untwist and become longer during a dry state of the air, and vice versa.
Cited in Inwards, p. 218.

cattle

484. A cold winter is revealed by cattle staying close together in the autumn.
Cited in Hyatt, p. 27.

485. Cattle go to the hills before rain.
Cited in Inwards, p. 180.

486. Cattle grouping under trees in the summer means a thunderstorm is coming.
Cited in Smith, p. 11.

487. Cattle with their tails to the northwest makes weather the best.
Cited in Alstad, p. 173.

488. Considerable lowing among cattle forecasts rain.
Cited in Hyatt, p. 27.

489. Expect rain when cattle low and gaze at the sky. Var.: When cattle low and gaze at the sky, expect rain.
Cited in Dunwoody, p. 30; Lee, p. 68.

490. Falling weather is imminent when cattle become capricious and fight each other.
     Cited in Hyatt, p. 27.
491. If cattle continue to feed peacefully, fine weather to continue.
     Cited in Alstad, p. 176.
492. If cattle or horses run and play in the evening, it will rain soon.
     Recorded in Bryant (Nebr.).
493. If cattle or sheep crowd together, rain will follow.
     Cited in Swainson, p. 231.
494. If cattle run around and collect together in the meadows, expect thunder.
     Cited in Dunwoody, p. 79.
495. If cattle turn up their nostrils and sniff the air, or if they lick their forefeet, or lie on their right side, it will rain. Vars.: (a) Cattle are said to foreshow rain when they lick their forefeet, or lie on the right side, or scratch themselves more than they usually do against posts or other objects. (b) When cattle lick their forefeet or scratch themselves more than usual against objects, it will rain.
     Cited in Swainson, p. 231; Dunwoody, p. 31; Lee, p. 68.
496. When a storm threatens, if cattle go under trees it will be a shower; if they continue to feed, it will probably be a continuous rain. Var.: When a storm is coming cattle will go under trees if it is to be a shower, but will continue to graze if it will be a long rain.
     Cited in Dunwoody, p. 30; Lee, p. 68.
497. When cattle are on hilltops, fine weather is coming.
     Cited in Lee, p. 68.
498. When cattle collect near the barn long before night and remain near the barn till late in the morning, expect a severe winter.
     Cited in Dunwoody, p. 30.
499. When cattle go out to pasture and lie down early in the day, it indicates early rain. Vars.: (a) When cattle lie around in the midday sun, it will rain. (b) When cattle lie down as they are put to pasture, rain is on its way.
     Cited in Dunwoody, p. 31; Hand, p. 22; Sloane 2, p. 33; Reeder, no pp.
500. When cattle graze in tight patterns, it will rain soon.
     Cited in Reeder, no pp.

501. When cattle lie down during light rain, it will soon pass.
Cited in Inwards, p. 180; Lee, p. 68.
502. When cattle stand with their backs to the wind, rain is coming.
Cited in Lee, p. 68.
503. When you see cattle sniffing the air and crowding together with their heads away from the wind, expect a storm.
Cited in Hyatt, p. 27.

cenizo bush
504. When the cenizo bush blooms, it will rain within three days.
Comment: Cenizo bush refers to a sage bush.
Cited in Reeder, no pp.

centipede
505. If centipedes have hairy backs, a bad winter is predicted.
Cited in Smith, p. 13.

chicken
506. A storm is approaching when chickens run about flapping their wings.
Cited in Hyatt, p. 24.
507. Chickens are said to be very noisy just before rain and cocks to crow at unusual hours.
Cited in Dunwoody, p. 35.
508. Chickens huddling together outside the henhouse instead of going to roost betoken rain.
Cited in Hyatt, p. 24.
509. Chickens puddling in the dust predict rain.
Cited in Freier, p. 67.
510. Chickens refusing to leave the henhouse in the morning warn you of rain.
Cited in Hyatt, p. 23.
511. Chickens standing with their tails to the wind are an omen of rain.
Cited in Hyatt, p. 24.
512. During rain if chickens pay no attention to it, you may expect continued rain; if they run to shelter, it won't last long. Vars.: (a) If chickens do not seek shelter during a rain, it will rain all day or at least for a few hours. (b) If chickens go out in the rain,

it will rain all day. (c) If chickens seek shelter at the beginning of a rain, it will be a shower only; if they do not seek shelter, the rain will last all day. (d) If it starts raining and the chickens run for shelter, it's not going to rain very long. (e) When chickens stay out in the rain, it will rain all day. (f) When the chickens run around in the rain, it will continue to rain.

Cited in Dunwoody, p. 35; Hyatt, p. 24; Smith, p. 5; Reeder, no pp. Recorded in Bryant (Wis.).

513. Heavy feathers on chickens indicate heavy weather during the winter. Var.: When chickens have a lot of feathers, thicker than usual, the winter will be long and cold.
Cited in Hyatt, p. 24; Smith, p. 14.

514. If chickens after dark sit on a fence and flap their wings, rain will fall before morning.
Cited in Hyatt, p. 24.

515. If chickens are roosting high, the following day will bring clear weather; if low, stormy weather.
Cited in Hyatt, p. 24.

516. If chickens fly up on something during a rain and preen their feathers, the rain will soon stop.
Cited in Hyatt, p. 24.

517. If chickens go to roost early, the weather next day will be good; if late, it will be bad.
Cited in Hyatt, p. 24.

518. If chickens moult in August, they prophesize a hard winter; if in October an open winter.
Cited in Hyatt, p. 24.

519. If chickens roll in the sand, rain is at hand.
Cited in Hyatt, p. 23.

520. If chickens sun themselves and oil themselves in the morning, there will be rain that day.
Recorded in Bryant (Miss.).

521. If during fair weather chickens begin to huddle together or search for sheltered places, rain is imminent.
Cited in Hyatt, p. 24.

522. If high roosts in winter are sought by chickens, colder weather is at hand.
Cited in Hyatt, p. 24.

523. If it is raining in the morning and chickens refuse to leave the henhouse, it will soon clear off; if they leave the henhouse, it will rain the entire day.
Cited in Hyatt, p. 24.

524. If the moulting of chickens starts in the front of their bodies, the first half of winter will be cold; if at the rear of their bodies, the second half of winter will be cold.
Cited in Hyatt, p. 24.

525. Predict rain when you notice chickens picking up little stones.
Cited in Hyatt, p. 23.

526. The singing of chickens in a rain is followed by fine weather.
Cited in Hyatt, p. 24.

527. When chickens come down from roost at night, rain will soon follow.
Cited in Dunwoody, p. 35.

528. When chickens crow before sundown, it is a sign of rain next day.
Cited in Dunwoody, p. 35.

529. When chickens light on fences during rain to plume themselves, it will soon clear.
Cited in Dunwoody, p. 35.

chicken breastbone

530. If in the autumn the soft end of a young chicken's breastbone is dark, look for a cold winter; the darker the bone, the cooler the winter.
Cited in Hyatt, p. 24.

531. If you can see through the breastbone of a freshly killed chicken, it signifies clear weather; if you cannot, foul weather.
Cited in Hyatt, p. 24.

chicken gizzard

532. If at anytime during the year a chicken gizzard is easy to clean good weather will follow; if difficult, bad weather.
Cited in Hyatt, p. 24.

533. If in the autumn the lining of a chicken gizzard is removed with difficulty, a severe winter may be prophesized; if with ease, a mild winter.
Cited in Hyatt, p. 24.

chickweed

534. Chickweed expands its leaves boldly and fully when fine weather is to follow; but if it should shut up, then the traveller is to put on his great coat. Var.: Chickweed expands its leaves boldly and fully when fine weather is to follow.
Cited in Swainson, p. 257; Dunwoody, p. 21; Inwards, p. 215.

535. Chickweeds close their leaves before a rain.
Cited in Freier, p. 39.

536. The chickweed expands its leaves boldly when fine weather is nigh. If it half opens its petals in a rain, it soon will be clear. If its blooms stay open all night, it probably will rain the next day.
Cited in Alstad, p. 95.

537. The flowers of the chickweed contract before rain.
Cited in Dunwoody, p. 65.

538. The half opening of the flowers of the chickweed is a sign that the wet will not last long.
Cited in Inwards, p. 215.

539. When chickweed opens its leaves fully, fine weather will follow; when they close up, expect a storm.
Cited in Lee, p. 72.

chimney

540. A singing chimney is warning you of a change in the weather.
Cited in Hyatt, p. 30.

541. If the wind sweeps down the chimney, cold weather will soon follow.
Cited in Hyatt, p. 30.

542. When chimneys smoke and soot falls, bad weather is at hand.
Cited in Mitchell, p. 230; Dunwoody, p. 18.

chimney swallow

543. When chimney swallows circle and call, a sign of rain.
Cited in Freier, p. 52.

chinquapin

544. It's always cool weather when the chinquapin blooms.
Cited in Mieder, p. 647. Recorded in Bryant (Miss.).

chipmunk

545. Chipmunks are seen as late as December before a mild winter.
Cited in Freier, p. 81.

546. In cold and early winters chipmunks are always housed for the winter in October. In short and mild winters they are seen until the first of December.
Cited in Dunwoody, p. 32; Garriott, p. 40; Hand, p. 40.

Christmas

547. A black Christmas makes a fat churchyard. Var.: They've always said a black Christmas fills the graveyards full.
Cited in Denham, p. 62; Marvin, p. 204; Whiting 2, p. 114.

548. A bright moonlight Christmas means light crops.
Recorded in Bryant (Miss.).

549. A clear and bright sun on Christmas Day foretelleth a peaceable year and plenty; but if the wind grow stormy before sunset, it betokeneth sickness in the spring and autumn quarters.
Cited in Inwards, p. 440.

550. A cold Christmas, a warm Easter.
Cited in Hyatt, p. 14.

551. A green Christmas, a white Easter. Vars.: (a) A green Christmas indicates a white Easter. (b) A green Christmas makes a white Easter, a white Christmas makes a green Easter. (c) If you have a green Christmas, you will have a white Easter.
Cited in Swainson, p. 161; Cheales, p. 20; Dunwoody, p. 99; Taylor 1, p. 115; Sloane 2, p. 34; Hyatt, p. 14; Smith, p. 13; Mieder, p. 99.

552. A green Christmas brings a heavy harvest.
Cited in Whitman, p. 40; Inwards, p. 67; Hand, p. 43; Wilshere, p. 15.

553. A green Christmas means a fat graveyard. Vars.: (a) A green Christmas, a full churchyard. (b) A green Christmas brings forth a fat graveyard. (c) A green Christmas fills the churchyard. (d) A green Yule (Christmas) makes a fat kirkyard. (e) Green Christmas and a fruitful graveyard. (f) Green Christmas bigger graveyards. (g) Green Christmas, fat graveyard. (h) Green Christmas makes a fat churchyard. Comment: It was believed that a mild winter followed by a severe spring was often fatal to old and weak people.

Cited in Mitchell, p. 230; Swainson, p. 155; Dunwoody, p. 101; Marvin, p. 204; Whitman, p. 40; Inwards, p. 67; Hand, p. 43; Sloane 2, p. 34; Smith, p. 14; Wilshere, p. 15; Whiting 2, p. 114; Mieder, p. 99. Recorded in Bryant (Utah).

554. A green Yule (Christmas) makes a fat churchyard.
   Cited in Simpson, p. 102.

555. A light Christmas, a heavy sheaf.
   Cited in Dunwoody, p. 101.

556. A warm Christmas, a cold Easter. Var.: A warm Christmas means a cold Easter.
   Cited in Swainson, p. 161; Dunwoody, p. 99; Hyatt, p. 14; Mieder, p. 99. Recorded in Bryant (Miss.).

557. A white Christmas, a green Easter. Var.: A white Christmas means a green Easter.
   Cited in Hyatt, p. 14; Smith, p. 10.

558. A white Christmas, a lean graveyard.
   Cited in Dunwoody, p. 76.

559. A windy Christmas and a calm Candlemas, are signs of a good year. Comment: Candlemas Day is February 2nd.
   Cited in Denham, p. 27; Swainson, p. 161; Inwards, p. 69.

560. At Christmas meadows green, at Easter covered with frost.
   Cited in Dunwoody, p. 100; Inwards, p. 68.

561. Christmas foretells the coming year.
   Cited in Lee, p. 151.

562. Christmas in mud, Easter in snow.
   Cited in Dunwoody, p. 91.

563. Christmas wet gives empty granary and barrel. Var.: Christmas wet, empty granary and barrel.
   Cited in Dunwoody, p. 100; Inwards, p. 68.

564. Days lengthen a cock's stride each day after Christmas.
   Cited in Wright, p. 128.

565. If at Christmas ice hangs on the willow, clover may be cut at Easter.
   Cited in Dunwoody, p. 91; Whitman, p. 40; Inwards, p. 68.

566. If Christmas Day on a Monday fall, a troublous winter we shall have all. Var.: If Christmas Day on Monday be, a great winter that year you'll see.
   Cited in Denham, p. 65; Cheales, p. 20; Wright, p. 126.

567. If Christmas Day on a Sunday fall, a troublous winter we shall have all.
Recorded in Bryant (Nebr.).

568. If Christmas Day on Thursday be, a windy winter you shall see; windy weather in each week, and hard tempests strong and thick; the summer shall be good and dry, corn and beast shall multiply.
Cited in Dunwoody, p. 101; Northall, p. 455.

569. If Christmas finds a bridge, he'll break it; if he finds none, he'll make one. Comment: Bridge refers to ice formation on a river or lake.
Cited in Dunwoody, p. 99; Garriott, p. 43; Inwards, p. 68; Hand, p. 43.

570. If Christmas night is clear and bright, it brings a blessed and rich year ahead.
Recorded in Bryant (Ohio.).

571. If it rain much during the twelve days after Christmas, it will be a wet year.
Cited in Swainson, p. 158; Inwards, p. 69.

572. If it rains on Christmas Day, there should be good grass but very little hay.
Cited in Wilshere, p. 4.

573. If it snows during Christmas night, the crops will do well.
Cited in Inwards, p. 68.

574. If it snows on Christmas night, we expect a good hop crop next year. Var.: Snow on Christmas night, good hop crop next year.
Cited in Dunwoody, p. 100; Inwards, p. 68.

575. If on Christmas night the wine ferments heavily in the barrels, a good year is to follow.
Cited in Whitman, p. 40; Inwards, p. 69.

576. If snow covers the apple trees at Christmas, the next autumn will see a bumper crop.
Cited in Wilshere, p. 15.

577. If that Christmas Day should fall upon Friday, know well all that winter season shall be easy, save great winds aloft shall fly.
Cited in Wright, p. 126; Inwards, p. 68.

578. If the sun shines through the apple trees on Christmas Day, there will be an abundent crop the following year. Vars.: (a) If the sun shines through the apple trees on Christmas Day, when autumn

comes, they will a load of fruit display. (b) Sun through the apple trees on Christmas Day, means a fine crop is on the way.
Cited in Swainson, p. 156; Dunwoody, p. 101; Whitman, p. 40; Inwards, p. 68; Page, p. 40; Wilshere, p. 15.

579. If the sun shines through the trees on Christmas Day, it's a good sign that the crops will be good in summer.
Cited in Smith, p. 11.

580. If windy on Christmas Day, trees will bring much fruit. Var.: If windy on Christmas, a fruitful harvest.
Cited in Dunwoody, p. 99; Inwards, p. 68; Sloane 2, p. 96.

581. The twelve days after Christmas indicate the kind of weather for the whole year. Each day in that order indicates the trend of weather for each month in regular order for the following year. Vars.: (a) The twelve days from Christmas to Epiphany are said to be the keys of the weather for the whole year. (b) The twelve days of Christmas determine the weather for each month of the year.
Cited in Inwards, p. 69; Sloane 2, p. 34; Freier, p. 88.

582. When on Christmas Eve at midnight, the wind waxes still, it betokens a fruitful year; when on the twelfth day afore day, it is somewhat windy, that betokens great plenty of oil.
Cited in Swainson, p. 158.

chrysalide

583. When the chrysalides are found suspended from the underside of rails, limbs, etc., as if to protect them from the rain, expect much rain. If they are found on slender branches, fair weather will last some time.
Cited in Dunwoody, p. 56; Inwards, p. 207.

chrysanthemum

584. There gay chrysanthemums repose, and when stern tempests lower, their silken fringes softly close against the shower.
Cited in Inwards, p. 216; Lee, p. 72.

cicada

585. One can tell the summer's weather by observing the cocoon of the cicada; if the head is upward, a dry summer; if the head is

downward, a wet summer.
Cited in Sloane 2, p. 125.

cigarette
586. If cigarettes taste like rope, it's a sign the weather outside is muggy.
Recorded in Bryant (N.Y.).

clam
587. Air bubbles over clam beds indicate rain. Var.: Air bubbles rise excessively over clam beds before a rain.
Cited in Dunwoody, p. 49; Garriott, p. 24; Inwards, p. 199; Alstad, p. 96; Hand, p. 24.

clothes
588. Clothes not taking starch on washing day is a portent of rain.
Cited in Hyatt, p. 32.

clothesline
589. A clothesline becoming taut foretells rain.
Cited in Hyatt, p. 31.

cloud
590. A bank of clouds in the west indicates rain. Var.: A bench or bank of clouds in the west means rain.
Cited in Swainson, p. 202; Inwards, p. 128.
591. A cloud with rounded top and flattened base carries rainfall in its face. Vars.: (a) A round-topped cloud and flattened base, carries rainfall in its face. (b) A round-topped cloud with flattened base, carries rain drops on its face. (c) A round-topped cloud, with flattened base carries rainfall in its face.
Cited in Dunwoody, p. 47; Whitman, p. 24; Inwards, p. 136; Page, p. 26; Freier, p. 43. Recorded in Bryant (N.Y.).
592. A long strip of clouds called a salmon, or Noah's Ark, stretching east and west, is a sign of stormy weather; but when it extends north and south, it is a sign of dry weather.
Cited in Steinmetz 2, p. 281; Swainson, p. 204; Dunwoody, p. 47.

593. A squall cloud that one sees through or under is not likely to bring or be accompanied by so much wind as a dark, continued cloud extending beyond the horizon.
Cited in Inwards, p. 128.

594. After clouds, a clear sun. Var.: After clouds, the sun shines.
Cited in Marvin, p. 204; Mieder, p. 572.

595. After clouds, clear weather. Var.: After clouds comes clear weather.
Cited in Marvin, p. 204; Mieder, p. 104.

596. Against much rain the clouds grow rapidly larger, especially before thunder. Var.: Against heavy rain every cloud rises bigger than the preceding, and all are in a growing state.
Cited in Dunwoody, p. 47; Inwards, p. 126.

597. All clouds bring not rain. Var.: Not every cloud brings rain.
Cited in Wilson, p. 128; Whiting 1, p. 75.

598. Anvil-shaped clouds are very likely to be followed by a gale of wind.
Cited in Dunwoody, p. 42.

599. Before eight o'clock clouds, after will bring sunshine.
Recorded in Bryant (Wis.).

600. Clouds are the storm signals of the sky.
Cited in Garriott, p. 11; Hand, p. 11.

601. Clouds being soft, undefined, and feathery, it will be fair.
Cited in Dunwoody, p. 44.

602. Clouds floating at different heights show different currents of air, and the upper one generally prevails. If this is northeast, fine weather may be expected; if southwest, rain.
Cited in Inwards, p. 127.

603. Clouds floating low enough to cast shadows on the ground are usually followed by rain.
Cited in Dunwoody, p. 45; Inwards, p. 126.

604. Clouds flying against the wind indicate unsettled weather. Var.: Clouds flying against the wind indicate rain.
Cited in Dunwoody, p. 43; Hand, p. 12.

605. Clouds in the east, obscuring the sun, indicate fair weather.
Cited in Dunwoody, p. 43.

606. Clouds in the morning, sailors take warning.
Recorded in Bryant (N.Y.).

607. Clouds large like rocks, great showers.
     Cited in Mitchell, p. 232.
608. Clouds moving in opposite directions indicate rain in about
     twelve hours.
     Cited in Sloane 2, p. 35.
609. Clouds on the setting sun's brow indicate rain.
     Cited in Freier, p. 32.
610. Clouds stretching north and south is the sign of drought, east and
     west the sign of blast. Var.: North and south the sign of drought,
     east and west the sign of blast.
     Cited in Dunwoody, p. 47; Inwards, p. 134; Lee, p. 93.
611. Clouds upon hills, if rising, do not bring rain; if falling, rain
     follows.
     Cited in Garriott, p. 13; Inwards, p. 140; Hand, p. 13.
612. Clouds without dew indicate rain.
     Cited in Dunwoody, p. 48.
613. Clouds without rain in summer indicate wind.
     Cited in Wright, p. 81.
614. Different kinds of clouds indicate rain.
     Cited in Dunwoody, p. 47.
615. Every cloud has a silver lining. Vars.: (a) Behind each cloud is
     a silver lining. (b) Look upon the brightest side of every cloud.
     (c) The darkest cloud has a silver lining. (d) The inner side of
     every cloud is bright and shining.
     Cited in Taylor 1, p. 111; Wilson, p. 128; Simpson, p. 38;
     Whiting 2, p. 121; Mieder, p. 104.
616. Every cloud is not a sign of a storm.
     Cited in Mieder, p. 104.
617. Few days pass without some clouds.
     Cited in Denham, p. 4.
618. From clouds with a golden glow at sunset pleasant weather is
     presaged.
     Cited in Hyatt, p. 5.
619. Hard-edged, oily-looking clouds, wind.
     Cited in Steinmetz 1, p. 123.
620. High upper clouds crossing the sun, moon, or stars, in a
     direction different from that of the lower clouds, or the wind

when felt below, foretell a change of wind towards their direction.
Cited in Steinmetz 1, p. 123; Inwards, p. 126.

621. If a layer of thin clouds drip up from the northwest, and under other clouds moving more to the south, expect fine weather.
Cited in Dunwoody, p. 45; Inwards, p. 126.

622. If at sunrise many clouds are seen in the west, and disappear, expect fine weather for a short time.
Cited in Dunwoody, p. 46.

623. If clouds appear suddenly in the south, expect rain.
Cited in Dunwoody, p. 46.

624. If clouds at the same height drive up with the wind and gradually become thinner and descend, expect fine weather.
Cited in Dunwoody, p. 44; Garriott, p. 13; Hand, p. 34.

625. If clouds be bright, 'twill clear tonight; if clouds be dark, 'twill rain, do you hark?
Cited in Dunwoody, p. 43; Marvin, p. 204; Inwards, p. 128; Lee, p. 94.

626. If clouds bunch together to form a tree, formerly described as a "cloud-baum" or "cloud-tree" by some of the old-time Germans, a rain is impending.
Cited in Inwards, p. 6.

627. If clouds drive up high from the south, expect a thaw.
Cited in Dunwoody, p. 107; Inwards, p. 126.

628. If clouds float at different heights and rates but generally in opposite directions, expect heavy rain.
Cited in Dunwoody, p. 44; Garriott, p. 13; Inwards, p. 127; Hand, p. 13.

629. If clouds fly to the west at sunrise, expect fine weather.
Cited in Dunwoody, p. 46.

630. If clouds form high in air in their white trains like locks of wool, they portend wind and probably rain. Var.: If clouds appear high in air in their white trains, wind, and probably rain, will follow.
Cited in Dunwoody, p. 44; Inwards, p. 133.

631. If clouds increase visibly, and the clear sky becomes less, it is a sign of rain.
Cited in Inwards, p. 125.

632. If clouds open and close, rain will continue.
Cited in Dunwoody, p. 45; Inwards, p. 136.

633. If, during dry weather, two layers of clouds appear moving in opposite directions, rain will follow.
Cited in Inwards, p. 127.
634. If large clouds decrease, fair weather.
Cited in Mitchell, p. 232; Steinmetz 1, p. 122.
635. If long strips of clouds drive at a slow rate high in the air, and gradually become larger, the sky having been previously clear, expect rain.
Cited in Dunwoody, p. 46; Inwards, p. 138.
636. If small clouds increase, much rain.
Cited in Mitchell, p. 232; Steinmetz 1, p. 122.
637. If the clouds, as they come forward, seem to diverge from a point in the horizon, a wind may be expected from that quarter, or the opposite.
Cited in Inwards, p. 126.
638. If the clouds be of different heights, the sky being grayish or dirty blue, with hardly any wind stirring, the wind, however changing from the west to south, or sometimes to southeast, without perceptibly increasing force, expect storm.
Cited in Dunwoody, p. 46.
639. If the clouds move against the wind, rain will follow.
Cited in Freier, p. 59.
640. If the upper current of clouds come from the northwest in the morning, a fine day will ensue.
Cited in Garriott, p. 13; Inwards, p. 126; Hand, p. 13; Freier, p. 46.
641. If there were no clouds, we should not enjoy the sun.
Recorded in Bryant (N.Y.).
642. If two strata of clouds appear in hot weather to move in different directions, they indicate thunder.
Cited in Garriott, p. 13; Inwards, p. 127; Hand, p. 13.
643. If you see clouds going cross wind, there is a storm in the air.
Cited in Dunwoody, p. 43; Inwards, p. 127.
644. In winter and in the North Atlantic a cloud rising from the northwest is an infallible forerunner of a great tempest.
Cited in Inwards, p. 126.
645. It never clouds up in a June night for a rain.
Cited in Garriott, p. 13; Hand, p. 13.

646. Large irregular masses of cloud, "like rocks and towers," are indicative of showery weather. If the barometer be low, rain is all the more probable.
Cited in Inwards, p. 137.

647. Light scuds (wind driven clouds) driving across heavy masses, show wind and rain; but if alone may indicate wind only.
Cited in Steinmetz 1, p. 123.

648. Long parallel bands of clouds in the direction of the wind indicate steady high winds to come.
Cited in Inwards, p. 129.

649. Many small clouds at northwest in the evening show that rain is gathering, and will suddenly fall.
Cited in Inwards, p. 124.

650. Misty clouds, forming or hanging on heights, show wind and rain coming, if they remain, increase or descend. If they rise or disperse, the weather will improve.
Cited in Inwards, p. 140.

651. Morning clouds opening before seven and closing afterward foretell rain before eleven.
Cited in Hyatt, p. 6.

652. No matter what the ground wind, if high clouds are moving from a westerly quadrant, fair weather will persist.
Cited in Sloane 2, p. 35.

653. One cloud is enough to eclipse all the sun.
Cited in Mieder, p. 104.

654. Rain-clouds appearing before moonrise will drift away as soon as the moon rises, but rain-clouds after the moon has risen always remain.
Cited in Hyatt, p. 5.

655. Small floating clouds over a bank of clouds, sign of rain.
Cited in Inwards, p. 135.

656. Small inky clouds foretell rain.
Cited in Dunwoody, p. 42; Inwards, p. 127.

657. Small scattering clouds flying high in the southwest foreshow whirlwinds.
Cited in Inwards, p. 126.

658. Soft-looking or delicate clouds foretell fine weather, with moderate or light breezes.
Cited in Steinmetz 1, p. 123; Dunwoody, p. 42.

659. Streaky clouds across the wind foreshow rain.
     Cited in Inwards, p. 130.
660. The fish-shaped cloud, if pointing east and west, indicates rain;
     if north and south, more fine weather.
     Cited in Inwards, p. 134.
661. The higher the clouds, the fairer the weather. Var.: The higher
     the clouds, the finer the weather.
     Cited in Humphreys 1, p. 441; Humphreys 2, p. 53; Whitman,
     p. 24; Inwards, p. 123; Alstad, p. 129; Sloane 2, p. 34; Lee, p.
     92; Freier, p. 45; Mieder, p. 104.
662. The pocky cloud, looking like festoons of drapery, forebodes a
     storm. Comment: Pock, a bag.
     Cited in Inwards, p. 138.
663. The rounded clouds called "water-wagons" which fly alone in
     the lower currents of wind forebode rain.
     Cited in Inwards, p. 137.
664. The softer clouds look, the less wind, but perhaps more rain may
     be expected; and the harder, more "greasy," rolled, tufted,
     ragged, the stronger the coming wind will prove.
     Cited in Steinmetz 1, p. 123.
665. Thin streaked clouds will eventually collect rain.
     Cited in Hyatt, p. 6.
666. Threatening clouds, without rain, in the old moon indicate
     drought.
     Cited in Dunwoody, p. 63.
667. Two currents of clouds indicate approaching rain, and in summer
     thunder.
     Cited in Dunwoody, p. 47.
668. When a heavy cloud comes up in the southwest and seems to
     settle back again, look out for a storm.
     Cited in Dunwoody, p. 87; Garriott, p. 13; Inwards, p. 125;
     Hand, p. 13.
669. When a plain sheet of the wane cloud is spread over a large
     surface at eventide, or when the sky gradually thickens with this
     cloud, a fall of steady rain is usually the consequence.
     Cited in Inwards, p. 133.
670. When clouds after a rain, disperse during the night, the weather
     will not remain clear.
     Cited in Dunwoody, p. 43; Marvin, p. 204; Inwards, p. 124.

671. When clouds appear like rocks and towers, the earth's refreshed with frequent showers.
Cited in Denham, p. 11; Steinmetz 1, p. 126; Steinmetz 2, p. 202; Swainson, p. 203; Northall, p. 460; Humphreys 1, p. 442; Humphreys 2, p. 53; Whitman, p. 24; Inwards, p. 137; Page, p. 26; Wilshere, p. 3; Freier, p. 42. Recorded in Bryant (Wis.).

672. When clouds are stationary and others accumulate by them, but the first remain still, it is a sign of a storm.
Cited in Inwards, p. 125.

673. When clouds bank the moon around no frost will cover the ground.
Cited in Alstad, p. 166.

674. When clouds gather around the mountains there will be rain.
Cited in Smith, p. 7.

675. When clouds sink below the hills, foul weather; when clouds rise above the hills, fair weather.
Cited in Freier, p. 43.

676. When e'er the clouds do weave, 'twill storm before they leave.
Comment: On the front side of a wide-spread storm the clouds of the humid winds come lower and lower until the rain begins. At times the lower clouds are so broken that the layer above may also be seen, the two streams crossing each other.
Cited in Humphreys 2, p. 51.

677. When high clouds and low, in different paths go, be sure that they show it'll soon rain and blow.
Cited in Alstad, p. 140.

678. When mountains and cliffs in the clouds appear, some sudden and violent showers are near.
Cited in Whitman, p. 24; Whitman, p. 64; Inwards, p. 137; Freier, p. 43.

679. When on clear days isolated clouds drive over the zenith from the rain-wind side, storm and rain follow within twenty-four hours.
Var.: When on clear days isolated clouds drive over the zenith from the rain-wind side, rain or snow will follow within twenty-four hours, more likely within a few hours.
Cited in Inwards, p. 123; Hand, p. 12.

680. When the ark is out, north and south, in the rain's mouth. Var.: When the ark is out, rain is about. Comment: The ark is a thin

strip of high feathery cloud that forms an archway across the sky. Cited in Page, p. 26.

681. When the carry goes west, good weather is past; when the carry goes east, good weather comes neist. Comment: Carry means the current of the clouds, neist means next.
Cited in Swainson, p. 227; Humphreys 2, p. 54; Inwards, p. 125.

682. When the clouds are upon the hills they'll come down by the mills. Vars.: (a) When it goes up in fops, it'll fall down in drops. (b) When the clouds go up the hill, they'll send down water to turn the mill. Comment: Fops are small clouds on hills.
Cited in Denham, p. 19; Swainson, p. 204; Dunwoody, p. 45; Northall, p. 460; Humphreys 1, p. 442; Humphreys 2, p. 53; Whitman, p. 24; Inwards, p. 140; Wilson, p. 128; Freier, p. 44.

683. When the clouds bank up the contrary way to the wind, there will be rain.
Cited in Inwards, p. 137.

684. When the clouds of the morn to the west fly away, you may conclude on a settled, fair day.
Cited in Marvin, p. 204; Inwards, p. 124.

685. When the clouds on the hill tops are thick and in motion, rain to the southwest is regarded as certain to follow.
Cited in Inwards, p. 140.

686. When the streamers point upward, the clouds are falling and rain is at hand. When streamers point downwards, the clouds are ascending amd drought is at hand. Vars.: (a) Clouds with streamers pointing upward carry rain. (b) When cloud streamers point upward, the clouds are descending, and rain is indicated; when cloud streamers point downward, the clouds are ascending, and dry weather is indicated.
Cited in Dunwoody, p. 46; Garriott, p. 12; Hand, p. 12; Hyatt, p. 6.

687. When you see two layers of clouds, it will rain.
Cited in Reeder, no pp.

688. Woolpack clouds on a quiet day mean fine weather is here to stay.
Cited in Alstad, p. 130.

cloud (black)

689. After black clouds, clear weather. Var.: After black clouds, fair weather.
Cited in Denham, p. 6; Inwards, p. 127; Wilson, p. 6; Freier, p. 43.
690. Black clouds in the north in winter indicate approaching snow.
Cited in Dunwoody, p. 42.
691. Black clouds thunder a great deal but rain little.
Cited in Mieder, p. 104.
692. If in the northwest before daylight ends there appears a company of small black clouds like flocks of sheep, it is a sure and certain sign of rain.
Cited in Inwards, p. 126.
693. Small black scuds (clouds), drifting from southwest, is a sign of rain.
Cited in Dunwoody, p. 43.
694. When the clouds build up to form a towering black anvil, a thunderstorm is almost certain.
Cited in Page, p. 26.
695. When the sheet (cloud) turns blackish gray, it soon will be a rainy day.
Cited in Alstad, p. 137.

cloud (brass)

696. Brassy-colored clouds in the west at sunset indicate wind.
Cited in Dunwoody, p. 47; Inwards, p. 127.

cloud (buttermilk)

697. Buttermilk clouds are rain-bearers.
Cited in Hyatt, p. 6.

cloud (cirrocumulus)

698. Before thunder, cirrocumulus clouds often appear in very dense and compact masses, in close contact.
Cited in Inwards, p. 134.
699. The cirrocumulus, when accompanied by the stratocumulus, is a sure indication of a coming storm.
Cited in Inwards, p. 135.

700. When cirrocumulus clouds appear in winter, expect warm and wet weather.
Cited in Dunwoody, p. 43; Garriott, p. 12; Hand, p. 12.
701. When cirrocumulus is seen overhead, if the fleeces gently merge into each other, and the edges are soft and transparent, settled weather prevails.
Cited in Inwards, p. 136.

cloud (cirrostratus)

702. If long lines of cirrostratus, extend along the horizon, and are slightly contracted in their center, expect heavy rain the following day.
Cited in Inwards, p. 134.
703. The waved cirrostratus cloud indicates heat and thunder.
Cited in Inwards, p. 134.
704. When, after a shower, the cirrostrati open up at the zenith, leaving broken or ragged edges pointing upwards, and settle down gloomily and compactly on the horizon, wind will follow, and will last for some time.
Cited in Mitchell, p. 221.

cloud (cirrus)

705. A large formation of murky white cirrus may merely indicate a backing of wind to an easterly quarter.
Cited in Inwards, p. 129.
706. After a long run of clear weather the appearance of light streaks of cirrus cloud at a great elevation is often the first sign of change. Var.: After a long run of clear weather, the appearance of cirrus clouds often is the first sign of change.
Cited in Inwards, p. 129; Alstad, p. 146.
707. Cirrus at right angles to the wind is a sign of rain.
Cited in Mitchell, p. 22; Dunwoody, p. 17.
708. Cirrus clouds announce the east wind; if their streaks point upward, they indicate rain; if downward, wind and dry weather. Var.: Cirrus clouds announce the east wind; if their under surface is level, and their streaks pointing upwards, they indicate rain; if downwards, wind and dry weather.
Cited in Garriott, p. 12; Inwards, p. 130; Hand, p. 12.

709. Cirrus of a long, straight, feathery kind, with soft edges and outlines, or with soft, delicate colors at sunrise and sunset, is a sign of fine weather.
Cited in Inwards, p. 129.

710. If cirrus clouds dissolve and appear to vanish, it is an indication of fine weather. Var.: If cirrus clouds dissolve and appear to evaporate, it is a sign of fine weather.
Cited in Dunwoody, p. 22; Inwards, p. 129; Alstad, p. 147; Hand, p. 12; Freier, p. 26.

711. If cirrus clouds form in fine weather with a falling barometer, it is almost sure to rain.
Cited in Dunwoody, p. 22; Garriott, p. 12; Hand, p. 12; Freier, p. 25.

712. If cirrus clouds get lower and denser to leeward, it presages bad weather from the opposite quarter.
Cited in Inwards, p. 130.

713. If cirrus mares' tails have ascending streaks or point upward, wind and storm are in the making; if they point downward, calm and dryness are in store.
Cited in Sloane 2, p. 35.

714. If the cirrus clouds appear to windward, and change to cirrostratus, it is a sign of rain.
Cited in Inwards, p. 129.

715. In unsettled weather sheet cirrus preceeds more wind or rain.
Cited in Inwards, p. 129; Alstad, p. 147.

716. The curdled cirrus cloud often indicates the approach of bad weather.
Cited in Inwards, p. 129.

717. When, after a clear frost, long streaks of cirrus are seen with their ends bending towards each other as they recede from the zenith, and when they point to the northeast, a thaw and southwest wind may be expected.
Cited in Mitchell, p. 221.

718. When cirri threads are brushed back from a southerly direction, expect rain and wind. Var.: When threads of cirrus clouds are brushed back from a southerly direction, expect rain and wind.
Cited in Dunwoody, p. 43; Garriott, p. 12; Inwards, p. 135.

719.   When cirrus clouds appear at lower elevations than usual, and with a denser character, expect a storm from the opposite quarter to the clouds.
Cited in Inwards, p. 132.

cloud (cumulus)

720.   Cumulus clouds, high up, are said to show that south and southwest winds are near at hand; and stratified clouds, low down, that east or north winds will prevail.
Cited in Mitchell, p. 22.

721.   If a fair day, with cumulus clouds, expect rain before night. Var.: Cumulus clouds in a clear blue sky, it will likely rain.
Cited in Dunwoody, p. 43; Freier, p. 44.

722.   If cumulus clouds are smaller at sunset than at noon, expect fair weather. Var.: If cumulus clouds are smaller at sunset than at noon, look for continued fair weather.
Cited in Dunwoody, p. 22; Inwards, p. 137; Hand, p. 13; Garriott, p. 13; Alstad, p. 128; Freier, p. 42.

723.   The formation of cumulus clouds to leeward during a strong wind indicates the approach of a calm with rain.
Cited in Inwards, p. 137.

724.   When at sea, if the stratocumulus cloud appear on the horizon, it is a sign that the weather is going to break up.
Cited in Inwards, p. 138.

725.   When cirrus merge into cirrostratus, and when cumulus increase toward evening and become lower, expect wet weather. Var.: When cirri merge into cirrostratus, and when cumulus increase towards evening and become lower clouds, expect wet weather.
Cited in Dunwoody, p. 43; Garriott, p. 12; Inwards, p. 130; Hand, p. 12.

726.   When cumulus clouds become heaped up to leeward during a strong wind at sunset, thunder may be expected during the night.
Cited in Garriott, p. 13; Hand, p. 13; Freier, p. 42.

727.   When cumulus (clouds) shrink with the passing day, another fine day is on the way.
Cited in Alstad, p. 128.

cloud (dark)

728. Dark clouds in the west at sunrise indicate rain that day. Var.: If at sunrise there are many dark clouds seen in the west and remain there, rain will fall on that day.
Cited in Dunwoody, p. 43.

729. If high, dark clouds are seen in spring, winter, or fall, expect cold weather. Var.: If high, dark clouds are seen in spring, winter, or autumn, expect cold weather.
Cited in Dunwoody, p. 44; Inwards, p. 126.

730. If the sun rises covered with a dark spotted cloud, expect rain on that day.
Cited in Dunwoody, p. 79.

731. If the sun sets in dark, heavy clouds expect rain next day.
Cited in Dunwoody, p. 76.

732. When small dark clouds (broken nimbus) appear against a patch of blue sky, there will be rain before sunset.
Cited in Inwards, p. 128.

cloud (fleece)

733. If clouds are fleecy like cotton, it will rain in twenty-four hours.
Cited in Wurtele, p. 293.

734. If, in winter, the clouds appear fleecy, with a very blue sky, expect cold rain or snow.
Cited in Dunwoody, p. 44.

735. If woolly fleeces spread the heavenly way, be sure no rain disturbs the summer day. Vars.: (a) If woollen fleeces spread the heavenly way, be sure no rain disturbs the summer day. (b) If woolly fleeces spread the heavenly way, no rain, be sure, disturbs the summer's day. (c) If wooly fleeces strew the heavenly way, be sure no rain disturbs the summer's day.
Cited in Swainson, p. 201; Dunwoody, p. 22; Northall, p. 460; Inwards, p. 135; Lee, p. 99; Page, p. 25; Freier, p. 42.

736. Light fleecy clouds produce rain only; heavy rough clouds, rain accompanied by wind.
Cited in Hyatt, p. 6.

737. When the clouds are formed like fleeces, but dense in the middle and bright toward the edge, with the sky bright, they are signs of a frost, with hail, snow, or rain.
Cited in Dunwoody, p. 44.

cloud (gray)

738. Clouds small and round, like a dapple gray with a north wind, fair weather for two or three days. Var.: Clouds small and round like a dapply-gray, with north wind, fair for a day.
Cited in Mitchell, p. 232; Dunwoody, p. 22; Freier, p. 45.

739. Large rounded masses or rolls of gray cloud, commonly arranged in parallel bands and often covering the whole sky, bring little or no rain.
Cited in Inwards, p. 138.

cloud (green)

740. A green cloud is a sign of hail.
Recorded in Bryant (Nebr.).

741. When you observe greenish-tinted masses of composite cloud collect in the southeast and remain there for several hours, expect a succession of heavy rains and gales.
Cited in Garriott, p. 13; Hand, p. 13.

cloud (hen)

742. Hen scarts and filly tails make lofty ships wear low sails. Comment: Hen scarts (scratches) are light clouds that resemble the scratches of hens on the ground, filly tails are clouds that resemble the tails of young mares.
Cited in Dunwoody, p. 44.

743. If clouds look as if scratched by a hen, get ready to reef your topsails then. Vars.: (a) Clouds look as if scratched by a hen, be ready to reef your topsails in. (b) Hen scratches and filly tails, get ready to reef your topsails.
Cited in Swainson, p. 203; Dunwoody, p. 15; Northall, p. 460; Marvin, p. 208; Whitman, p. 23; Inwards, p. 132; Lee, p. 95; Page, p. 25. Recorded in Bryant (Wis.).

cloud (mackerel)

744. A mackerel sky, not twenty-four hours dry. Comment: Mackerel and mackerel scales refer to the dark bands on the mackerel's back.
Cited in Hand, p. 12.

745. Mackerel clouds in sky, expect more wet than dry.

Cited in Dunwoody, p. 45; Garriott, p. 12; Marvin, p. 204; Whitman, p. 23; Inwards, p. 134; Hand, p. 12; Lee, p. 95.

746. Mackerel scales and mares' tails make lofty ships carry low sails. Vars.: (a) Mackerel backs and mare's tails make lofty ships carry low sails. (b) Mare's tails and mackerel scales, make tall ships take in their sails. Comment: Mackerel scales and mare's tails refers to cirrocumulus clouds.
Cited in Steinmetz 1, p. 125; Steinmetz 2, p. 280; Cheales, p. 26; Dunwoody, p. 45; Garriott, p. 12; Humphreys 2, p. 51; Whitman, p. 23; Sloane 1, p. 28; Alstad, p. 145; Hand, p. 12; Sloane 2, p. 54; Smith, p. 14.

747. Mackerel scales, furl your sails.
Cited in Dunwoody, p. 45; Inwards, p. 134; Lee, p. 95.

748. The mackerel clouds always indicate a storm if they first appear about 15 degrees north of west.
Cited in Dunwoody, p. 45; Garriott, p. 12; Hand, p. 12.

cloud (Monday)

749. Monday clouds portend cloudy weather two more days that week.
Cited in Hyatt, p. 6.

cloud (mountain)

750. In the morning mountains, in the evening fountains. Comment: Towering thunder-clouds that look like mountains early on a summer's day usually portend intense storm activating during the heat of the afternoon, and often herald torrent of rain in the evening.
Cited in Northall, p. 465; Alstad, p. 132; Humphreys 2, p. 53; Whitman, p. 23; Inwards, p. 137; Wilshere, p. 3.

751. The mountain is the mother of weather in the valley.
Cited in Alstad, p. 36.

752. When mountains and cliffs with clouds appear, some sudden and violent showers are near.
Recorded in Bryant (N.Y.).

cloud (nimbus)

753. When scattered patches or streaks of nimbus come driving up from the southwest, they are called by the sailors "Prophet

Clouds," and indicate wind.
Cited in Swainson, p. 202; Inwards, p. 139.

cloud (pink)
754. If clouds look slightly pink and completely cover the sky as a blanket in the winter, it will snow.
Cited in Smith, p. 9.

cloud (red)
755. If there be red clouds in the west at sunset, it will be fair; if the clouds have a tint of purple or if red bordered with black in the southeast, it will be very fine.
Cited in Dunwoody, p. 45.
756. Narrow, horizontal red clouds after sunset in the west indicate rain before thirty-six hours.
Cited in Dunwoody, p. 44; Inwards, p. 127.
757. Red clouds at sunrise indicate rain on the following day. Var.: If the clouds at sunrise be red, there will be rain the following day.
Cited in Dunwoody, p. 46.
758. Red clouds at sunrise indicate storm.
Cited in Dunwoody, p. 46.
759. Red clouds in the east, rain the next day.
Cited in Inwards, p. 127.
760. Red clouds in the morning, a traveler's warning; red clouds at night, a traveler's delight.
Recorded in Bryant (Ill.).

cloud (rosy)
761. When clouds are gathered toward the sun at setting, with a rosy hue, they foretell rain.
Cited in Dunwoody, p. 45.

cloud (silver)
762. Dusky or tarnished silver colored clouds indicate hail.
Cited in Dunwoody, p. 43; Inwards, p. 127.

cloud (stratus)
763. A stratus at night, with a generally diffused fog the next morning, is usually followed by a fine day, if the barometer be high

and steady. If the barometer be low, the fog will probably turn to rain. Comment: Stratus are a formless layer of gray clouds, resembling fog.
Cited in Inwards, p. 139.

cloud (tail)

764. A change in the weather rides the sky, with mares' tails streaming on high.
Cited in Alstad, p. 147.
765. Clouds resembling a mare's tail presage rain.
Cited in Hyatt, p. 6.

766. Horses' tails and fishes' scales, make sailors spread their sails.
Cited in Hyatt, p. 6. Recorded in Bryant (La.).
767. Mare's tail clouds signify rain within three days.
Recorded in Bryant (Nebr.).
768. Mare's tails and mackerel sky, not long wet, nor not long dry.
Cited in Northall, p. 459; Wright, p. 80; Inwards, p. 134; Lee, p. 96.
769. Mares' tails, mares' tails, make lofty ships carry low sails. Var.: Fishes' scales and mare's tails make lofty ships carry low sails. Comment: Horse's tails and fishes' scales refer to cumulus clouds.
Cited in Lee, p. 95. Recorded in Bryant (La.).
770. When the tails of clouds are turned downwards, fair weather or slight showers often follow.
Cited in Inwards, p. 129.
771. When you see wispy (mare tail clouds), it will rain within seventy-two hours.
Cited in Reeder, no pp.

cloud (white)

772. A small increasing white cloud about the size of a hand to windward is a sure precursor of a storm.
Cited in Inwards, p. 128.
773. Heavy, white, rolling clouds in front of a storm denote high wind.
Cited in Dunwoody, p. 88.

774. If on a fair day in winter a white bank of clouds arise in the south, expect snow.
Cited in Dunwoody, p. 47.

775. If small white clouds are seen to collect together, their edges appearing rough, expect wind.
Cited in Dunwoody, p. 47; Inwards, p. 134.

776. Small white clouds indicate rain within three days.
Cited in Hyatt, p. 6.

777. Small white clouds, like a flock of sheep, driving northwest, indicate continued fine weather.
Cited in Inwards, p. 135.

778. White drift-clouds often called "sheep" are a rain warning.
Cited in Hyatt, p. 6.

cloud (wisp)

779. Curly wisps and blown-back pieces (of clouds), are not a bad sign.
Cited in Inwards, p. 129.

cloudiness

780. When a general cloudiness covers the sky and small, black fragments of clouds fly underneath, they indicate rain, and probably it will be lasting.
Cited in Dunwoody, p. 44.

cloudless

781. A small cloudless place in the northeast horizon is regarded both by seamen and landsmen as a certain precursor of fine weather or a clearing up.
Cited in Dunwoody, p. 14.

cloudy

782. Cloudy mornings turn to clear evenings. Vars.: (a) A cloudy morning bodes a fair afternoon. (b) Cloudy mornings may turn to clear evenings. (c) Cloudy mornings turn to clear afternoons (evenings).
Cited in Denham, p. 6; Marvin, p. 203; Inwards, p. 124; Wilson, p. 128; Wilshere, p. 3; Mieder, p. 104.

783. If cloudy and it soon decreases, certain fair weather.

Cited in Freier, p. 43.

784. If there be a cloudy sky, with dark clouds driving fast under high clouds, expect violent gusts of wind.
Cited in Inwards, p. 127.

785. It takes three cloudy days to bring a heavy snow.
Cited in Dunwoody, p. 76.

clover

786. Clover contracts its leaves at the approach of a storm. Var.: Clover contracts its leaves before a storm.
Cited in Swainson, p. 258; Inwards, p. 214; Sloane 2, p. 112; Lee, p. 73.

787. Clover leaves turned up so as to show light underside indicate approaching rain. Var.: Clover leaves show their bottom sides before rain.
Cited in Dunwoody, p. 65; Garriott, p. 25; Hand, p. 25; Freier, p. 57.

788. Clover stems bristle before a rain.
Cited in Alstad, p. 94.

789. Clovers contract at the close of a storm.
Cited in Dunwoody, p. 65.

790. If the trefoil contracts its leaves, expect heavy rains. Comment: The trefoil is a clover.
Cited in Dunwoody, p. 68.

791. The trefoil swells in the stalk against rain, so that it stands up very stiff, but the leaves droop and hang down. Var.: The stalk of the trefoil swells before rain.
Cited in Steinmetz 1, p. 112; Inwards, p. 214; Lee, p. 73.

792. Trefoil or clovergrass against stormy and tempestuous weather will seem rough, and the leaves of it stare and rise up, as if afraid of an assault.
Cited in Swainson, p. 258.

793. When clover grass is rough to the touch stormy weather is at hand.
Cited in Inwards, p. 214.

794. You will always find clover blossoms closed just before a rain.
Cited in Hyatt, p. 14.

coal

795. Coals becoming alternately bright and dim indicate approaching storms. Var.: Coals become alternately bright and dim before storms.
Cited in Dunwoody, p. 107; Inwards, p. 221; Lee, p. 75.

796. Coals covered with thick white ashes indicate snow in winter and rain in summer.
Cited in Dunwoody, p. 107; Inwards, p. 221; Freier, p. 64.

797. If burning coals stick to the bottom of a pot, it is a sign of a tempest.
Cited in Freier, p. 27.

coat

798. A wise man caries his cloak (coat) in fair weather, and a fool wants his in rain.
Cited in Inwards, p. 23.

799. Button to the chin till May be in.
Cited in Wilshere, p. 11.

800. Cast not a clout till May be out. Vars.: (a) Cast ne'er a clout 'ere May be out. (b) Change not a clout till May be out. (c) Change not a clout till May be out, if you change in June 'twill be too soon. (d) Don't toss aside a clout till May is out. (e) Ne'er cast a clout before May is out. (f) Take off neither rag nor clout till May is out. (g) Till May is out, cast not a clout. (h) What you put off and what you put on, never change till May be gone.
Cited in Denham, p. 43; Swainson, p. 89; Cheales, p. 23; Dunwoody, p. 97; Northall, p. 439; Inwards, p. 54; Wilson, p. 106; Page, p. 10; Wilshere, p. 11; Simpson, p. 32; Mieder, p. 408. Recorded in Bryant (Minn., N.Y., Vt.).

801. Don't have thy cloak (coat) to make when it begins to rain. Var.: Have not thy cloak to make, when it begins to rain.
Cited in Denham, p. 4; Wilson, p. 127.

802. It never rains when you wear an overcoat.
Recorded in Bryant (Wash.).

803. One should turn his coat according to the weather.
Cited in Mieder, p. 104.

804. Though the sun shines, leave not your coat at home.
Cited in Denham, p. 11.

805. Who doffs his coat on a winter's day, will gladly put it on in May. Vars.: (a) He that drops a coat on a winter day will gladly put it on in May. (b) Who doffs his coat on a winter day gladly puts it on in the month of May.
Cited in Mitchell, p. 230; Swainson, p. 89; Dunwoody, p. 88; Garriott, p. 45; Whitman, p. 45; Hand, p. 45; Freier, p. 78.

cobweb

806. Cobwebs on the grass are a sign of fair weather. Var.: Cobwebs on the grass, a sign of fair weather.
Cited in Whitman, p. 50; Alstad, p. 115.
807. Cobwebs on the grass are a sign of frost.
Cited in Sloane 2, p. 39.
808. If, early in the morning, you see cobwebs formed by dew on the grass, it is sure to rain.
Cited in Boughton, p. 124.
809. When cobwebs are webbing up tight the first week of September, you can look for a very cold and long winter. For a mild winter they should web up in the first week of October.
Cited in Hyatt, p. 19.

cock

810. Cock crow before two in the morning, of two days wet it is a warning.
Cited in Sloane 2, p. 52.
811. Cocks are said to clap their wings in an unusual manner before rain, and hens to rub in the dust and seem very uneasy. Vars.: (a) Cocks are said to clap their wings in an unusual way, and to crow more than ususal and at an earlier hour, just before rain. (b) Cocks clap their wings excessively before a storm.
Cited in Dunwoody, p. 35; Marvin, p. 209; Alstad, p. 81.
812. If a cock bobs his head after crowing, fair weather will follow.
Cited in Lee, p. 62.
813. If cocks crow at unusual hours, but especially when a hen and chickens crowd into the house, these are sure signs of rain.
Cited in Steinmetz 1, p. 111.
814. If cocks crow during a downpour it will be fine before night.
Cited in Inwards, p. 187; Sloane 2, p. 129.

815. If cocks crow late and early, clapping their wings unusually, rain is expected. Vars.: (a) If cocks crow late and early, clapping their wings occasionally, rain is expected. (b) If the cock crows more than usual or earlier, expect rain.
Cited in Swainson, p. 237; Dunwoody, p. 35; Garriott, p. 23; Inwards, p. 187; Hand, p. 23.

816. If the cock crows on going to bed, he's sure to rise with a watery head. Vars.: (a) If the cock crows going to bed, he wakens with a watery head. (b) If the cock goes crowing to bed, he'll certainly rise with a watery head.
Cited in Denham, p. 18; Mitchell, p. 227; Garriott, p. 23; Marvin, p. 209; Whitman, p. 48; Inwards, p. 187; Alstad, p. 77; Hand, p. 23; Sloane 2, p. 52; Hyatt, p. 24; Lee, p. 62; Page, p. 19; Wilshere, p. 22.

817. If the cock moult before the hen, we shall have weather thick and thin; but if the hen moult before the cock, we shall have weather hard as a rock. Var.: If the cock moult before the hen, we shall have weather thick and thin; but if the hen moult before the cock, we shall have weather as hard as a block.
Cited in Denham, p. 10; Swainson, p. 238; Dunwoody, p. 35; Marvin, p. 209; Inwards, p. 187; Freier, p. 80. Recorded in Bryant (Wis.).

818. When a cock crows after a shower, clear weather is coming.
Cited in Lee, p. 63.

819. When cocks crow and then drink, rain and thunder are on the brink.
Cited in Lee, p. 62.

820. When the cock crows at unusual times, wet or snowy weather is expected.
Cited in Mitchell, p. 227.

cockle

821. Cockles and most shellfish are observed against a tempest to have gravel sticking hard unto their shells to help them down, if raised from the bottom by the surges. Vars.: (a) Cockles and most shell fish are observed against a tempest to have gravel sticking hard to their shells, as a providence of nature to stay or poise themselves, and to help weight them down if raised from the bottom by surges. (b) Cockles and most shellfish have gravel sticking to

their shells during a storm, which is nature's way to help weigh them down and protect them from being tossed around in the surging water. (c) Cockles have more gravel sticking to their shells before a tempest.

Cited in Swainson, p. 248; Dunwoody, p. 49; Inwards, p. 199; Lee, p. 60.

cocklebur

822. Frost has never been known to catch the cockle or blackberry bloom. Vars.: (a) After the burs of a cocklebur bush have started forming, you need not worry about frost for six weeks. (b) As long as the top bur on a cocklebur bush stays green, so long will there be no frost.
Cited in Dunwoody, p. 66; Hyatt, p. 14.

823. The ripening of the very top bur on a cocklebur bush signifies an exceedingly bad winter.
Cited in Hyatt, p. 14.

824. When cockleburs mature brown, it indicates frost.
Cited in Dunwoody, p. 65; Inwards, p. 212.

825. When the top burs of the cocklebur bush have ripened, winter is at hand.
Cited in Hyatt, p. 14.

cockroach

826. When cockroaches fly it is a sign of approaching rain. Vars.: (a) Cockroaches are more active before a storm. (b) When cockroaches fly, rain will come.
Cited in Dunwoody, p. 56; Lee, p. 57; Freier, p. 28.

codfish

827. The cod is said to take in ballast (swallow stones) before a storm.
Cited in Dunwoody, p. 49.

828. When the codfish's eyes are bloated, wind is coming.
Cited in Lee, p. 59.

cold

829. An hour's cold will suck out seven year's heat. Vars.: (a) An hour's cold can suck out seven year's warmth. (b) One hour's cold will spoil seven years warming.

Cited in Denham, p. 3; Wilson, p. 389; Whiting 2, p. 328.

830. Cold and cunning come from the north, but cunning sans wisdom is nothing worth.
Cited in Mieder, p. 106.

831. Cold is the night, when the stars shine bright.
Cited in Lee, p. 30.

832. The colder the day, the hotter the sun.
Cited in Humphreys 2, p. 121.

833. Very cold weather is generally followed by rain or snow.
Cited in Steinmetz 2, p. 287.

834. When there are three days cold, expect three days colder.
Cited in Dunwoody, p. 100; Wright, p. 122; Inwards, p. 74.

comet

835. Comets bring cold weather.
Cited in Dunwoody, p. 73; Marvin, p. 207.

Connecticut

836. When you can see the Connecticut shore from Long Island, with great clarity, it will rain in about twelve to fifteen hours.
Cited in Sloane 2, p. 14.

convolvulus

837. The convolvulus folds up its petals at the approach of rain. Var.: If the convolvulus does not open its petals by seven in the morning, it will rain or thunder that day; it also closes before a storm.
Cited in Swainson, p. 258; Dunwoody, p. 64; Garriott, p. 25; Inwards, p. 215; Hand, p. 25.

cordage

838. Sailors note the tightening of the cordage on ships as a sign of coming rain.
Cited in Mitchell, p. 230; Garriott, p. 21; Inwards, p. 218; Hand, p. 21.

cormorant

839. Cormorants swiftly returning from sea to land, making a great noise indicates wind.

Cited in Mitchell, p. 233.

840. When cormorants fly from the sea and sea fowls seek their prey in pools or ponds, expect wind.
Cited in Dunwoody, p. 35.

corn

841. A coming storm your shooting corns presage, and aches will throb, your hollow tooth will rage. Vars.: (a) Corns giving trouble indicate bad weather. (b) With coming storm your aching corns presage, and aches will throb, your hollow tooth will rage.
Cited in Dunwoody, p. 14; Whitman, p. 51; Inwards, p. 217; Sloane 1, p. 139; Sloane 2, p. 62; Lee, p. 40.

842. If corns, wounds, and sores itch or ache more than ususal, rain is to fall shortly. Vars.: (a) If a persons corns hurt, it is going to rain. (b) If corns, wounds, and sores itch or ache more than usual; rain is likely to fall shortly. (c) If you have a corn on your left foot, it will always hurt just before a rain. (c) Painful corns mean rain or a thunderstorm. (d) When corns ache rain follows.
Cited in Dunwoody, p. 14; Garriott, p. 21; Inwards, p. 217; Hand, p. 21; Hyatt, p. 30; Smith, p. 7.

843. If your corns hurt, it will snow.
Cited in Smith, p. 10.

corn (grain)

844. Corn blades twined about the stalk mean rain. Var.: Twisted-up corn blades denote rain.
Cited in Hyatt, p. 14.

845. Corn fodder dry and crisp indicates fair weather; but damp and limp, rain. Vars.: (a) Corn fodder turns limp before a rain. (b) If corn fodder is dry and crisp it indicates fair weather, if damp and limp it indicates rain. (c) When corn fodder is crisp, fair weather; when corn fodder is limp, rain is coming.
Cited in Dunwoody, p. 65; Garriott, p. 25; Alstad, p. 93; Hand, p. 25; Freier, p. 39.

846. Corn must be knee high by the fourth of July.
Cited in Mieder, p. 116.

847. Corn should be planted in the light of the moon when the oak leaves are as big as squirrels ears; potatoes should be planted

when the moon is dark. Var.: Plant your corn when the leaves of the oak tree are the size of a mouse's ears.
Recorded in Bryant (N.Y., Vt.).

848. Don't hurry with your corn, don't hurry with your harrows; snow lies behind the dike, more may come and fill the furrows.
Cited in Inwards, p. 30.

849. Don't plant corn till the dogwood is in bloom.
Recorded in Bryant (Wis.).

850. If maize (corn) is hard to husk, expect a severe winter.
Cited in Sloane 2, p. 131.

851. If while cutting corn in the autumn the ears fall to the ground as soon as you hit the stalk, deep snows will fall during winter.
Cited in Hyatt, p. 15.

852. It's time to plant corn when the toads trill.
Recorded in Bryant (Vt.).

853. Look at your corn in May, and you'll come weeping away; look at the same in June, and you'll come home in another tune.
Cited in Denham, p. 46; Swainson, p. 93; Cheales, p. 22; Northall, p. 487.

854. Many hunks of corn are a sign of a cold winter.
Recorded in Bryant (Nebr.).

855. Red corn is followed by a rigorous winter.
Cited in Hyatt, p. 15.

856. When corn fodder stands all dry and crisp, go on your outing, there's no great risk.
Cited in Freier, p. 37.

857. When ears of corn are close to the ground, it means a severe winter.
Cited in Smith, p. 13.

858. When the corn is above the crow's back the frost is over.
Cited in Inwards, p. 30.

corn husk

859. Ears of corn are covered with thicker and stronger husks in cold winters. Vars.: (a) If corn has thick husks, there'll be a hard winter. (b) If corn husks are thicker than usual, a cold winter is ahead. (c) If the corn has thick shucks, it will be a hard winter. (d) Thick and tight corn husks predict a hard winter. (e) Thick, heavy corn shucks foretell a cold winter. (f) When corn shucks

are plentiful, a severe winter is ahead. (g) When corn shucks are thick, it is a sign of a cold winter.

Cited in Dunwoody, p. 65; Inwards, p. 213; Sloane 2, p. 35; Smith, p. 13; Wurtele, p. 299; Freier, p. 80. Recorded in Bryant (Miss., Vt.).

860. If corn husks are long, a long winter is approaching; if short, a short winter.

Cited in Hyatt, p. 14.

861. If corn husks are pointed, it foretells a hard winter; if blunt, a good winter.

Cited in Hyatt, p. 14.

862. If corn husks are thin, a moderate winter may be expected; if thick, a harsh winter.

Cited in Hyatt, p. 14.

863. If the corn husk entirely conceals the ear, predict a closed winter; if the tip of the ear protrudes through the corn husk, predict an open winter. Var.: If the corn husk tightly enfolds the ear, winter will be severe.

Cited in Hyatt, p. 14.

corn silk

864. If corn silk has a light texture, look for a light winter; if a heavy texture, a heavy winter.

Cited in Hyatt, p. 14.

865. If corn silk is abundant, a cold winter is portended; if scanty, a warm winter.

Cited in Hyatt, p. 14.

corncob

866. If corncobs have scattered grains, prepare for an uneven winter; if corncobs are full-grained, prepare for a normal winter.

Cited in Hyatt, p. 14.

867. If the kernels on corncobs are in crooked rows, an irregular winter will follow; if in straight rows, a regular winter.

Cited in Hyatt, p. 14.

Corpus Christi Day

868. Corpus Christi Day clear, gives a good year. Comment: Corpus Christi Day is a religious festival day; the first Thursday after Trinity Sunday.
Cited in Dunwoody, p. 102; Inwards, p. 72; Lee, p. 152.

869. If it rains on Corpus Christi Day, the rye granary will be light.
Cited in Dunwoody, p. 102; Inwards, p. 72.

cottonwood

870. Cottonwood and quaking aspen trees turn up their leaves before rain. Vars.: (a) Cottonwood leaves turn over before a rain. (b) Cottonwood turn up their leaves before a rain.
Cited in Dunwoody, p. 65; Garriott, p. 25; Inwards, p. 212; Hand, p. 25; Sloane 2, p. 112; Freier, p. 57.

cow

871. A cow with its tail to the west, makes weather the best; a cow with its tail to the east, makes weather the least.
Cited in Sloane 2, p. 33.

872. An unexpected returning of cows from the pasture portends a rainstorm in summer or a snowstorm in winter.
Cited in Hyatt, p. 27.

873. Cows lowing constantly is a sign of bad weather.
Recorded in Bryant (Miss.).

874. Cows lying down in the barnyard or pasture during the morning foretell rain before night.
Cited in Hyatt, p. 27.

875. Cows lying down in the morning or standing with their backs to the wind mean rain within a few hours.
Cited in Wilshere, p. 22.

876. Cows remaining near the stable in November warn you of a hard winter.
Cited in Hyatt, p. 27.

877. If a cow kicks backwards while being milked in the morning, rain may be expected.
Cited in Hyatt, p. 27.

878. If a cow raises her tail over her back and runs, a storm within twenty-four hours is indicated.
Cited in Hyatt, p. 27.

879. If cows and sheep sniff the air, a sign of rain.
     Cited in Freier, p. 67.
880. If cows huddle its going to rain.
     Cited in Freier, p. 50.
881. If cows lie down before 9:00 a.m. in the morning, it will rain that day.
     Cited in Smith, p. 5.
882. If cows lie down early in the morning, it will rain before night.
     Cited in Lee, p. 69.
883. If cows look up in the air, and sniff it, a sure sign of rain. Var.: If cows sniff, or stretch out their necks and sniff, or raise their heads and look up into the air, a storm is brewing.
     Cited in Steinmetz 1, p. 111; Hyatt, p. 27.
884. If the cows all graze in the same direction, there will be falling weather.
     Cited in Smith, p. 1.
885. In colder weather cows will lie down and chew the cud when rain approaches, or they will huddle together in a field corner, standing with their tails to the wind.
     Cited in Page, p. 15.
886. Kine (cow), when they assemble at one end of a field with their tails to windward, often indicate rain or wind. Comment: Kine is the archaic plural of cow.
     Cited in Garriott, p. 23; Hand, p. 23.
887. The huddling of cows during day is an indication of a storm.
     Cited in Hyatt, p. 27.
888. When a cow stops and shakes her foot, it indicates that there is bad weather behind her.
     Cited in Dunwoody, p. 30.
889. When a cow tries to scratch its ear, it means a shower is very near; when it thumps its ribs with angry tail, look out for thunder, lightning, hail. Vars.: (a) When a cow endeavors to scratch its ear, it means a shower is very near, when it thumps its ribs with angry tail, look out for thunder, lightning, hail. (b) When a cow tries to scratch her ear, it's the sign that rain is very near. (c) When a cow will scratch her ear, it will mean a rain is near.

Cited in Whitman, p. 47; Inwards, p. 181; Sloane 2, p. 121; Hyatt, p. 27; Lee, p. 44; Page, p. 14; Freier, p. 50. Recorded in Bryant (Wis.).

890. When a herd of cows is lying down, it is going to rain.
Cited in Smith, p. 5.

891. When an old cow raises her head high and sniffs the air, soon a change to nasty weather will come.
Cited in Freier, p. 51.

892. When cows bellow in the evening, expect snow that night.
Cited in Dunwoody, p. 20; Inwards, p. 180.

893. When cows fail their milk, expect stormy and cold weather. Var.: When cows don't give milk, expect stormy and cold weather.
Cited in Dunwoody, p. 30; Inwards, p. 180; Lee, p. 69.

894. When cows lie down, this is often a sign of rain.
Recorded in Bryant (Vt.).

895. When cows refuse to go to pasture in the morning, it will rain before night. Vars.: (a) If cows refuse to go to pasture, expect a storm. (b) If cows refuse to go to the pasture when they are loosed in the morning, it signifies rain.
Cited in Dunwoody, p. 30; Sloane 2, p. 121; Hyatt, p. 27; Lee, p. 68.

896. When kine (cows) and horses lie with their heads upon the ground, it is a sign of rain.
Cited in Inwards, p. 181.

897. When the cows come home with hay pieces dropping out of their mouths, then rain will come.
Cited in Freier, p. 51.

898. When the cows lie around on a hill it means rain.
Cited in Smith, p. 5.

899. When the cows lie down in the field, it is a sign of rain.
Cited in Boughton, p. 124.

cow's dropping

900. If a cow's droppings are frozen in September, they will thaw in October.
Cited in Freier, p. 83.

cowslip

901. The cowslip stalks being short are said to foreshow a dry summer.
Cited in Inwards, p. 216.

crab

902. Before the storm the crab his briny home sidelong forsakes, and strives on land to roam.
Cited in Whitman, p. 49; Lee, p. 59; Freier, p. 23.

903. The appearance of crabs and lobsters indicates that spring has come, and that there will be no more freezing weather.
Cited in Dunwoody, p. 49.

crane

904. Cranes flying towards the source of a river indicate rain.
Cited in Inwards, p. 196.

905. Cranes follow the last frost.
Cited in Dunwoody, p. 35.

906. Cranes forsaking the valleys means rain.
Cited in Mitchell, p. 233.

907. Cranes, soaring aloft and quietly in the air, foreshows fair weather; but if they make much noise, it foreshows a storm that's near at hand.
Cited in Swainson, p. 235; Dunwoody, p. 20; Garriott, p. 19; Inwards, p. 196; Hand, p. 19; Freier, p. 53.

908. If cranes appear in autumn early, a severe winter is expected.
Vars.: (a) If cranes appear early in the autumn expect a severe winter. (b) If cranes come early in autumn, expect a severe winter.
Cited in Swainson, p. 234; Dunwoody, p. 35; Garriott, p. 40; Inwards, p. 196; Hand, p. 40.

909. If cranes place their bills under their wings, expect rain.
Cited in Dunwoody, p. 35.

910. One crane the sign of rain, two cranes the sign of dry weather.
Recorded in Bryant (Ill., Ind.).

911. There will be no rain the day the crane flies down the creek.
Cited in Dunwoody, p. 35.

912. When cranes make a great noise or scream, expect rain.

Cited in Dunwoody, p. 35; Garriott, p. 23; Hand, p. 23; Freier, p. 52.

913. When the cranes early (in October) fly southward, it indicates a cold winter.
Cited in Dunwoody, p. 35; Garriott, p. 40; Hand, p. 40.

crawfish

914. A great number of crawfish quitting the water for land is a rain sign.
Cited in Hyatt, p. 19.

cream

915. Cream makes most freely with a north wind.
Cited in Inwards, p. 115.

916. Cream rises to the top of milk most freely with a north wind.
Cited in Lee, p. 73.

917. When cream and milk turn sour in the night, there are thunderstorms nearby and they will probably arrive.
Cited in Dunwoody, p. 107; Inwards, p. 218; Lee, p. 73.

creek

918. In dry weather, when creeks and springs that have gone dry become moist, or, as we may say, begin to sweat, it indicates approaching rain. Many springs that have gone dry will give a good flow of water just before a rain.
Cited in Dunwoody, p. 107; Garriott, p. 22; Hand, p. 22.

crepe myrtle

919. When crepe myrtle begins to bloom, rain will come very soon.
Cited in Smith, p. 14.

cricket

920. Count the number of cricket chirps in fourteen seconds, add forty, and you have the temperature in degrees Fahrenheit.
Cited in Freier, p. 27.

921. Crickets are accurate thermometers; they chirp faster when warm and slower when cold.
Cited in Sloane 2, p. 42.

922. Crickets are the housewife's barometer, foretelling her when it will rain.
     Cited in Swainson, p. 254.
923. Crickets chirp faster as the temperature rises.
     Cited in Smith, p. 11.
924. Expect frost six weeks from the time the cricket first sings.
     Recorded in Bryant (Nebr.).
925. If crickets chirp in the house, the weather will become colder.
     Cited in Hyatt, p. 18.
926. If you hear crickets chirping, it is an omen of rain.
     Cited in Hyatt, p. 18.
927. If you spit on a cricket and bury it on its back, rain will come.
     Cited in Smith, p. 14.
928. If you step on a cricket it will surely rain.
     Cited in Smith, p. 5.
929. Kill a cricket and it will rain within three days.
     Cited in Hyatt, p. 18.
930. The killing of a cricket brings rain that day or the following day.
     Cited in Hyatt, p. 18.
931. To tell the temperature, take 72 as the number of cricket chirps per minite at 60 degrees F. For every four chirps extra, add 1 degree F.; for every four chirps less, subtract 1 degree F.
     Cited in Sloane 2, p. 125.
932. When crickets chirp unusually, rain is expected.
     Cited in Swainson, p. 254; Inwards, p. 208; Sloane 2, p. 125.
933. When crickets crawl up the wall, it will rain; the higher they crawl, the more it will rain.
     Cited in Reeder, no pp.
934. When crickets sing louder than usual, expect rain. Var.: If crickets chirp louder than usual, or loudly at night, they are informing you of rain.
     Cited in Dunwoody, p. 56; Hyatt, p. 18.
935. When you hear crickets calling, bad weather will follow.
     Cited in Smith, p. 1.

crocus
936. Open crocus, warm weather; closed crocus, cold weather.
     Cited in Freier, p. 28.

937. Spring is just around the corner when the crocus blooms.
Cited in Smith, p. 11.

crow

938. A large flock of crows signifies a change of weather; in summer,
a rain; in winter, a snow.
Cited in Hyatt, p. 22.

939. If a crow cries, it is going to rain.
Cited in Smith, p. 5.

940. If a crow hollers in the morning, look for rain by night.
Cited in Smith, p. 5; Lee, p. 61.

941. If crows feed busily, and hurry over the ground in one direction
and in a compact body, a storm will soon follow.
Cited in Mitchell, p. 226.

942. If crows fly south, a severe winter may be expected; if they fly
north, the reverse. Var.: If crows fly south, we'll have a severe
winter.
Cited in Dunwoody, p. 35; Garriott, p. 40; Hand, p. 40; Freier,
p. 79.

943. If crows gather together in large flocks, it will rain.
Recorded in Bryant (Nebr.).

944. If crows make much noise and fly round and round, expect rain.
Var.: If crows make much noise and fly round and round, a sign
of rain.
Cited in Dunwoody, p. 36; Garriott, p. 23; Hand, p. 23; Freier,
p. 52.

945. If the caws of crows are remarkably loud and incessant anytime
during the day, rain is near.
Cited in Hyatt, p. 22.

946. If the crow hath any interruption in her note like hiccough, or
croak with a kind of swallowing, it signifies wind and rain.
Cited in Inwards, p. 191.

947. If, when crows are flying high, they suddenly dart down and
wheel about in circles, wind is prognosticated.
Cited in Mitchell, p. 226.

948. In autumn and winter, if crows, after feeding in the morning,
return to the rookery and hang about it, rain is expected soon.
Cited in Mitchell, p. 227.

949. One crow flying alone is a sign of foul weather, but if crows fly in pairs, expect fine weather.
   Cited in Dunwoody, p. 35; Hand, p. 23.

950. One flying crow presages a bad storm; two flying crows presage a mild storm.
   Cited in Hyatt, p. 22.

951. The cawing of a crow early in the morning foretells fair weather for the day.
   Cited in Hyatt, p. 21.

952. The continual prating of the crow, chiefly twice or thrice quick calling, indicates rain and stormy weather.
   Cited in Inwards, p. 191.

953. The hoarse crow croaks before rain.
   Cited in Page, p. 19.

954. The low flight of crows indicates coming rain.
   Cited in Mitchell, p. 226.

955. The wicked crow aloud, foul weather threats.
   Cited in Inwards, p. 191.

956. Three crows are an omen of rain.
   Cited in Hyatt, p. 22.

957. To obtain frost, kill a crow and spread its fat on the salt water.
   Cited in Inwards, p. 191.

958. When crows are gathered around on the ground, it is a sign that rain will soon come down.
   Cited in Smith, p. 14.

959. When crows are going home to roost, if they fly high, the next day will be fair and vice versa.
   Cited in Mitchell, p. 226.

960. When crows go to the water and beat it with their wings, expect foul weather.
   Cited in Sloane 2, p. 130.

961. When crows sit in rows on dykes or palings, wind is looked for.
   Cited in Mitchell, p. 226.

962. When the crow or raven gapeth, against the sun, in summer, heat follows.
   Cited in Inwards, p. 191.

cuckold

963. In rain and sunshine, cuckolds go to heaven.

Cited in Denham, p. 3.

cuckoo

964. Cuckoo oats and woodcock hay make a farmer run away.
Comment: If oats cannot be sown till the cuckoo is heard, or the
after-crop of hay cannot be gathered till the woodcocks come
over (about October 20th), the farmer is sure to lose.
Cited in Inwards, p. 192.

965. Cuckoos halloing in low lands indicate rain; on high lands
indicate fair weather.
Cited in Dunwoody, p. 36.

966. If a cuckoo can be penned up in an enclosure of hedges and
trees, to prevent it from flying away, then the summer will never
end.
Cited in Page, p. 22.

967. If the cuckoo does not cease singing at midsummer, corn will be
dear.
Cited in Inwards, p. 192.

968. If the cuckoo sings when the hedge is brown, sell thy horse and
buy thy corn. Comment: You will not be able to afford horse
corn.
Cited in Inwards, p. 191.

969. If the cuckoo sings when the hedge is green, keep thy horse and
sell thy corn. Comment: It will be so plentiful that you will have
enough and to spare.
Cited in Inwards, p. 192; Freier, p. 83.

970. In April, the cuckoo shews his bill; in May, he sings both night
and day; in June, he altereth his tune; in July, away he'll fly; in
August, go he must. Var.: April fourteenth, first cuckoo day.
Cited in Denham, p. 42; Wright, p. 40.

971. The cuckoo comes in April, sings a song in May; then in June
another tune, and then she flies away. Var.: The cuckoo comes
in April, he sings his song in May, in the middle of June he
changes his tune, and in July he flies away.
Cited in Taylor 1, p. 110; Page, p. 22.

972. The cuckoo comes of mid-March, and cucks of mid-April, and
goes away Midsummer month, when the corn begins to fill.
Cited in Denham, p. 38.

973. The cuckoo of summer is the scold crow of winter.

Cited in Mieder, p. 129.

974. When the cuckoo comes to the bare thorn, sell your cow and buy your corn; but when she comes to the full bit, sell your corn and buy your sheep. Comment: A late spring is bad for cattle, and an early spring is bad for corn.
Cited in Inwards, p. 30.

975. When the cuckoo is heard in low lands, it indicates rain; when on high lands, fair weather. Var.: When the cuckoo is heard in low lands, it indicates rain; on high lands, fair weather.
Cited in Garriott, p. 18; Inwards, p. 192; Hand, p. 18.

976. When the cuckoo sings in the sunny sky, all the roads will soon be dry.
Cited in Inwards, p. 192; Lee, p. 65.

curl

977. Curls that kink and cords that bind; signs of rain and heavy wind.
Cited in Humphreys 2, p. 76; Whitman, p. 29; Sloane 2, p. 24; Hyatt, p. 30.

current

978. If there is more than one current in the air, the barometer fluctuates and the weather is uncertain.
Cited in Alstad, p. 41.

cuttlefish

979. Cuttles with their many legs swimming on the top of the water, and striving to be above the waves, do presage a storm. Vars.: (a) Cuttlefish swimming on the surface of water indicate the approach of a storm. (b) Cuttlefish swimming on the surface portend a storm. (c) Cuttlefish swimming on the surface presage a storm.
Cited in Swainson, p. 249; Dunwoody, p. 50; Inwards, p. 199; Lee, p. 59.

daisy

980. Spring is here when you can tread on nine daisies at once on the village green. Vars.: (a) It is not spring until you can plant your foot upon twelve daisies. (b) Spring has come when a maid can set her foot on seven daisies at once.
Cited in Wright, p. 34; Wurtele, p. 298; Simpson, p. 211.

981. The daisy shuts its eye before rain. Var.: Daisies close before rain.
Cited in Inwards, p. 215; Alstad, p. 94; Freier, p. 29.

dandelion

982. Dandelions and daisies tend to shut when bad weather comes.
Cited in Page, p. 10.

983. If the down flies off dandelions and thistles, when there is no wind, it is a sign of rain. Var.: If the down flies off colt's foot, dandelion, and thistles, when there is no wind, it is a sign of rain.
Cited in Swainson, p. 258; Wright, p. 78; Inwards, p. 214.

984. The dandelions close their blossoms before a storm.
Cited in Dunwoody, p. 65; Inwards, p. 214.

985. The flowers of the dandelion and daisy close before rain. Vars.: (a) Dandelion blossoms close before a rain. (b) Dandelion blossoms shut just before a rain. (c) Dandelions close before a rain.
Cited in Dunwoody, p. 65; Alstad, p. 95; Hyatt, p. 15; Freier, p. 38.

986. To have dandelions bloom in January is an omen of clement weather for the rest of the winter.
Cited in Hyatt, p. 15.

987. When the dandelions bloom early in the spring, there will be a short season. When they bloom late, expect a dry summer.
Cited in Dunwoody, p. 107; Inwards, p. 214.

988. When the down of the dandelion contracts, it is a sign of rain. Var.: When the down of the dandelion closes up, it is a sign of rain.
Cited in Swainson, p. 258; Inwards, p. 214; Sloane 2, p. 112; Lee, p. 72.

daw

989. When three daws are seen on St. Peter's vane together, then we're sure to have bad weather. Comment: A daw or jackdaw is a member of the crow family.
Cited in Swainson, p. 240.

dawn

990. A low dawn indicates foul weather; a high dawn indicates wind.
Cited in Dunwoody, p. 78.

991. The darkest hour is nearest the dawn. Vars.: (a) It's always darkest before dawn. (b) It's always darkest before the sun shines. (c) The darkest hour is just before dawn. (d) When the night is darkest, dawn is nearest.
Cited in Denham, p. 4; Taylor 1, p. 111; Wilson, p. 168; Whiting 1, p. 225; Simpson, p. 48; Mieder, p. 134. Recorded in Bryant (Ill.).

day

992. A bad day has a good night.
Cited in Dunwoody, p. 100; Inwards, p. 74.

993. A calm and fair day followed by absence of dew indicates rain.
Cited in Dunwoody, p. 48.

994. A cold day or two does not make winter.
Cited in Whiting 1, p. 95.

995. A muggy day without the slightest trace of a breeze is a token of a thunderstorm.
Cited in Hyatt, p. 12.

996. After a wet day the whole sky often clears at night.
Cited in Dunwoody, p. 116.

997. As the day lengthens, so the cold strengthens. Vars.: (a) As the days grow longer, the cold grows stronger. (b) The longer the day the colder the night (c) When the day is lengthened, cold strengthens. (d) When the days begin to lengthen, the cold begins to strengthen.
Cited in Denham, p. 6; Swainson, p. 20; Dunwoody, p. 100; Northall, p. 476; Garriott, p. 41; Marvin, p. 218; Taylor 1, p. 112; Whitman, p. 36; Inwards, p. 37; Taylor 2, p. 93; Wilson, p. 169; Wurtele, p. 298; Lee, p. 30; Page, p. 44; Wilshere, p. 17; Simpson, p. 48; Whiting 2, p. 153. Recorded in Bryant (Ohio).

998. As the days begin to shorten, the heat begins to scorch them.
Cited in Dunwoody, p. 76; Garriott, p. 41; Marvin, p. 218;
Whitman, p. 37; Inwards, p. 32; Hand, p. 41.

999. As the days grow longer, the storms grow stronger.
Cited in Denham, p. 15; Northall, p. 476; Inwards, p. 30.

1000. Between the hours of ten and two, will show you what the day
will do. Var.: Between ten and two, you'll see what the day will
do.
Cited in Dunwoody, p. 79; Northall, p. 477; Smith, p. 4.

1001. Call not the day a fine one in the morning.
Cited in Inwards, p. 74.

1002. Dark days and storms heighten the appreciation of sunshine.
Cited in Mieder, p. 135.

1003. Do not wait for a rainy day to fix your roof.
Cited in Mieder, p. 135.

1004. Due to the fact that one day we have rain, it does not mean that
we will have it the next day.
Recorded in Bryant (Wis.).

1005. Each day dawns but once.
Recorded in Bryant (Ga., Ohio).

1006. Fine and unusually warm days during the colder months are
called weather breeders.
Cited in Hand, p. 41.

1007. From twelve to two see what the day will do.
Cited in Wright, p. 120.

1008. It is day still, while the sun shines.
Cited in Wilson, p. 170.

1009. It is never a bad day, that has a good night.
Cited in Wilson, p. 25.

1010. Moving day is always a rainy day.
Cited in Hyatt, p. 32.

1011. Never bargain with the devil on a dark day.
Recorded in Bryant (Ohio).

1012. No day is over until the sun has set.
Cited in Mieder, p. 136.

1013. No day so clear, but has dark clouds.
Cited in Wilson, p. 169.

1014. On a hot day muffle yourself to move.
Cited in Mieder, p. 136.

1015. Praise a fair day at night. Vars.: (a) In the evening one may
praise the day. (b) Praise not the day before the night.

Cited in Denham, p. 11; Cheales, p. 19; Whitman, p. 51; Inwards, p. 74; Wilson, p. 643; Mieder, p. 136.

1016. Say nothing of the day 'till the sun is set.
Cited in Denham, p. 6.

1017. Some days are darker than others.
Cited in Mieder, p. 136.

1018. The darkest hour is just before the day.
Cited in Taylor 2, p. 193.

1019. The day never becomes brighter by finding fault with the sun.
Cited in Mieder, p. 136.

1020. The first three days of any season rule the weather of that season.
Cited in Dunwoody, p. 100; Lee, p. 145.

1021. The longer the day, the colder the night.
Recorded in Bryant (Ohio).

1022. The longest day must have an end. Vars.: (a) Be the day never so long, at length comes the evensong. (b) Long as the day may be, the night comes last. (c) The longest day will come to an end. (d) The longest day will have an end.
Cited in Wilson, p. 482; Mieder, p. 177.

1023. When the days get shorter, comes the winter.
Cited in Dunwoody, p. 91.

1024. You can tell by two what a day is going to do.
Cited in Whitman, p. 69.

1025. You can tell the day by the morning.
Cited in Mieder, p. 137.

December

1026. December changeable and mild, the whole winter will remain a child.
Cited in Dunwoody, p. 92.

1027. December cold, with snow, brings rye everywhere. Var.: December cold with snow, good for rye.
Cited in Dunwoody, p. 100; Garriott, p. 45; Whitman, p. 44; Inwards, p. 76; Hand, p. 45; Freier, p. 83.

1028. December's frost and January's flood, never boded the husband-man's good.
Cited in Swainson, p. 150; Northall, p. 443; Wright, p. 122; Inwards, p. 67.

1029. If it rains on Sunday in December, before mass, it will rain the whole week.

Cited in Dunwoody, p. 99.

1030. The direction of the wind on the first three days of December shows whence the wind will blow during the three following months.
Cited in Hyatt, p. 9.

1031. The twelve days commencing December 25th and ending January 5th are said to be the keys of the weather of the year.
Cited in Dunwoody, p. 100.

1032. The weather on the first three days of December regulates the weather for the three winter months.
Cited in Hyatt, p. 14.

deer

1033. Deer and elk come down from the mountains at least two days before a storm.
Cited in Dunwoody, p. 33; Lee, p. 67.

1034. Deer move to lower wooded areas before a rain.
Cited in Freier, p. 50.

1035. When deer are in gray coat in October, expect a severe winter.
Cited in Dunwoody, p. 31.

dew

1036. A heavy dew in the middle latitudes is said to indicate southerly winds.
Cited in Dunwoody, p. 48.

1037. A heavy dew with a south to east wind, fair; with a northwest wind, rain.
Cited in Dunwoody, p. 48.

1038. A light dew in the morning will be followed by rain; a heavy dew, by splendid weather.
Cited in Hyatt, p. 12.

1039. Dew in the night, next day will be bright.
Cited in Page, p. 5.

1040. Dew is an indication of fine weather; so is fog.
Cited in Inwards, p. 146.

1041. During summer a heavy dew is sometimes followed by a southerly wind in the afternoon.
Cited in Dunwoody, p. 48; Inwards, p. 146.

1042. Heavy dew indicates fair weather.
Cited in Dunwoody, p. 48; Inwards, p. 146; Wilshere, p. 21.

1043. Heavy dews in hot weather indicate a continuation of fair weather.
Cited in Mitchell, p. 231; Dunwoody, p. 20.

1044. Heavy dews in March; heavy fogs in August.
Cited in Hyatt, p. 13.

1045. If in clear summer nights there is no dew, expect rain next day.
Cited in Inwards, p. 146.

1046. If nights three dewless there be, 'twill rain, you're sure to see.
Cited in Dunwoody, p. 48; Wright, p. 78; Inwards, p. 147.

1047. If the dew drips off the house three days in a row, it will rain.
Cited in Reeder, no pp.

1048. If the dew lies on the grass plentifully after a fair day, it indicates that the following day will be fair.
Cited in Dunwoody, p. 48.

1049. If there is a heavy dew and it soon dries, expect fine weather; if it remains long on the grass, expect rain in twenty-four hours.
Cited in Dunwoody, p. 48.

1050. If there is a heavy dew it indicates fair weather; no dew indicates rain.
Cited in Dunwoody, p. 48; Inwards, p. 146.

1051. If there is a profuse dew in summer, it is about seven to one that the weather will be fine.
Cited in Inwards, p. 146.

1052. If there is dew on the grass in the morning, fair weather.
Cited in Freier, p. 36.

1053. If there is no dew and no wind after a fair day, rain will follow.
Cited in Dunwoody, p. 48.

1054. If your feet you wet with dew in the morning, you may keep them dry for the rest of the day.
Cited in Dunwoody, p. 48.

1055. Much dew after a fair day indicates another fair day.
Cited in Dunwoody, p. 48.

1056. No dew after a hot day fortells rain.
Cited in Mitchell, p. 231; Dunwoody, p. 20.

1057. No dew in the morning is a forecast of rain.
Cited in Hyatt, p. 12.

1058. No dew indicates rain.
Cited in Dunwoody, p. 48; Inwards, p. 146.

1059. The absence of dew for three days indicates rain.
Cited in Dunwoody, p. 48.

1060. The dews of the evening industriously shun, they're the tears of the sky for the loss of the sun.
Cited in Humphreys 2, p. 56; Whitman, p. 34; Alstad, p. 116.

1061. The lack of dew for three mornings brings rain, whereas others say rain is brought by three dewy mornings.
Cited in Hyatt, p. 12.

1062. The number of dews before Easter will indicate the number of hoar frosts to occur after Easter, and the number of dews to occur in August.
Cited in Dunwoody, p. 48.

1063. When the dew is on the grass, rain will never come to pass.
Cited in Humphreys 2, p. 59; Whitman, p. 34; Alstad, p. 116; Sloane 2, p. 36; Hyatt, p. 12; Smith, p. 4; Lee, p. 28; Wilshere, p. 21; Freier, p. 36. Recorded in Bryant (Ohio).

1064. When there is no dew at such times as usually there is, it foreshoweth rain.
Cited in Inwards, p. 147.

1065. With dew before midnight, the next day will sure be bright.
Cited in Dunwoody, p. 48; Inwards, p. 146; Alstad, p. 115.

digestion

1066. In persons of weak and irritable constitution, the digestive powers are much influenced by the weather; before storms such persons are uneasy.
Cited in Inwards, p. 217.

dirt bird

1067. The dirt bird sings, and we shall have rain.
Cited in Swainson, p. 247; Marvin, p. 216; Inwards, p. 194; Wilson, p. 189.

ditch

1068. When the ditch and the pond offend the nose, then look for rain and stormy blows. Vars.: (a) When ditches and cellars smell most, a long rain is near. (b) When the ditch and pond affect the nose, then look out for rain and storm blows.
Cited in Whitman, p. 50; Sloane 2, p. 55; Lee, p. 36; Freier, p. 25. Recorded in Bryant (Wis.).

dog

1069. A dog chewing or chasing his tail presages rain.

Cited in Hyatt, p. 27.
1070. A dog rolling on the ground is a sign of violent wind.
Cited in Lee, p. 70.
1071. A dog that becomes sportive and darts about all day is foretelling a windstorm.
Cited in Hyatt, p. 27.
1072. After a dog rolling on the ground has turned over three times, there will be a storm.
Cited in Hyatt, p. 27.
1073. Barking dogs, cats sneezing, washing behind their ears or generally restless, may indicate weather change.
Cited in Wilshere, p. 22.
1074. Dog's tails straighten when rain is near.
Cited in Lee, p. 70.
1075. Dogs digging or making deep holes in the ground are said to indicate rain thereby. Vars.: (a) Dogs making holes in the ground, eating grass in the morning, or refusing meat are said to indicate coming rain. (b) Dogs making holes in the ground indicate coming rain.
Cited in Dunwoody, p. 31; Garriott, p. 23; Inwards, p. 178; Hand, p. 23.
1076. Dogs hunt better before a rain.
Cited in Freier, p. 66.
1077. Dogs refusing meat is an indication of rain.
Cited in Dunwoody, p. 31; Garriott, p. 23; Inwards, p. 178.
1078. Dogs shedding their hair in autumn is a sign of an open winter.
Cited in Hyatt, p. 27.
1079. If a dog howls and looks down to the ground, we will have a rain.
Cited in Hyatt, p. 28.
1080. If a dog howls at the moon in summer, a rain is foretold; if in winter, a snow.
Cited in Hyatt, p. 28.
1081. If a dog howls when someone leaves the house, it indicates rain. Vars.: (a) Dogs howling when anyone goes out indicates rain. (b) If a dog howls while company is leaving your house, it indicates rain.
Cited in Dunwoody, p. 31; Inwards, p. 178.
1082. If a dog lies on his back with feet up in the air, prepare for stormy weather; moreover, the storm will come from the

direction toward which his nose points.
Cited in Hyatt, p. 27.

1083. If a dog lies on its back, it is a sign of rain.
Recorded in Bryant (Nebr.).

1084. If a dog pulls his feet up high while walking, a change in the weather is coming.
Cited in Freier, p. 30.

1085. If a dog, while eating grass, changes his position frequently, rain is in the air.
Cited in Hyatt, p. 27.

1086. If dogs and horses sniff the air, a summer shower will soon be there.
Cited in Freier, p. 66.

1087. If dogs howl, expect a storm.
Cited in Sloane 2, p. 121.

1088. If dogs roll on the ground and scratch, or become drowsy and stupid, it is a sign of rain.
Cited in Swainson, p. 231; Inwards, p. 178.

1089. If dogs roll on the ground and scratch, or eat grass and refuse meat, a sign of rain.
Cited in Sloane 2, p. 120.

1090. If it is raining and a dog washes itself before seven, it will clear by eleven.
Cited in Hyatt, p. 27.

1091. If your dog has an exceptionally thick coat of fur, watch out for the coming winter, it will be snowy, cold and damp.
Cited in Smith, p. 14.

1092. Stepping on a dog's tail will cause a rain.
Cited in Hyatt, p. 27.

1093. The unusual howling of dogs portends a storm.
Cited in Inwards, p. 178; Sloane 2, p. 120.

1094. To see a dog lying in a draught is a sign of warmer weather.
Cited in Hyatt, p. 28.

1095. Watch for rain, if you can smell a dog's skin.
Cited in Hyatt, p. 27.

1096. When a dog acts restless and unusually alert, rain is coming.
Cited in Smith, p. 6.

1097. When a dog or cat eats grass in the morning, it will certainly rain before night.
Cited in Dunwoody, p. 31.

1098. When a dog puts its nose in the snow, another snowstorm is brewing.
Cited in Smith, p. 10.

1099. When a dog rolls on his back, it will soon rain. Var.: Dogs rolling on their back, expect rain.
Cited in Freier, p. 49.

1100. When dogs chew on grass, it is going to rain.
Recorded in Bryant (Nebr.).

1101. When dogs dig holes, howl when anyone goes out, eat grass, or refuse meat, it is a sign of rain.
Cited in Lee, p. 70.

1102. When dogs eat grass it will be rainy. Vars.: (a) Expect rain when dogs eat grass. (b) If you see a dog eating grass, it will rain soon. (c) When a dog eats grass, it is a sign of rain.
Cited in Swainson, p. 231; Dunwoody, p. 29; Garriott, p. 23; Inwards, p. 178; Sloane 2, p. 36; Hyatt, p. 27.

1103. When dogs rub themselves in winter, it will thaw soon.
Cited in Lee, p. 70.

1104. When dogs sleep on their back in the yard, it is a sign of rain.
Cited in Reeder, no pp.

dog days

1105. As the dog days commence so they end. Var.: As the dog days begin, so they will end. Comment: The dog days cover the hottest period of the summer, from July 3rd to August 11th, when the weather is sultry and the air stifling. In Roman times it was thought that Sirius, the dog star, added its heat to the heat of the sun during this period.
Cited in Dunwoody, p. 91; Garriott, p. 42; Whitman, p. 39; Inwards, p. 58; Hand, p. 42.

1106. Dog days are when dogs go mad. Comment: As drought time was often plague time too, people believed that sickness and madness came with Sirius, the dog star.
Cited in Sloane 2, p. 36.

1107. Dog days bright and clear indicate a good year; but when accompanied by rain, we hope for better times in vain. Var.: Dog days bright and clear indicate a happy year; but when accompanied by rain, for better times our hopes are vain.
Cited in Dunwoody, p. 98; Garriott, p. 42; Whitman, p. 39; Inwards, p. 58; Hand, p. 42.

1108. Rain on first dog day, it will rain for forty days after. Vars.: (a) If it rains on the first dog day, it will rain for forty days.
Cited in Dunwoody, p. 102; Whitman, p. 39; Inwards, p. 58; Hyatt, p. 10.

dogrose

1109. If many dogroses are seen, expect a severe winter.
Cited in Sloane 2, p. 131.

dogwood

1110. Frost will not occur after the dogwood blossoms.
Cited in Dunwoody, p. 65; Inwards, p. 212; Sloane 2, p. 131.

1111. When the blooms of the dogwood tree are full, expect a cold winter. When blooms of the same are light, expect a warm winter.
Cited in Dunwoody, p. 65; Inwards, p. 212.

dolphin

1112. Dolphins, as well as porpoises, when they come about a ship, and sport and gambol on the surface of the water, betoken a storm, hence they are regarded as unlucky omens by sailors.
Cited in Dunwoody, p. 50; Inwards, p. 198.

1113. Dolphins in fair and calm weather pursuing one another, as one of their waterish pastimes, foreshows wind, but if they play thus when the seas are rough and troubled, it is a sign of fair and calm weather to ensue. Var.: Dolphins pursuing one another in calm weather foreshow wind, and from that part whence they fetch their frisks; but if they play in rough weather, it is a sign of coming calm.
Cited in Swainson, p. 249; Inwards, p. 198.

1114. Dolphins sporting in a calm sea are thought to prognosticate wind from that quarter whence they come; but if they play in a rough sea, and throw water about, it will be fine.
Cited in Inwards, p. 198.

1115. If dolphins are seen to leap and toss, fine weather may be expected, and the wind will blow from the quarter in which they are seen.
Cited in Inwards, p. 198.

1116. When dolphins and porpoises play near a ship, it is a sign of a storm.
Cited in Sloane 2, p. 122.

donkey

1117. It is time to stack your hay and corn, when the old donkey blows his horn. Var.: When the donkey blows his horn, 'tis time to house your hay and corn. Comment: "Donkey blows his horn" refers to thunder.
Cited in Swainson, p. 229; Dunwoody, p. 31; Inwards, p. 180; Lee, p. 71. Recorded in Bryant (N.Y.).

door

1118. A door sticking to the jamb means rain.
Cited in Hyatt, p. 30.
1119. Doors and drawers stick before a rain.
Cited in Freier, p. 37.
1120. Doors and windows are hard to shut when rain impends. Var.: Doors and windows are harder to open and shut in damp weather.
Cited in Steinmetz 1, p. 113; Garriott, p. 21; Inwards, p. 1616; Hand, p. 21; Lee, p. 74.
1121. When doors and windows start to stick, it will probably rain.
Cited in Freier, p. 37.

dotterel

1122. When dotterel do first appear, it shows that frost is very near; but when the dotterel do go, then you may look for heavy snow. Comment: A dotterel is a plover.
Cited in Mitchell, p. 227; Inwards, p. 196.

dove

1123. A dove cooing in a tree is a rain omen.
Cited in Hyatt, p. 22.
1124. Doves come later home to the cote before a storm.
Cited in Alstad, p. 78.
1125. Doves flying in bunches is a sign of rain.
Cited in Reeder, no pp.
1126. Doves that coo constantly and are more restless than usual warn you of rain.
Cited in Hyatt, p. 22.
1127. If a dove mourns in February, there will be no more cold weather.
Cited in Wurtele, p. 298.

1128. The cooing of a dove is a sure sign of rain. Var.: Rain doves coo before a rain.
Cited in Smith, p. 6; Freier, p. 53.
1129. There will be neither freeze nor frost, after doves have cooed in the spring.
Cited in Hyatt, p. 22.
1130. When the morning dove calls "shoo-oo", rain is coming.
Cited in Smith, p. 6.

dream
1131. If a person dreams about blood, it foretells a snowstorm.
Cited in Smith, p. 10.
1132. When you dream of seeing the dead, it is a sign of rain. Var.: If you dream of a dead person, it is a sign of rain.
Cited in Smith, p. 8. Recorded in Bryant (Miss.).

drop
1133. Little drops produce a shower.
Cited in Mieder, p. 169.
1134. Many drops make a shower.
Cited in Denham, p. 11; Swainson, p. 212.

drought
1135. After great droughts come great rains.
Cited in Inwards, p. 158.
1136. Drought never bred dearth in England.
Cited in Denham, p. 2; Inwards, p. 31.
1137. In Texas and the southwest when the wind shifts during a drought, expect rain.
Cited in Dunwoody, p. 87.

duck
1138. Ducks and geese flying north indicate warm weather; flying south, cold weather.
Recorded in Bryant (Nebr.).
1139. Ducks and geese quack and cackle more loudly when rain approaches.
Cited in Page, p. 19.
1140. Ducks, geese, and other aquatic birds are more noisy and active than usual before rain.
Cited in Steinmetz 2, p. 290.

1141. If a duck flaps its wings continually, rain is in the air.
      Cited in Hyatt, p. 25.
1142. If ducks fly backwards and forwards, and continually plunge in
      water and wash themselves incessantly, wet weather will ensue.
      Var.: If ducks and geese fly back and forth and duck into the
      water to wash themselves, wet weather will ensue.
      Cited in Swainson, p. 236; Inwards, p. 188; Sloane 2, p. 129.
1143. If ducks or drakes do shake and flutter their wings when they
      rise, it is a sign of ensuing water.
      Cited in Inwards, p. 187.
1144. If wild ducks or wild geese start south in early autumn, an early
      winter is betokened; if in late autumn, a late winter.
      Cited in Hyatt, p. 22.
1145. Very cold weather and long hard winters are shown by the early
      arrival of winter migrants such as ducks, geese and swans
      together with much larger than ususal flocks of fieldfares and
      redwings.
      Cited in Page, p. 20.
1146. When ducks are driving through the burn, that night the weather
      takes a turn. Comment: Burn means water.
      Cited in Swainson, p. 235; Northall, p. 473; Inwards, p. 187;
      Lee, p. 64.
1147. When ducks quack loudly, it's a sign of rain.
      Cited in Freier, p. 22.
1148. When ducks quack profusely, they are said to be calling for rain.
      Cited in Lee, p. 63.
1149. When the ducks go south early, there will be a hard winter.
      Recorded in Bryant (Nebr.).
1150. Wild ducks scattered around the lakes near Lake Superior form
      in large flocks and go south one month earlier in cold or early
      winters than in mild or pleasant winters.
      Cited in Dunwoody, p. 41; Garriott, p. 40; Hand, p. 40.

duck breastbone
1151. If a breastbone of a duck be red, it signifieth a long winter; if
      white, the contrary.
      Cited in Inwards, p. 187.

dust
1152. Dust in March brings grass and foliage.
      Cited in Garriott, p. 42.

1153. Dust rising from the road in dry weather, when there is little wind, predicts change.
Cited in Mitchell, p. 229.

1154. Dust rising in dry weather is a sign of approaching change.
Cited in Dunwoody, p. 107; Inwards, p. 220.

1155. If dust whirls round in eddies when being blown about by the wind, it is a sign of rain.
Cited in Inwards, p. 220.

ear

1156. A ringing in your ear is a rain sign.
    Cited in Hyatt, p. 30.
1157. A singing in the ears sometimes indicates a change of weather, generally an increase of pressure or rise in the barometer.
    Cited in Inwards, p. 217.
1158. Ringing in the ear at night indicates a change of wind.
    Cited in Dunwoody, p. 107; Inwards, p. 217.

earthworm

1159. If many earthworms appear, it presages rain. Vars.: (a) If many earthworms appear, rain will follow. (b) The earthworm appearing in large numbers on the surface indicates rain.
    Cited in Mitchell, p. 228; Swainson, p. 252; Inwards, p. 201.
1160. When common garden worms form many "casts," rain or frost will appear according to the season of the year.
    Cited in Inwards, p. 201.
1161. When earthworms appear in the daytime, expect rain; but when early in the evening, it indicates a mild night with heavy dew and two days' fine weather.
    Cited in Inwards, p. 201.
1162. Worms descend to a great depth before either a long drought or a severe frost.
    Cited in Inwards, p. 201.

Easter

1163. A cloudy Easter will be followed by seven weeks of cloudy weather.
    Cited in Hyatt, p. 14.
1164. A good deal of rain on Easter Day, gives a good crop of corn, but little hay.
    Cited in Northall, p. 451; Inwards, p. 71.
1165. A late Easter means a late spring. Var.: A late Easter, a late spring.
    Cited in Boughton, p. 124; Hyatt, p. 14.
1166. An early Easter means an early spring. Var.: An early Easter, and early spring.
    Cited in Boughton, p. 124; Hyatt, p. 14.
1167. An Easter rain means rainy weather for the next six Sundays and ten Mondays.

Cited in Hyatt, p. 10.

1168. An Easter rain means wet weather for the next six Sundays.
Cited in Hyatt, p. 10.

1169. Before Easter, winter is not to be trusted.
Cited in Sloane 2, p. 36.

1170. Easter in snow, Christmas in mud; Christmas in snow, Easter in mud. Var.: Easter in snow, Christmas in mud, Christmas in mud, Easter in snow.
Cited in Dunwoody, p. 102; Whitman, p. 40; Inwards, p. 68; Page, p. 37; Wilshere, p. 16; Freier, p. 78.

1171. From whatever direction the wind blows on Easter, it will blow for the next six weeks.
Cited in Hyatt, p. 9.

1172. If fair weather from Easter to Whitsuntide, the butter will be cheap. Comment: Whitsuntide is the week beginning with Whitsunday (seventh Sunday after Easter) or Pentecost.
Cited in Dunwoody, p. 102; Inwards, p. 72.

1173. If it rains on an Easter falling on April Fool's Day, it will not rain for seven Sundays.
Cited in Hyatt, p. 10.

1174. If it rains on Easter Day, plenty of grass and little hay. Var.: Rain on Easter Day, plenty of grass, but little good hay.
Cited in Cheales, p. 21; Northall, p. 451.

1175. If it rains on Easter Day, there shall be good grass but very bad hay.
Cited in Wilson, p. 213; Page, p. 37.

1176. If it rains on Easter Sunday, it will rain seven Sundays following. Vars.: (a) If it rains on Easter Sunday, it will rain seven Sundays in a row. (b) Rain on Easter Sunday means rain for the next seven Sundays. (c) Rainy weather on Easter will not cease for seven days.
Cited in Boughton, p. 123; Hyatt, p. 10; Smith, p. 7; Freier, p. 91; Reeder, no pp.

1177. If the sun shine clear on Easter Day or Palm Sunday, or both of them, there will be fine weather, plenty of corn and other fruits of the earth.
Cited in Swainson, p. 71.

1178. If the sun shines on Easter Day, it shines on Whitsunday. Vars.: (a) If the sun shines on Easter Day, it shines on Whitsunday likewise. (b) If the sun shines on Easter Day, it will shine on Whitsunday. Comment: Whitsunday is the seventh Sunday after Easter.

Cited in Denham, p. 33; Swainson, p. 71; Dunwoody, p. 77; Inwards, p. 70.

1179. If you can wet a handkerchief with rain on Easter, look for a fine crop year.
Cited in Hyatt, p. 34.

1180. Late Easter, long, cold spring.
Cited in Inwards, p. 70.

1181. Past Easter frost, fruit not lost.
Cited in Inwards, p. 71.

1182. Pleasant Easter weather means a bountiful crop and harvest.
Cited in Lee, p. 152.

1183. Rain on Easter gives slim fodder. Var.: Rain at Easter gives slim fodder.
Cited in Dunwoody, p. 102; Inwards, p. 70.

1184. Such weather as there is on Easter Day there will be at harvest. Var.: The weather on Easter Day is supposed to indicate the weather at harvest.
Cited in Inwards, p. 71; Lee, p. 152.

1185. The direction the wind blows on Easter will be its direction during the forty days that follow.
Cited in Hyatt, p. 9.

1186. There will not be any grapes the year it rains on Easter.
Cited in Hyatt, p. 34.

1187. Wet weather on Easter remains until Ascension Day forty days later.
Cited in Hyatt, p. 10.

1188. Wind in the northeast about six o'clock on Easter morning foretells seven weeks of rain.
Cited in Hyatt, p. 9.

eel

1189. Eels in greater numbers than usual during the spring betoken high water.
Cited in Hyatt, p. 18.

1190. If eels are very lively, it is a sign of rain.
Cited in Dunwoody, p. 50; Inwards, p. 199; Lee, p. 59.

1191. The catching of an eel warns you of a rise in the river, because eels are caught only before high water.
Cited in Hyatt, p. 19.

1192. They are nought but eels, that never will appear till that tempestuous winds or thunder tear their slimy beds.
Cited in Inwards, p. 199; Lee, p. 60.

Ember Days

1193. If the weather on Ember Days is fair, three months of good weather will follow; if the weather on these three days is rainy, the following three months will be wet. Comment: Ember Days are a Wednesday, Friday, or Saturday in any of the four weeks commencing on the first Sunday in Lent or on Whitsuntide or including or immediately following September 14th or December 13th set apart for fasting and prayer and observed especially hv the Anglican and Roman Catholic churches.
Cited in Hyatt, p. 14.

1194. The direction of the wind on the Ember Days of September (Wednesday, Friday and Saturday after the fourteenth) determines the weather for winter: if it blows from the north, expect a closed winter; if from the south, an open winter.
Cited in Hyatt, p. 9.

English summer

1195. An English summer, two fine days and a thunderstorm. Vars.: (a) An English summer, three hot days and a thunderstorm. (b) An English summer, two hot days and a thunderstorm.
Cited in Denham, p. 48; Inwards, p. 31; Wilson, p. 223.

equinoctial gale

1196. The vernal equinoctial gales are stronger than the autumnal.
Cited in Dunwoody, p. 43; Inwards, p. 70.

equinoctial storm

1197. As clear off the line, or equinoctial storm, so will all storms clear for six months. Var.: As the equinoctial storms clear, so will all storms clear for six months.
Cited in Dunwoody, p. 89; Inwards, p. 69.

1198. In equinoctial storms fish bite the best before the sun crosses the line.
Cited in Dunwoody, p. 50.

equinox
1199. As the wind and weather is at the time of the equinox, so will be the wind and weather generally during the following three months. Var.: As the wind and weather at the equinoxes, so will they be for the next three months.
Cited in Dunwoody, p. 89; Whitman, p. 63; Inwards, p. 69.
1200. If near the time of the equinox it blows in the day, it generally hushes towards evening.
Cited in Inwards, p. 70.
1201. If the wind is northeast at vernal equinox, it will be a good season for wheat and a poor one for other kinds of corn; but if south or southwest, it will be good for other corn, but bad for wheat.
Cited in Inwards, p. 70.
1202. Wind northeast or north at noon of the vernal equinox, no fine weather before midsummer. If westerly or southwesterly, fine weather till midsummer.
Cited in Inwards, p. 69.

evening
1203. An evening red and morning gray, will set the traveler on his way, but if the evening's gray, and the morning red, put on your hat or you'll wet your head. Vars.: (a) A red evening and a gray morning set the pilgram a-walking. (b) An evening gray and morning red will send the shepherd wet to bed. (c) An evening red and morning gray are sure signs of a fine day; evening gray and morning red, put on your hat or you'll wet your head. (d) An evening red and morning gray is a token of a bonnie day. (e) An evening red and morning gray make the pilgram sing. (f) Evening gray and morning red, put on your hat or you'll wet your head. (g) Evening gray and morning red send the traveler back to bed. (h) Evening gray and morning red will make the shepherd hang his head. (i) Evening gray and morning red will pour rain on the pilgram's head. (j) Evening red and morning gray, two sure signs of a fair day. (k) Evening red and morning

gray will send the sailor on his way, but evening gray and morning red will bring rain down upon his head. (l) If the evening is red and the morning gray, it is the sign of a bonnie day; if the evening's gray and the morning red, the lamb and ewe go wet to bed. (m) The evening red and the morning gray is the sign of a bright and cheery day; the evening gray and the morning red, put on your hat or you'll wet your head. (n) When the evening is gray and the morning red the sailor is sure to have a wet head.

Cited in Denham, p. 19; Mitchell, p. 221; Steinmetz 1, p. 119; Steinmetz 2, p. 278; Swainson, p. 181; Dunwoody, p. 68; Northall, p. 460; Garriott, p. 26; Humphreys 1, p. 437; Marvin, p. 205; Taylor 1, p. 112; Whitman, p. 32; , no pp.; Brunt, p. 69; Inwards, p. 83; Taylor 2, p. 122; Hand, p. 26; Sloane 2, p. 54; Hyatt, p. 4; Smith, p. 4; Wilson, p. 227; Wurtele, p. 296; Lee, p. 80; Page, p. 5; Wilshere, p. 3; Freier, p. 33; Mieder, p. 183. Recorded in Bryant (Vt., Wis.).

fall down

1204. If a person falls down while walking, it means a storm; a large storm, if the person is large; a small storm, if the person is small.
Cited in Hyatt, p. 32.

feather

1205. Heavy feathers on the Thanksgiving turkey means hard winter.
Cited in Smith, p. 13.

February

1206. A February spring is not worth a pin.
Cited in Marvin, p. 211; Inwards, p. 40; Lee, p. 135.

1207. A February spring is worth nothing. Var.: A February spring is nothing worth.
Cited in Northall, p. 433; Humphreys 2, p. 10; Whitman, p. 42; Wilshere, p. 8.

1208. February, and ye be fair, the hoggs'll mend, and nothing pair. Comment: Hoggs are sheep which have not been shorn; pair means lessen.
Cited in Inwards, p. 39.

1209. February, and ye be foul, the hoggs'll die in ilka pool. Comment: Hoggs are sheep which have not been shorn; ilka means every.
Cited in Inwards, p. 39.

1210. February builds a bridge (of ice), and March breaks it down. Var.: February makes a bridge and March breaks it.
Cited in Denham, p. 27; Swainson, p. 42; Dunwoody, p. 94; Inwards, p. 41; Wilson, p. 252; Page, p. 44; Wilshere, p. 8.

1211. February doth cut and shear.
Cited in Dunwoody, p. 94; Inwards, p. 39; Wilshere, p. 8.

1212. February fill ditch, black or white, don't care which; if it be white, it's better to like. Comment: Black or white refers to rain or snow.
Cited in Swainson, p. 40; Northall, p. 433; Whitman, p. 19; Inwards, p. 40.

1213. February fills the dyke, either with black or white. Vars.: (a) February fill dyke, be it black or be it white, but if it be white, 'tis better to like. (b) February fill dyke, fill with either black or

white; March muck it out, with besom and clout. (c) February
fill the dyke, weather either black or white, but if it be white, it's
better to like. (d) February fill dyke with what thou dost like.
Cited in Denham, p. 31; Cheales, p. 21; Dunwoody, p. 94;
Northall, p. 431; Marvin, p. 211; Inwards, p. 40; Page, p. 44;
Whiting 1, p. 425; Whiting 2, p. 218; Wilshere, p. 7.

1214. February rain is only good to fill ditches.
Cited in Dunwoody, p. 94; Garriott, p. 44; Whitman, p. 42;
Hand, p. 44; Freier, p. 80.

1215. February sun is dearly won.
Recorded in Bryant (Can. [Ont.]).

1216. Fog in February means frosts in May.
Cited in Inwards, p. 41.

1217. For every thunder with rain in February, there will be a cold
spell in May.
Cited in Dunwoody, p. 94; Inwards, p. 41.

1218. Heavy north winds in February forbode a fruitful year.
Cited in Dunwoody, p. 94.

1219. If bees get out in February, the next day will be windy and rainy.
Cited in Inwards, p. 40.

1220. If February gives much snow, a fine summer it doth foreshow.
Cited in Dunwoody, p. 94; Northall, p. 434; Garriott, p. 44;
Wright, p. 22; Whitman, p. 42; Hand, p. 44; Freier, p. 82.

1221. If in February there be no rain, 'tis neither good for hay nor
grain.
Cited in Swainson, p. 39; Wilshere, p. 8; Simpson, p. 78.

1222. If the last eight days of February and the first twenty days of
March are for the most part rainy, then the spring and summer
quarters will be so too.
Cited in Steinmetz 2, p. 193.

1223. If the last eighteen days of February and ten days of March be,
for the most part, rainy, then the spring and summer quarters are
likely to be so too. Var.: If the eighteen last days of February be
wet, and the first ten of March, you'll see that the summer too,
will prove too wet and danger will ensue.
Cited in Steinmetz 1, p. 136; Inwards, p. 44.

1224. In February, if thou hearest thunder, thou wilt see a summer's
wonder.

Cited in Denham, p. 30; Northall, p. 434; Inwards, p. 41; Wilshere, p. 8.

1225. Of all the months in the year, curse a fair February. Vars.: (a) All the months of the year curse a fair February. (b) All the months of the year fear a fair February.
Cited in Denham, p. 6; Swainson, p. 39; Cheales, p. 21; Dunwoody, p. 94; Northall, p. 433; Humphreys 1, p. 430; Wright, p. 18; Marvin, p. 211; Humphreys 2, p. 10; Whitman, p. 42; Inwards, p. 39; Wilson, p. 541; Wilshere, p. 7.

1226. Rain and sunshine together in February means a bountiful harvest for the year.
Cited in Hyatt, p. 34.

1227. Sleet in February is followed by a fine apple crop.
Cited in Hyatt, p. 34.

1228. The cold days of February will be the warm days of March; the warm days of February will be the cold days of March.
Cited in Hyatt, p. 13.

1229. There is always one fine week in February.
Cited in Dunwoody, p. 94; Garriott, p. 44; Whitman, p. 42; Inwards, p. 39; Hand, p. 44; Freier, p. 82.

1230. Thunder in February or March, poor sugar maple year.
Cited in Inwards, p. 41.

1231. Violent north winds in February herald a fertile year.
Cited in Dunwoody, p. 94; Inwards, p. 41.

1232. Warm February bad hay crop; cold February good hay crop.
Cited in Marvin, p. 211; Inwards, p. 39.

1233. When gnats dance in February, the husbandman becomes a beggar.
Cited in Swainson, p. 38; Inwards, p. 39.

1234. When in February it is mild, the spring brings the frost by night.
Cited in Dunwoody, p. 91.

1235. When it rains in February, all the year suffers.
Cited in Inwards, p. 40.

1236. When the cat in February lies in the sun, she will creep behind the stove in March.
Cited in Inwards, p. 41.

1237. When the north wind does not blow in February, it will surely come in March.
Cited in Inwards, p. 41.

1238. Whenever the latter part of February and beginning of March are dry, there will be a deficiency of rain up to Midsummer day.
Cited in Inwards, p. 41.

1239. Winter's back breaks about the middle of February.
Cited in Inwards, p. 44.

February 2

1240. February 2nd bright and clear gives a good flax year.
Cited in Dunwoody, p. 94.

1241. Good weather on February 2nd indicates a long continuance of winter, and a bad crop; on the contrary, if foul it is a good omen.
Cited in Inwards, p. 42; Freier, p. 74.

1242. If a storm on February 2nd, spring is near; but if that day be bright and clear, the spring will be late.
Cited in Garriott, p. 42; Inwards, p. 43; Hand, p. 42.

1243. If bear comes out on the second of February, and if he sees his shadow he returns for six weeks.
Cited in Hand, p. 39.

1244. If it snows on the second of February, only so much as may be seen on a black ox, then summer will come soon.
Cited in Inwards, p. 43.

1245. If it storms on February 2nd, then spring is not very far; but when bright and clear, then the spring will be late.
Cited in Dunwoody, p. 90.

1246. If on the second of February the goose finds it wet, then the sheep will have grass on March 25th.
Cited in Dunwoody, p. 94; Inwards, p. 43.

1247. If the groundhog is sunning himself on the second of February, he will return for four weeks to his winter quarters again. Var.: If the groundhog is sunning himself on February 2nd, he will return to his winter quarters for six more weeks of winter.
Cited in Inwards, p. 43; Hand, p. 42; Freier, p. 75.

1248. If the groundhog sees his shadow on February 2nd, there will be six more weeks of winter. Var.: If the groundhog sees his shadow on February 2nd, there will be another month of winter.
Cited in Freier, p. 74; Arora, p. 6.

1249. If the groundhog sees its shadow on February 2nd, the fruit that year will not be wormy.

Cited in Hyatt, p. 34.

1250. Sunshine on February 2nd fills the barns and cellars.
Cited in Hyatt, p. 34.

1251. When drops hang on the fence on the second of February, icicles
will hang there on the twenty-fifth of March.
Cited in Dunwoody, p. 94; Inwards, p. 43.

February 22

1252. A freeze on February 22nd is a sign of forty more freezes.
Cited in Hyatt, p. 11.

feet

1253. When your feet hurt, it will rain soon.
Cited in Hyatt, p. 30.

1254. With some persons, the feet tend to go cold before snow.
Cited in Inwards, p. 218.

fennel

1255. When fennel blooms, frost follows.
Cited in Dunwoody, p. 66.

field

1256. A field requires three things; fair weather, sound seed, and a
good husbandman. Var.: A field has three needs, good weather,
good seed, and a good husbandman.
Cited in Denham, p. 3; Mieder, p. 207.

1257. Happy are the fields that receive summer rain.
Cited in Swainson, p. 15.

field mouse

1258. It is a token of a cold winter when field mice store a large
quantity of corn in their burrows.
Cited in Hyatt, p. 29.

1259. When the field mouse makes its burrow with the opening to the
south, it expects a severe winter; when to the north, it appre-
hends much rain.
Cited in Inwards, p. 184.

fire

1260. A fire failing to burn presages a change of weather.
Cited in Hyatt, p. 31.

1261. A fire hard to kindle indicates bad weather.
Cited in Dunwoody, p. 109; Inwards, p. 221; Freier, p. 64.

1262. A wood fire in winter pops more before snow.
Cited in Dunwoody, p. 75.

1263. Fire before a snow crackles, sizzles and spits; coals or embers pop out of the fireplace or stove; and the stove itself cracks.
Cited in Hyatt, p. 31.

1264. Fire is said to burn brighter and throw out more heat just before a storm.
Cited in Inwards, p. 221.

1265. Fires burn brightly and fiercely before and during frosty clear weather.
Cited in Wilshere, p. 23.

1266. Fires burning paler than usual and murmuring within are significant of storms.
Cited in Freier, p. 27.

1267. If during the summer a wood fire simmers, it will rain; if during winter a wood fire flutters or sighs, it will turn colder.
Cited in Hyatt, p. 31.

1268. If the fire burns unusually fierce and bright in winter, there will be frost and clear weather, and it is hotter during a storm.
Cited in Dunwoody, p. 109; Inwards, p. 221.

1269. If the fire sparks and spits in the fireplace, snow is coming.
Cited in Smith, p. 10.

1270. When fires burn faster than usual, and with a blue flame, frosty weather may be expected.
Cited in Mitchell, p. 230; Dunwoody, p. 21.

1271. When the fire crackles and crackles lightly, it is said to be treading snow.
Cited in Dunwoody, p. 109; Inwards, p. 221.

1272. When wood fires crackle it is a sign of snow.
Cited in Smith, p. 10.

1273. When, in winter or spring, during rough, sleety, and rainy weather, you hear the fire cracking, and feel it throwing out more heat, the weather will probably soon clear up without frost

or frosty air.
Cited in Steinmetz 1, p. 120.
1274. Wood fire coals frequently snuffing during the winter months
foretell snow.
Cited in Dunwoody, p. 119.

firefly
1275. Fireflies are out before a rain.
Cited in Freier, p. 62.
1276. Fireflies fly very low before a rain.
Cited in Lee, p. 57.
1277. Fireflies in great numbers indicate fair weather.
Cited in Dunwoody, p. 56.

firelight
1278. Firelight reflected on the woodwork of a room is a warning of
cold weather.
Cited in Hyatt, p. 31.

fish
1279. Black fish in schools indicate an approaching gale.
Cited in Dunwoody, p. 49; Garriott, p. 24.
1280. Blue fish, pike, and other fish jump with head towards the point
where a storm is forming.
Cited in Dunwoody, p. 49.
1281. Fish become inactive just before thunder showers, silent, and
won't bite.
Cited in Dunwoody, p. 49; Alstad, p. 97.
1282. Fish bite best as the moon grows full.
Cited in Whitman, p. 68.
1283. Fish bite best before a rain.
Cited in Freier, p. 23.
1284. Fish bite best when the moon is in the tail.
Cited in Dunwoody, p. 50.
1285. Fish bite least with wind in the east.
Cited in Wurtele, p. 300.
1286. Fish can forecast hot weather, for if temperatures are likely to be
high they keep to the cooler, lower water, and are reluctant to

take a hook, but when rain is imminent they gambol about near the surface, causing ripples and splashes.
Cited in Page, p. 15.

1287. Fish come to the surface when rain is imminent.
Cited in Wilshere, p. 23.

1288. Fish swim upstream and catfish jump out of the water before rain. Var.: Fish swim upstream and catfish jump from the water before a rain.
Cited in Inwards, p. 199; Sloane 2, p. 122.

1289. Fish swim upstream before rain.
Cited in Dunwoody, p. 49.

1290. Fishes rise more than usual at the approach of a storm.
Cited in Swainson, p. 248; Inwards, p. 198.

1291. Fishes sport most and bite most eagerly against rain than at any other time.
Cited in Dunwoody, p. 49; Garriott, p. 24; Hand, p. 24.

1292. If fish flop up in the water, you can look for rain; the more the flopping, the larger the storm.
Cited in Hyatt, p. 19.

1293. Near the surface, quick to bite, catch your fish when rain's in sight.
Cited in Alstad, p. 96; Sloane 2, p. 37.

1294. The appearance of a great number of fish on the west coast of the Gulf of Mexico indicates bad weather and easterly winds.
Cited in Dunwoody, p. 51; Garriott, p. 24; Hand, p. 24.

1295. When fish bite readily and swim near the surface, rain may be expected. Vars.: (a) If fish swim near the surface of water, rain is signified. (b) When fish bite readily and stay near the surface of the water, wet weather is near.
Cited in Dunwoody, p. 49; Garriott, p. 24; Inwards, p. 198; Hand, p. 24; Sloane 2, p. 122; Hyatt, p. 19; Smith, p. 6.

1296. When fish break water and bite eagerly, expect rain.
Cited in Freier, p. 23.

1297. When fish hook well, and they are hard to haul up, it is a sign of wind.
Cited in Lee, p. 59.

1298. When fish jump up after flies, expect rain.
Cited in Dunwoody, p. 50.

flea

1299. When eager bites the thirsty flea, clouds and rain you sure shall see.
Cited in Dunwoody, p. 56; Whitman, p. 49; Inwards, p. 207; Sloane 2, p. 124; Lee, p. 57; Freier, p. 62.

1300. When fleas do very many grow, then 'twill surely rain or snow.
Cited in Dunwoody, p. 56; Inwards, p. 207; Lee, p. 57.

flood

1301. A flood in the river means fine weather.
Cited in Marvin, p. 203.

1302. What comes with the flood goes with the wind.
Recorded in Bryant (Minn., N.C., N.Y.).

floor

1303. As soon as an oiled floor begins to sweat, you will know that a rain is imminent.
Cited in Hyatt, p. 30.

1304. Floors saturated with oil become very damp just before rain.
Var.: Oiled floors become damp before rain.
Cited in Dunwoody, p. 109; Garriott, p. 22; Inwards, p. 218; Hand, p. 22.

1305. Oily floors quite slippery get, before the rain makes everything wet.
Cited in Freier, p. 35.

floor matting

1306. If the matting on the floor is shrinking, dry weather may be expected; when matting expands, expect wet weather. Var.: When the matting on the floor shrinks, expect dry weather; when it expands expect wet weather.
Cited in Dunwoody, p. 116; Garriott, p. 21; Inwards, p. 218; Hand, p. 21; Lee, p. 74.

flower

1307. Flowers in bloom late in autumn indicate a bad winter.
Cited in Inwards, p. 65.

1308. Flowers remaining open all night and having a stronger fragrance than usual forecast rain.

Cited in Hyatt, p. 15.

1309. If a flower blooms twice during the year, a sharp winter is revealed.
Cited in Hyatt, p. 15.

1310. If the flowers keep open all night, the weather will be wet next day.
Cited in Inwards, p. 215; Lee, p. 71.

1311. Just before rain, flowers smell stronger and sweeter. Vars.: (a) Flowers just before a rain are always more fragrant. (b) Flowers smell best just before a rain.
Cited in Steinmetz 1, p. 112; Hyatt, p. 15; Freier, p. 29.

1312. No summer flowers are half so sweet as those of early spring.
Cited in Wright, p. 30.

1313. The blooming of flowers late in the autumn presages a bitter winter say some; a mild winter say others.
Cited in Hyatt, p. 15.

1314. The odor of flowers is more apparent just before a shower. Var.: The odor of flowers is more apparent just before a rain.
Cited in Garriott, p. 25; Inwards, p. 74; Alstad, p. 108; Hand, p. 25.

1315. When the perfume of flowers is unusually perceptible rain may be expected. Var.: When the perfume of flowers is unusually strong, expect rain.
Cited in Garriott, p. 21; Lee, p. 71; Freier, p. 29.

fly

1316. A fly on your nose you slap and it goes, if it comes back again it will bring a good rain.
Cited in Dunwoody, p. 56; Inwards, p. 107; Sloane 2, p. 124; Lee, p. 57.

1317. A large number of flies about the house denotes rain.
Cited in Hyatt, p. 18.

1318. Flies sting and are more troublesome than usual when the humidity increases before rain.
Cited in Garriott, p. 21; Hand, p. 21.

1319. Flies tend to collect in swirling swarms before rain.
Cited in Sloane 2, p. 124.

1320. House flies coming into the house in large numbers indicate rain.
Cited in Dunwoody, p. 57; Inwards, p. 207.

1321. If a fly lands on you and seems to stick, it is going to rain.
Cited in Smith, p. 4.
1322. If a fly lands on your nose, swat it till it goes, if the fly then lands again, it will bring back heavy rain.
Cited in Page, p. 20.
1323. If flies come in great swarms, rain follows soon.
Cited in Freier, p. 62.
1324. If flies in the spring or summer grow busier or blinder than at other times, or are seen to shroud themselves in warm places, expect either hail, cold storms of rain, or much wet weather.
Cited in Inwards, p. 207.
1325. If flies sting and are more troublesome than usual, a change approaches.
Cited in Inwards, p. 207.
1326. If flies try to get in, it is a sign of rain.
Recorded in Bryant (Nebr.).
1327. If in autumn the flies repair unto their winter quarters, presages frosty mornings, cold storms and the approach of winter.
Cited in Inwards, p. 207.
1328. If little flies or gnats are seen to hover together about the beams of the sun before it sets and fly together, making, as it were, the form of a pillar, it is a sure token of fair weather.
Cited in Inwards, p. 208.
1329. If the flies sting when they sit on your arms, it is going to rain.
Cited in Reeder, no pp.
1330. If you see little yellow flies hopping from one flower to another in the fall, that is a sign of an open winter.
Cited in Hyatt, p. 18.
1331. Small flies swarming together and sporting in the sunbeams give omen of fair weather.
Cited in Inwards, p. 207.
1332. When eager bites the thirsty fly, clouds and rain you will surely spy.
Cited in Alstad, p. 99.
1333. When flies bite greedily, expect rain. Vars.: (a) Expect rain soon, when flies begin biting, or bite harder or oftener than usual. (b) Flies bite greedily before a rain. (c) Flies bite more before a rain. (d) When flies bite and sting more than usual, it is a sign of rain.

Cited in Dunwoody, p. 56; Garriott, p. 24; Hand, p. 24; Sloane 2, p. 37; Hyatt, p. 18; Smith, p. 4; Freier, p. 62.

1334. When flies cling to the ceiling, or disappear rain may be expected.
Cited in Steinmetz 1, p. 111; Swainson, p. 255; Inwards, p. 207; Sloane 2, p. 124.

1335. When flies congregate in swarms, rain follows soon.
Cited in Dunwoody, p. 56; Garriott, p. 24; Hand, p. 24.

1336. When flies keep near the ground it is a sign of rain.
Cited in Mitchell, p. 228.

1337. When flies start to drop from the ceiling, autumn and cold weather are approaching.
Cited in Hyatt, p. 18.

1338. When harvest flies sing, warm weather will follow. Var.: When harvest flies hum, warm weather to come.
Cited in Dunwoody, p. 57; Inwards, p. 207.

1339. When you see a lot of flies in the house or on the screen, it will soon rain.
Cited in Reeder, no pp.

flying squirrel

1340. When the flying squirrels sing in midwinter, it indicates an early spring.
Cited in Dunwoody, p. 31.

foam

1341. A foam swimming on the waves is said to forebode in calm weather a continuance of the same for some days.
Cited in Inwards, p. 154.

1342. Foam on the water in a river or creek signifies rain; therefore this is also a sign of high water.
Cited in Hyatt, p. 13.

1343. Look for foam on the river before a rain.
Cited in Freier, p. 24.

1344. Much foam in a river foretells a storm. Var.: A lot of foam on a river foretells a storm.
Cited in Inwards, p. 154; Lee, p. 74.

1345. When there is a lot of foam at the whirlpools in a river, it will rain.

Cited in Reeder, no pp.

fog

1346. A fog and a small moon bring an easterly wind soon.
Cited in Northall, p. 460; Inwards, p. 88.

1347. A fog cannot be dispelled with a fan.
Cited in Denham, p. 11.

1348. A fog from the sea brings honey to the bee; a fog from the hills brings corn to the mills.
Cited in Inwards, p. 147.

1349. A fog in August indicates a severe winter and plenty of snow.
Cited in Dunwoody, p. 51; Inwards, p. 61.

1350. A fog in February indicates a frost in the following May.
Cited in Dunwoody, p. 52.

1351. A foggy morning will fade away; a foggy afternoon will stay.
Cited in Hyatt, p. 12.

1352. A ground fog foretells fine weather.
Cited in Wurtele, p. 293.

1353. A large number of foggy days in October indicates a hard winter.
Cited in Page, p. 50.

1354. A morning fog lifting early is an omen of rain; lifting late, a clear day.
Cited in Hyatt, p. 12.

1355. A rising fog indicates fair weather; if the fog settles down, expect stormy weather. Var.: A rising fog indicates fair weather; if fog settles down expect rain.
Cited in Dunwoody, p. 53; Garriott, p. 21; Hand, p. 21.

1356. A summer fog for fair, a winter fog for rain; a fact most everywhere, in valley and on plain.
Cited in Humphreys 2, p. 48; Whitman, p. 34; Sloane 2, p. 38; Lee, p. 29.

1357. A summer fog is a good indication of fair weather. Var.: A summer fog is for fair weather.
Cited in Dunwoody, p. 53; Inwards, p. 32.

1358. A summer's fog will roast a hog.
Cited in Mieder, p. 218.

1359. A winter's fog will freeze a dog.
Cited in Dunwoody, p. 53; Inwards, p. 35; Mieder, p. 218.

1360. As many days of fog in March, so many days of frost in May, on corresponding days.
Cited in Wright, p. 35.

1361. As many foggy mornings as there are in January, so many frosty mornings will there be in May.
Cited in Hyatt, p. 12.

1362. As many fogs as there are in March, as many fogs there will be in June.
Recorded in Bryant (Vt.).

1363. As much fog as plagues you in March, so many thunderstorms after one hundred days.
Cited in Dunwoody, p. 91.

1364. As much fog in March, so much rain in summer.
Cited in Dunwoody, p. 91.

1365. Count sixty days from the first fog in August to the first frost in the fall.
Recorded in Bryant (Miss.).

1366. Count the number of foggy mornings in the spring, summer and fall, and that will be the number of snowfalls in winter.
Cited in Smith, p. 9.

1367. Fog after hard frosts and fog after mild weather indicate a change in the weather.
Cited in Dunwoody, p. 52.

1368. Fog and mist raise higher seas than wind.
Cited in Dunwoody, p. 84.

1369. Fog from seaward in Maine, don't expect the rain; fog from land in warm, batten down for storm.
Cited in Sloane 2, p. 14.

1370. Fog from seaward, fair weather; fog from landward, rain. Var.: Fog from seaward, weather fair; fog from land brings rainy air.
Cited in Dunwoody, p. 53; Garriott, p. 21; Inwards, p. 147; Hand, p. 21; Sloane 2, p. 38.

1371. Fog in April foretells a failure of the wheat crop next year.
Cited in Dunwoody, p. 52.

1372. Fog in February, frost in May. Var.: For every fog in February, a frost in May.
Recorded in Bryant (Ohio).

1373. Fog in January makes a wet spring.
Cited in Dunwoody, p. 90.

1374. Fog in March, thunder in July.
Cited in Inwards, p. 48.

1375. Fog in October suggests a cold winter.
Cited in Page, p. 51.

1376. Fog in the hollow, fine day to follow.
Recorded in Bryant (N.Y., S.C.).

1377. Fog in the morning, clear afternoon.
Recorded in Bryant (N.Y.).

1378. Fog in the morning, sailor take warning; fog in the night, sailor's delight.
Cited in Sloane 2, p. 38.

1379. Fog on the meadow means fair weather.
Recorded in Bryant (Vt.).

1380. Fogs are signs of change.
Cited in Inwards, p. 147.

1381. Fogs in March, frost in May.
Cited in Inwards, p. 47; Wilson, p. 511; Wilshere, p. 9.

1382. For every fog in October there will be a snow during the winter; for each heavy fog a heavy snow, and for each light fog a light snow.
Cited in Dunwoody, p. 53.

1383. Frozen smog makes an ice fog.
Recorded in Bryant (N.Y.).

1384. He that would have a bad day must go out in the fog after a frost.
Cited in Swainson, p. 208; Dunwoody, p. 52; Whitman, p. 34.

1385. Heavy fog in winter, when it hangs below the trees, is followed by rain.
Cited in Dunwoody, p. 52; Inwards, p. 147.

1386. High fog in the sky brings bad weather bye 'n' bye.
Cited in Alstad, p. 138.

1387. If a fog in the morning fades away without rising, the weather will be good; if the fog rises, the weather will be bad.
Cited in Hyatt, p. 12.

1388. If fog clears from the bottom up, the weather will be fair.
Cited in Freier, p. 63.

1389. If fog clears from the top down, expect some rain.
Cited in Freier, p. 63.

1390. If fog goes up before seven, it will rain before eleven. Vars.: (a) If fog goes up before seven, it will come down before eleven. (b) If the weather is foggy at 7:00 a.m. it will clear at 11:00 a.m. (c) Fog at seven, clear at eleven.
Recorded in Bryant (Ill., Miss. Can. [Ont.]).

1391. If fogs in August are light, a light winter may be expected; if they are heavy, a heavy winter.
Cited in Hyatt, p. 12.

1392. If fogs or mists rise in low ground and soon vanish, expect fair weather.
Cited in Alstad, p. 123.

1393. If the first three days of April be foggy, there will be a flood in June.
Cited in Dunwoody, p. 52.

1394. If the fog goes up in the morning, it will rain before night.
Recorded in Bryant (Ill., Miss.).

1395. If the fog on the hillside goes hunting, it will clear up.
Recorded in Bryant (Vt.).

1396. If there be a damp fog or mist, accompanied by wind, expect rain.
Cited in Dunwoody, p. 52; Inwards, p. 147.

1397. If there be continued fog, expect frost.
Cited in Dunwoody, p. 52; Inwards, p. 147.

1398. In summer a fog from the south, warm weather; from the north, rain.
Cited in Inwards, p. 32.

1399. In summer, when fog comes with a southerly wind, it indicates warm weather; when it comes with a northerly wind, it is a sign of heavy rain.
Cited in Dunwoody, p. 53.

1400. In the Mississippi Valley, when fogs occur in August, expect fever and ague in the following fall.
Cited in Dunwoody, p. 51; Inwards, p. 147.

1401. Light fog passing under sun from south to north in the morning indicates rain in twenty-four or forty-eight hours.
Cited in Dunwoody, p. 52; Inwards, p. 147.

1402. Much fog in autumn, much snow in winter. Var.: Much autumn fog; much winter snow.

Cited in Dunwoody, p. 92; Inwards, p. 33; Sloane 2, p. 38; Hyatt, p. 12.

1403. Observe on what day in August the first heavy fog occurs, and you may expect a hard frost on the same day in October.
Cited in Dunwoody, p. 51; Inwards, p. 61.

1404. So many August fogs, so many winter mists.
Cited in Inwards, p. 61.

1405. So many fogs in August we see, so many snows that year will be. Vars.: (a) As many fogs as you have in August is the same number of snows you will have in winter. (b) For every fog in August, there will be one snow the next winter. (c) The number of fogs in August determines the number of snows in winter.
Cited in Dunwoody, p. 52; Hyatt, p. 12; Smith, p. 9.

1406. Summer fog from seaward, weather fair. Summer fog from land, brings rainy air.
Cited in Alstad, p. 125.

1407. The day on which a fog occurs in January will be the date of a frost in May.
Cited in Hyatt, p. 12.

1408. The number of August fogs indicates the number of winter mists.
Cited in Dunwoody, p. 51.

1409. Three foggy mornings will bring a rain three times harder than usual.
Cited in Sloane 2, p. 38.

1410. Three foggy or misty mornings indicate rain.
Cited in Garriott, p. 21; Marvin, p. 203; Inwards, p. 145; Hand, p. 21.

1411. Three fogs in January mean three frosts in May.
Cited in Smith, p. 2.

1412. When a morning fog turns into clouds of different layers, the clouds increasing in size, expect a rain.
Cited in Dunwoody, p. 53.

1413. When fog comes from the north, rain will continue.
Cited in Smith, p. 8.

1414. When fog falls fair weather follows; when it rises rain follows.
Var.: When the fog falls, fair weather follows; when it rises, rain ensues.
Cited in Dunwoody, p. 52; Inwards, p. 147.

1415. When fog goes up, the rain is o'er; when fog comes down, 'twill rain some more.
Cited in Humphreys 2, p. 49; Lee, p. 29.

1416. When light fog clouds on evenings are observed to rise from the valleys and hang around the summits of mountains, rain follows.
Cited in Dunwoody, p. 52.

1417. When the fog climbs the mountain, more rain will fall.
Recorded in Bryant (Vt.).

1418. When the fog goes up skipping, the rain comes down dripping.
Recorded in Bryant (Vt.).

1419. When the fog goes up the hill the rain comes down the mill.
Var.: Fog on the hills, more water for the mills.
Cited in Dunwoody, p. 52. Recorded in Bryant (Vt.).

1420. When the fog goes up the mountain, you may go hunting; when it comes down the mountain, you may go fishing. Comment: In the former event it will be fair, in the latter it will rain.
Cited in Dunwoody, p. 52; Inwards, p. 147.

1421. When there is fog there is little or no rain.
Cited in Whitman, p. 35.

1422. When with hanging fog smoke rises vertically, rain follows.
Cited in Dunwoody, p. 109; Inwards, p. 147.

foliage

1423. Heavy foliage, heavy winter; meagre foliage, meagre winter.
Cited in Hyatt, p. 16.

food

1424. If all the food at the table is consumed or none of it is wasted, weather next day will be fair.
Cited in Hyatt, p. 31.

1425. If food dries quickly while being cooked, beans and potatoes especially, a rain is impending.
Cited in Hyatt, p. 32.

foot

1426. An itching on the sole of your foot signifies rain in summer and snow in winter.
Cited in Hyatt, p. 30.

1427. If frostbitten feet itch, snow may be expected.
Cited in Hyatt, p. 30.

forest

1428. When the forest murmurs, and the mountain roars, then close your windows and shut your doors.
Cited in Humphreys 2, p. 68; Whitman, p. 35; Inwards, p. 151; Lee, p. 37.

fork

1429. If you drop a fork, then a spoon, and the latter lies across the former, it is going to storm.
Cited in Hyatt, p. 31.

fowl

1430. Domestic fowls dress their feathers when the storm is about to cease. Var.: If fowl dress their feathers during a storm, it is about to cease.
Cited in Dunwoody, p. 36; Inwards, p. 187.

1431. Domestic fowls look towards the sky before rain.
Cited in Dunwoody, p. 36.

1432. Domestic fowls stand on one leg before cold weather. Var.: Fowl standing on one leg is considered a sign of cold weather.
Cited in Dunwoody, p. 36; Inwards, p. 187.

1433. Fowl will run to shelter and stay there if they think the weather will clear; but if they see it is to be wet all day, they come out and face it.
Cited in Inwards, p. 187.

1434. If fowls grub in the dust, it indicates coming rain. Var.: If fowls grub in the dust and clap their wings, or if their wings droop, or crowd into a house, it indicates rain.
Cited in Swainson, p. 236; Inwards, p. 187.

1435. If fowls roll in the sand, rain is at hand. Var.: If fowls roll in the sand, foul weather is at hand.
Cited in Swainson, p. 236; Northall, p. 473; Garriott, p. 23; Inwards, p. 187; Hand, p. 23; Sloane 2, p. 129; Lee, p. 60.

1436. If fowls' wings droop, rain is at hand.
Cited in Swainson, p. 237.

1437. If the fowls huddle together outside the henhouse instead of going to roost, there will be wet weather.
Cited in Inwards, p. 187; Sloane 2, p. 129.

1438. If water fowl make more noise than usual, expect frost soon.
Cited in Dunwoody, p. 40.

1439. The fowl sings, we shall have rain.
Recorded in Bryant (Wis.).

1440. When fowls collect together and pick or straighten their feathers, expect a change of weather.
Cited in Dunwoody, p. 36; Inwards, p. 187.

1441. When fowls roost in daytime, expect rain. Var.: When fowls look towards the sky, or roost in the daytime, expect rain.
Cited in Dunwoody, p. 36; Inwards, p. 187.

fox

1442. Foxes barking at night indicates storm. Var.: When foxes bark at night, a storm is on the way.
Cited in Dunwoody, p. 31; Lee, p. 66.

1443. When foxes bark and utter shrill cries, expect a violent tempest of wind and rain within three days.
Cited in Inwards, p. 183.

foxfire

1444. Foxfire seen at night indicates cold.
Cited in Dunwoody, p. 66.

Friday

1445. A rainy Friday, a rainy Sunday.
Cited in Hyatt, p. 10.

1446. As the Friday so the Sunday, as the Sunday so the week. Var.: As the Friday, so the Sunday.
Cited in Cheales, p. 19; Dunwoody, p. 101; Whitman, p. 41; Inwards, p. 73.

1447. Friday dawns clear as a bell, rain on Sunday, sure as hell.
Cited in Lee, p. 146.

1448. If on Friday it rains, 'twill on Sunday again; if Friday be clear, have a Sunday no fear.
Cited in Dunwoody, p. 100; Wright, p. 120; Whitman, p. 41; Inwards, p. 73; Hand, p. 43. Recorded in Bryant (Wis.).

1449. If the sun sets clear on Friday, generally expect rain before Monday.
Cited in Dunwoody, p. 101; Inwards, p. 73.
1450. If the sun sets clear on Friday, it will blow before Sunday night.
Cited in Dunwoody, p. 100; Inwards, p. 73.
1451. On Friday the weather changes.
Cited in Whiting 2, p. 242.
1452. The first Friday of each month is an almanac index for the trend of weather the rest of the month.
Cited in Sloane 2, p. 38.
1453. Wet Friday, wet Sunday, wet week.
Cited in Wilshere, p. 4.
1454. Whatever the weather is on Friday, that will be the weather until the following Friday.
Cited in Hyatt, p. 13.
1455. Whatever the weather is on Friday, that will be the weather all next week.
Cited in Hyatt, p. 13.

frog
1456. A small frog clinging to the well chain is a rain portent.
Cited in Hyatt, p. 20.
1457. Abundance of yellow frogs are accounted a good sign in a hay field, probably as indicating fine weather.
Cited in Dunwoody, p. 72.
1458. As soon as you hear the male frog croaking, you will know that spring is here.
Cited in Hyatt, p. 20.
1459. Croaking frogs in spring will be three times frozen in. Vars.: (a) After the first croaking of frogs in spring, you will look through glass three times (see three more freezes) before spring. (b) After the first frog croaks, you will look through glass (a thin sheet of ice) before spring. (c) Frogs must be frozen three times in spring after they begin to croak. (d) Peepers (peep frogs) have to freeze up three times before spring comes.
Cited in Dunwoody, p. 72; Boughton, p. 125; Hyatt, p. 20.
1460. Frogs croak before a rain; but in the sun are quiet again.
Cited in Freier, p. 41.

1461. Frogs croak importunately before rain.
       Cited in Steinmetz 1, p. 112.
1462. Frogs croak more noisily and come abroad in the evening in
       large numbers, before rain.
       Cited in Dunwoody, p. 72.
1463. Frogs croaking during the day are calling for rain; it will soon
       come.
       Cited in Hyatt, p. 20.
1464. Frogs croaking in the lagoon, means that rain will come real
       soon.
       Cited in Freier, p. 40.
1465. Frogs heard in March are an omen of an early spring.
       Cited in Hyatt, p. 20.
1466. Frogs singing in the evening indicate fair weather for next day.
       Cited in Dunwoody, p. 72.
1467. If frogs croak long and loudly at night, rain is at hand.
       Cited in Hyatt, p. 20.
1468. If frogs make a noise in the time of cold rain, warm dry weather
       will follow.
       Cited in Denham, p. 3; Swainson, p. 250; Inwards, p. 202;
       Freier, p. 106.
1469. If frogs, instead of yellow, appear russet-green, it will presently
       rain. Var.: The color of a frog changing from yellow to reddish
       indicates rain.
       Cited in Swainson, p. 251; Dunwoody, p. 72; Inwards, p. 202.
1470. Never plant anything until the frogs have croaked three different
       times, because there will be killing frost until they do.
       Cited in Hyatt, p. 20.
1471. Spring will come after the frogs are frozen three times.
       Recorded in Bryant (Wis.).
1472. The croaking of frogs means rain.
       Cited in Smith, p. 4.
1473. The louder the frogs, the more's the rain. Vars.: (a) The louder
       the frog, the more the rain. (b) The louder the frog, the nearer
       the rain.
       Cited in Dunwoody, p. 72; Whitman, p. 48; Inwards, p. 202;
       Alstad, p. 107; Sloane 2, p. 123.
1474. When a frog hollers in the spring, it is a sign that winter has
       ended.

Cited in Smith, p. 12.

1475. When a frog hollers, rain will soon foller.
Cited in Smith, p. 4.

1476. When frogs croak at night, it will rain.
Cited in Reeder, no pp.

1477. When frogs croak much it is a sign of rain.
Cited in Swainson, p. 250; Inwards, p. 202.

1478. When frogs croak three times, it indicates that winter has broken.
Cited in Dunwoody, p. 72.

1479. When frogs croak, winters broke.
Cited in Wurtele, p. 298.

1480. When frogs jump across the road, they are looking for rain.
Cited in Freier, p. 41.

1481. When frogs spawn in the middle of the water it is a sign of drought; and when at the side, it foretells a wet summer. Var.: When the frog spawns in the middle of the water, it is a sign of drought; and when at the side, it indicates a wet summer.
Cited in Mitchell, p. 228; Swainson, p. 251; Inwards, p. 203; Sloane 2, p. 123.

1482. When frogs warble, they herald rain.
Cited in Sloane 2, p. 123.

1483. When the color of frogs is observed to be dark, wet weather is thought to be very close.
Cited in Mitchell, p. 228.

1484. When there are many peeping frogs in the spring, there will be a dry year.
Cited in Boughton, p. 124.

1485. When you hear the first frogs in the spring, the frost is out of the ground.
Cited in Sloane 2, p. 39.

1486. You cause a rain by killing a frog or toad.
Cited in Hyatt, p. 20.

frog fish

1487. Frog fish crawling indicate rain.
Cited in Dunwoody, p. 50.

frost

1488. A frost clinging to trees late in the morning foretells snow.

Cited in Hyatt, p. 11.

1489. A hoar frost; third day crost; the fourth lost. Farewell frost, nothing got is nothing lost. Var.: A hoar frost, third day crost, the fourth lost.
Cited in Northall, p. 460; Inwards, p. 163.

1490. A long frost is a black frost.
Cited in Dunwoody, p. 16.

1491. A very heavy white frost in winter is followed by a thaw.
Cited in Dunwoody, p. 55.

1492. A white frost is a rain omen.
Cited in Hyatt, p. 11.

1493. A white frost never lasts more than three days.
Cited in Dunwoody, p. 16; Inwards, p. 163; Wilson, p. 292.

1494. After March 15th a frost never damages plants.
Cited in Hyatt, p. 12.

1495. Bearded frost, forerunner of snow. Var.: Bearded frost is a forerunner of snow.
Cited in Swainson, p. 209; Dunwoody, p. 53; Inwards, p. 946.

1496. Black frost indicates dry cold weather.
Cited in Dunwoody, p. 54.

1497. Black frost, long frost, hoar frost three days, then rain.
Cited in Wilshere, p. 16.

1498. Early frosts are generally followed by a long hard winter.
Cited in Dunwoody, p. 54.

1499. Frost occurring in the dark of the moon kills fruit, buds, and blossoms; but frost in the light of the moon will not. Var.: Frost during the light of the moon does not nip plants or fruit tree blossoms; during the dark of the moon it does.
Cited in Dunwoody, p. 53; Hyatt, p. 11.

1500. Frost on December 21st, the shortest day, is said to indicate a severe winter.
Cited in Inwards, p. 67.

1501. Frost suddenly following heavy rain seldom lasts long.
Cited in Mitchell, p. 229; Dunwoody, p. 16.

1502. Frost will probably occur when the temperature is forty degrees and the wind northwest.
Cited in Dunwoody, p. 55.

1503. Frosts end in foul weather.
Cited in Dunwoody, p. 54.

1504. Heavy frost is a sign of bad weather.
 Cited in Smith, p. 3.
1505. Heavy frosts are generally followed by fine, clear weather. Var.:
 Heavy frosts generally are followed by fine clear weather.
 Cited in Dunwoody, p. 54; Humphreys 2, p. 3; Whitman, p. 75.
 Recorded in Bryant (Utah, Wis.).
1506. Heavy frosts bring heavy rains; no frost, no rain.
 Cited in Dunwoody, p. 55; Garriott, p. 21; Hand, p. 21.
1507. Heavy white frost indicates warmer weather.
 Cited in Dunwoody, p. 54.
1508. Hoar frost and gypsies never stay nine days in the same place.
 Cited in Inwards, p. 163; Alstad, p. 118.
1509. Hoar frost and no snow is hurtful to fields, trees and grain.
 Cited in Dunwoody, p. 93; Inwards, p. 37.
1510. Hoar frost indicates rain.
 Cited in Dunwoody, p. 54; Garriott, p. 21; Hand, p. 21.
1511. Hoar frost on May 1st indicates a good harvest.
 Cited in Garriott, p. 42.
1512. Hoar frosts in spring and autumn are generally followed by rain
 in twenty-four hours.
 Cited in Steinmetz 2, p. 83.
1513. If a frost occurs three mornings in a row, it will rain.
 Cited in Smith, p. 2.
1514. If any day of September is cold but without frost, we will have
 no frost until the same date in October.
 Cited in Hyatt, p. 12.
1515. If hoar frost comes on morning twain, the third day surely will
 have rain. Comment: Twain in an archaic variant of two.
 Cited in Whitman, p. 34; Inwards, p. 163.
1516. If October bring heavy frosts and winds, then will January and
 February be mild.
 Cited in Garriott, p. 45.
1517. If the first frost occurs late, the following winter will be mild but
 variable. If first frost occurs early, it indicates a severe winter.
 Cited in Dunwoody, p. 54.
1518. If there be an abundance of hoar frost, expect rain.
 Cited in Dunwoody, p. 54.
1519. If there is no killing frost in September, there will be none until
 after October 15th.

Cited in Hyatt, p. 12.

1520. In winter if the fences and trees are covered with white frost, expect a thaw.
Cited in Dunwoody, p. 54.

1521. Light or white frosts are always followed by wet weather, either the same day or three days after.
Cited in Dunwoody, p. 54.

1522. Moonlight nights have the hardest frosts.
Cited in Dunwoody, p. 54; Inwards, p. 88; Alstad, p. 121.

1523. Past the Easter frost and fruit is safe.
Cited in Dunwoody, p. 54.

1524. Plants are not harmed by frost while the wind is in the north, but they are harmed by frost while the wind is in the south.
Cited in Hyatt, p. 12.

1525. Six months from last frost to next frost.
Cited in Dunwoody, p. 55.

1526. So many frosts in March, so many in May.
Cited in Dunwoody, p. 95; Wilson, p. 292.

1527. The first and last frost are the worst.
Cited in Wilson, p. 292.

1528. The frost hurts not the weeds.
Cited in Mieder, p. 242.

1529. The frost will bring home the pig.
Cited in Mieder, p. 242.

1530. There will be as many frosts in June as there are fogs in February.
Cited in Dunwoody, p. 54; Inwards, p. 41.

1531. They must hunger in frost that will not work in heat.
Cited in Mieder, p. 242.

1532. Three frosts in succession are a sign of rain.
Cited in Dunwoody, p. 55.

1533. Three rimy frosts, and then it rains. Comment: Rimy means covered with rime, an accumulation of granular ice tufts on the windward side of exposed objects.
Cited in Page, p. 7.

1534. Three successive mornings of hoar frost in the autumn and spring are generally followed by a continued rain. Comment: Hoar means gray or white.
Cited in Steinmetz 1, p. 119.

1535. Three white frosts and a rain.
Recorded in Bryant (Oreg.).

1536. Three white frosts and then a storm.
Cited in Dunwoody, p. 55.

1537. Three white frosts will be followed by a black one.
Cited in Wilshere, p. 16.

1538. Two white frosts and the hoot of an owl bring rain.
Recorded in Bryant (Can. [Ont.]).

1539. When the frost gets into the air, it will rain.
Cited in Dunwoody, p. 16.

1540. When the hoar frost is first accompanied by east wind, it indicates that the cold will continue a long time.
Cited in Inwards, p. 117.

1541. White frost on three successive nights indicates a thaw or rain.
Var.: Three successive white frosts mean a rain.
Cited in Dunwoody, p. 55; Hyatt, p. 11.

frosty

1542. Frosty nights and hot summer days set the cornfields all in a blaze.
Cited in Swainson, p. 17; Dunwoody, p. 89.

1543. If the trees are frosty and the sun takes it away before noon, it is a sign of rain.
Cited in Dunwoody, p. 54.

1544. In frosty weather, the stars appear clearest and most sparkling.
Cited in Sloane 2, p. 114.

fruit tree

1545. If a fruit tree blooms twice the same year, it means a harsh winter. Var.: If a fruit tree has two crops the same season, it means a harsh winter.
Cited in Hyatt, p. 16.

1546. When fruit trees and berry bushes are loaded, there is going to be a hard winter.
Cited in Smith, p. 12.

gale

1547. A gale moderating at sunset will increase before midnight, but if it moderates after midnight, the weather will improve.
Cited in Dunwoody, p. 84.
1548. Easterly gales, without rain, during the spring equinox, foretell a dry summer.
Cited in Mitchell, p. 229; Inwards, p. 117.
1549. Hoist up the sail while the gale doth last.
Recorded in Bryant (Can. [Ont.]).
1550. If, during the absence of wind, the surface of the sea becomes agitated by a long, rolling swell, a gale may be expected.
Cited in Dunwoody, p. 109; Alstad, p. 109.
1551. Leave not the harbor in a gale.
Cited in Whiting 1, p. 198.
1552. When after a stiff breeze there ensue a dead calm and drizzling rain, with a fall in the barometer, expect a gale from the southwest.
Cited in Inwards, p. 122.

gate

1553. If a gate opens and slams incessantly, cold weather is denoted.
Cited in Hyatt, p. 30.
1554. To keep shutting and opening a gate will bring rain.
Cited in Hyatt, p. 32.

gentian

1555. The gentian closes up both flowers and leaves before rain.
Cited in Inwards, p. 216; Sloane 2, p. 112.

glassware

1556. The sweating of glassware, a water pitcher in particular, is a presage of rain.
Cited in Hyatt, p. 31.

glowworm

1557. Before rain: glowworms numerous, clear, and bright, illuminate the dewy hills at night. Var.: Glowworms numerous and bright, indicate rain.
Cited in Dunwoody, p. 56; Alstad, p. 102.

1558. If glowworms shine much, it will rain.
       Cited in Swainson, p. 251; Dunwoody, p. 14; Inwards, p. 201.
1559. When the glowworm glows, dry hot weather follows.
       Cited in Dunwoody, p. 56.
1560. When the glowworm lights her lamp, the air is always damp.
       Cited in Swainson, p. 251; Northall, p. 473; Whitman, p. 49;
       Inwards, p. 201; Lee, p. 78.

gnat

1561. Gnats appear in swarms just before warmer weather with rain.
       Cited in Hyatt, p. 18.
1562. Gnats flying in a vortex in the beams of the sun, fair weather will
       follow; when they frisk about more wildly increasing heat is
       indicated; when they seek the shade and bite more frequently, the
       signs are of coming rain.
       Cited in Dunwoody, p. 57.
1563. Gnats gather in clouds to play in beams of sunlight in fair
       weather, but hide in bushes and grass and bite excessively before
       a rain.
       Cited in Alstad, p. 100.
1564. Gnats in October are a sign of long, fair weather. Var.: Gnats in
       October are a sign of protracted fair weather.
       Cited in Dunwoody, p. 57; Inwards, p. 208.
1565. If gnats bite sharper than usual, expect rain.
       Cited in Inwards, p. 208.
1566. If gnats collect in the evening before sunset, and form a vortex
       or column, fine weather.
       Cited in Inwards, p. 208.
1567. If gnats, flies, etc., bite sharper than usual, expect rain.
       Cited in Dunwoody, p. 57.
1568. If gnats fly in compact bodies in the beams of the setting sun,
       expect fine weather. Var.: If many gnats are seen flying in
       compact bodies in the beams of the setting sun, there will be fine
       weather.
       Cited in Dunwoody, p. 57; Inwards, p. 208; Lee, p. 57.
1569. If gnats fly in large numbers, the weather will be fine.
       Cited in Dunwoody, p. 57; Inwards, p. 208.
1570. If gnats play up and down, it is a sign of heat; but if in the
       shade, it presages mild showers.

Cited in Inwards, p. 208.

1571. If gnats sting much, it is held to be an unfailing indication of rain.
Cited in Inwards, p. 208.

1572. Many gnats in spring indicate that the autumn will be warm.
Cited in Dunwoody, p. 57; Inwards, p. 208.

1573. The gnats bite and I scratch in vain, because they know it is going to rain.
Cited in Lee, p. 51.

1574. When gnats are swarming, it will rain soon.
Cited in Reeder, no pp.

1575. When gnats bite keenly look for rain and wind. Var.: When gnats bite keenly and when fleas keep near the ground, we look for wind and rain.
Cited in Mitchell, p. 228; Dunwoody, p. 14.

1576. When gnats bite keenly, rain is near.
Cited in Smith, p. 4.

1577. When gnats dance in February the husbandmen becomes a beggar. Comment: Husbandmen means farmers.
Cited in Dunwoody, p. 57.

goat

1578. Flocks of goats graze down the mountains before the approach of a storm, and upwards before fair weather.
Cited in Inwards, p. 182.

1579. Goats graze down the mountain before a rain and up the mountain for fair weather.
Cited in Sloane 2, p. 120.

1580. Goats leave the high and exposed grounds and seek shelter in a recess before a storm. Var.: Goats leave high ground and seek shelter before a storm.
Cited in Mitchell, p. 227; Swainson, p. 231; Dunwoody, p. 14; Inwards, p. 182; Sloane 2, p. 120; Lee, p. 69.

1581. If goats and sheep quit their pastures with reluctance, it will rain the next day.
Cited in Swainson, p. 231; Inwards, p. 40.

1582. If goats "bawl," it is going to rain.
Cited in Lee, p. 69.

1583. If goats leave their homes during a rain, it will soon clear.
Cited in Lee, p. 69.
1584. Mountain goats come to lower ground before a rain.
Cited in Freier, p. 50.
1585. The goat will utter her peculiar cry before rain.
Cited in Dunwoody, p. 31; Inwards, p. 182; Sloane 2, p. 120.

goat's beard
1586. Goat's beard keeps its flowers closed in damp weather.
Cited in Inwards, p. 213; Sloane 2, p. 112.
1587. When goat's beard closes its petals at midday, expect rain.
Cited in Dunwoody, p. 66.

God
1588. God tempers the storm to the shorn lamb.
Cited in Denham, p. 7; Taylor 2, p. 154; Wilson, p. 312;
Whiting 1, p. 179; Simpson, p. 95; Whiting 2, p. 260; Mieder,
p. 255.
1589. When God wills all winds bring rain.
Cited in Cheales, p. 25.

goldenrod
1590. Frost may be expected six weeks from the time the goldenrod
first blooms.
Recorded in Bryant (Nebr.).

goldfish
1591. Goldfish in a bowl wash their faces or suck air when rain is
coming.
Cited in Hyatt, p. 19.

Good Friday
1592. A Good Friday rain is worthless.
Cited in Hyatt, p. 34.
1593. A wet Good Friday and a wet Easter Day, makes plenty of grass
but very little hay. Vars.: (a) A wet Good Friday and Easter Day
brings plenty of grass, but little good hay. (b) A wet Good
Friday and Saturday, brings plenty of grass but little hay. (c)

Rain on Good Friday or Easter Day, a good crop of grass, but a bad one of hay.
Cited in Dunwoody, p. 101; Northall, p. 450; Wright, p. 38; Inwards, p. 70; Lee, p. 152.

1594. Either a north or a northwest wind on Good Friday will be succeeded by six weeks of inclement weather.
Cited in Hyatt, p. 9.

1595. Good Friday rain brings a fertile year.
Cited in Dunwoody, p. 102.

1596. If it freezes in Christ's grave (while Christ was in the grave, Good Friday to Easter), it can freeze anytime during the forty days that follow.
Cited in Hyatt, p. 11.

1597. If it rains on Good Friday, it will rain the seven following Sundays.
Cited in Hyatt, p. 10.

1598. If it rains on Good Friday, summer will be hot and dry.
Cited in Hyatt, p. 10.

1599. If it rains on Good Friday, there will not be much rain the remainder of the year.
Cited in Hyatt, p. 10.

1600. Rain on Good Friday forebodes a fruitful year. Var.: Rain on Good Friday foreshows a fruitful year.
Cited in Dunwoody, p. 101; Inwards, p. 70.

1601. The direction of the wind on Good Friday will prevail during the next forty days.
Cited in Hyatt, p. 9.

1602. When it rains on Good Friday, the rest of the spring won't do the crops any good.
Cited in Boughton, p. 124.

goose

1603. A very heavy plumage of geese in fall indicates an approaching cold winter.
Cited in Dunwoody, p. 36.

1604. Everything is lovely and the goose hawks high. Vars.: (a) All is well and the goose hangs high. (b) Everything is lovely when the goose hangs high. (c) Everything is lovely when the goose honks high. Comment: Birds fly high when the barometer is high and low when the barometer is low. When the barometer is high the air is heavier and has more sustaining capacity, thus allowing birds to fly high with less effort.

Cited in Dunwoody, p. 36; Garriott, p. 18; Hand, p. 18; Freier, p. 21; Mieder, p. 261.

1605. Geese call down the rain with their cackling.
Cited in Sloane 2, p. 129.

1606. Geese flying high is a sign of fair weather. Var.: Geese fly higher in fair weather than in foul.
Cited in Dunwoody, p. 36; Sloane 2, p. 32.

1607. Geese flying north signify spring.
Cited in Boughton, p. 125.

1608. Geese flying out to sea and free-range chickens scrapping contentedly are signs of good weather.
Cited in Page, p. 18.

1609. Geese wash and sparrows fly in flocks before rain.
Cited in Dunwoody, p. 37.

1610. If a goose after it has dusted itself gets up and flaps its wings, a rain is not far away.
Cited in Hyatt, p. 25.

1611. If domestic geese walk east and fly west, expect cold weather.
Cited in Dunwoody, p. 37.

1612. If geese raise up and flap their wings while swimming, a rain will arrive soon.
Cited in Hyatt, p. 25.

1613. If the goose honks high, fair weather; if the goose honks low, foul weather.
Cited in Freier, p. 20.

1614. If the wild geese fly south early, it will be a hard winter.
Cited in Smith, p. 12.

1615. If wild geese along the river are restive in daytime and clamorous at night, they are preparing to go south because winter is at hand.
Cited in Hyatt, p. 22.

1616. If wild geese go to the south in October, but return in great numbers in a few days, the ensuing winter will be mild.
Cited in Inwards, p. 188.

1617. Inspect your geese after they have gone to roost and the direction toward which their heads are pointed will be the quarter of the

wind next day.
Cited in Hyatt, p. 25.

1618. The goose and the gander begin to meander; the matter is plain, they are dancing for rain.
Cited in Inwards, p. 188; Sloane 2, p. 130; Lee, p. 62.

1619. Unquiet geese portend rain.
Cited in Hyatt, p. 25.

1620. When geese and ducks go into the water and flap their wings throwing the water over their backs, rain is approaching.
Cited in Dunwoody, p. 37.

1621. When geese cackle, it will rain.
Cited in Lee, p. 62; Freier, p. 22.

1622. When geese fly at ten o'clock or in the first part of the night, it is a sign of cold weather.
Cited in Dunwoody, p. 37.

1623. When geese or ducks stand on one leg, expect cold weather.
Cited in Dunwoody, p. 37.

1624. When tame geese fly, it will rain.
Cited in Lee, p. 62.

1625. When the barnyard goose walks south to north, rain will surely soon break forth.
Cited in Lee, p. 62.

1626. When the geese fly in a "V" shape, it will be a very cold winter.
Cited in Smith, p. 13.

1627. When wild geese fly southward early, winter is coming soon.
Cited in Smith, p. 14.

1628. When wild geese fly to the southeast in the fall, in Kansas, expect a blizzard. Var.: When wild geese fly to the southeast in Kansas, expect a blizzard.
Cited in Dunwoody, p. 41; Garriott, p. 40; Inwards, p. 188; Hand, p. 40; Sloane 2, p. 14.

1629. When you see geese in water washing themselves, expect rain.
Cited in Dunwoody, p. 37.

1630. Wild geese fly high in pleasant weather and low in bad weather.
Cited in Dunwoody, p. 36; Garriott, p. 18; Hand, p. 18; Freier, p. 21.

1631. Wild geese flying directly south and very high indicates a very cold winter. When flying low and remaining along the river, they indicate a warm winter in Idaho. For spring, just the reverse when flying north.

Cited in Dunwoody, p. 41; Garriott, p. 40; Inwards, p. 188; Hand, p. 40.

1632. Wild geese flying north are a sign of warm weather.

Cited in Whitman, p. 48.

1633. Wild geese flying over in great numbers indicates approaching storm.

Cited in Dunwoody, p. 41.

1634. Wild geese flying past large bodies of water indicates change of weather; going south cold, going north warm.

Cited in Dunwoody, p. 41; Garriott, p. 23; Inwards, p. 188; Hand, p. 23.

1635. Wild geese moving south indicates approaching cold weather, moving north, indicates that most of winter is over.

Cited in Dunwoody, p. 41; Garriott, p. 40; Inwards, p. 188; Hand, p. 40.

1636. Wild geese, wild geese, ganging to the sea, good weather it will be; wild geese, wild geese, ganging to the hill, the weather it will spill. Vars.: (a) If the wild geese gang out to sea, good weather there will surely be. (b) Wild geese, wild geese going out to sea, all fine weather it will be.

Cited in Swainson, p. 247; Dunwoody, p. 41; Garriott, p. 23; Whitman, p. 48; Inwards, p. 188; Hand, p. 23; Sloane 2, p. 130; Lee, p. 47; Freier, p. 53.

goose bone

1637. If the November goose bone be thick, so will the winter weather be; if the November goose bone be thin, so will the winter weather be.

Cited in Dunwoody, p. 36; Garriott, p. 40; Whitman, p. 48; Hand, p. 40; Freier, p. 79.

1638. The whiteness of a goose bone indicates the amount of snow.

Cited in Freier, p. 82.

1639. When the goose bone exposed to air retains its color, expect clear weather.

Cited in Dunwoody, p. 114.

1640. When the goose bone exposed to air turns blue, it indicates rain.
Var.: When the goose bone, exposed to air, turns blue, it
indicates rain; when it retains its color, expect clear weather.
Cited in Dunwoody, p. 114; Inwards, p. 188.

goose breastbone

1641. If in autumn a goose has a white breastbone, we will have a mild
winter; if a dark breastbone, a cold winter.
Cited in Hyatt, p. 25.

1642. If the breastbone of a goose is dark colored after cooking, no
genial spring, and vice versa.
Cited in Inwards, p. 188.

1643. If the breastbone of a goose is red, or has many red spots, expect
a cold and stormy winter; but if only a few spots are visible, the
winter will be mild. Var.: If the breastbone of the Thanksgiving
goose is red or has many red spots, expect a cold and stormy
winter; but, if only a few spots are visible, we will have a mild
winter.
Cited in Dunwoody, p. 36; Garriott, p. 40; Hand, p. 40; Freier,
p. 79.

1644. If the breastbone of a November goose is thick, expect a thick
winter; if thin, a thin winter.
Cited in Hyatt, p. 25.

1645. If you find a long breastbone in an autumnal goose, a long winter
is denoted; if a short breastbone, a short winter.
Cited in Hyatt, p. 25.

1646. The whiteness of a goose's breastbone indicates the amount of
snow during a winter.
Cited in Dunwoody, p. 36; Garriott, p. 40; Inwards, p. 188;
Hand, p. 40.

gossamer

1647. Gossamer floating over fields and hedges in the early morning
means fine weather.
Cited in Wilshere, p. 23.

1648. Gossamer is said when abundant in the air to afford a sign of a
fine autumn. Comment: Gossamer is a fine filmy substance
consisting of cobwebs.
Cited in Dunwoody, p. 57.

1649. When gossamer is seen flying without apparent wind, rain is on the way.
    Cited in Alstad, p. 68.
1650. When you see gossamer flying, be sure the air is drying.
    Cited in Swainson, p. 256; Northall, p. 474; Inwards, p. 207.

grape
1651. Grapes are ruined by rain on July 4th.
    Cited in Hyatt, p. 34.

grass
1652. Grass never grows when the wind blows.
    Cited in Denham, p. 7.
1653. Grass that remains green into late autumn is a portent of a warm winter.
    Cited in Hyatt, p. 15.
1654. Grasses of all kinds are loaded with seeds before a severe winter.
    Cited in Dunwoody, p. 66.
1655. When the grass is dry at morning light, look for rain before the night. Vars.: (a) If the grass be dry at morning light, look out for rain before the night. (b) When the grass is dry at night, look for rain before light.
    Cited in Humphreys 2, p. 59; Whitman, p. 34; Smith, p. 8; Lee, p. 28; Wilshere, p. 21. Recorded in Bryant (Ohio).
1656. You must look for grass in April on the top of an oak. Comment: Because the grass seldom springs well before the oak begins to put forth.
    Cited in Swainson, p. 81.

green ray
1657. Glimpse you ever the green ray, count the morrow a fine day. Comment: Green ray refers to the refraction of the sun's rays as it goes below the horizon at sea.
    Cited in Humphreys 2, p. 16; Whitman, p. 32; Sloane 2, p. 126; Lee, p. 77.

groundhog
1658. If a groundhog lays aside little food in autumn, a mild winter may be prophesized; if much food, a cold winter.

Cited in Hyatt, p. 28.

1659. If the groundhog is sunning himself on the second of February, he will return for four weeks to his winter quarters again.
Cited in Dunwoody, p. 94; Garriott, p. 42.

1660. If the groundhog sees its shadow on February 2nd, there will be six more weeks of winter. Vars.: (a) If a groundhog sees his shadow on February 2nd, it foreshadows rain for the seven following Sundays. (b) If a groundhog sees his shadow on February 2nd, seven weeks of rain are foreshadowed. (c) If a groundhog sees his shadow on February 2nd, spring is at a distance; four weeks, or six weeks, or eight weeks; if he does not see his shadow, spring is very near. (d) If the groundhog sees his shadow on February 2nd, there will be six weeks of bad weather. (e) When the groundhog sees his shadow on February 2nd, the second half of the winter will be severe.
Cited in Hyatt, p. 28; Smith, p. 1. Recorded in Bryant (Wis. Can. [Ont.]).

Groundhog Day

1661. February, Groundhog Day; half your meat and half your hay. Comment: Groundhog Day is February 2nd, half way through the winter.
Cited in Whiting 1, p. 218; Whiting 2, p. 218.

grouse

1662. The gathering of grouse into large flocks indicates snow.
Cited in Mitchell, p. 227; Inwards, p. 190.

1663. The winter will be severe if the roughed grouse have heavier than usual fringes on their toes.
Cited in Freier, p. 69.

1664. When grouse approach the farmyard, it is a sign of severe weather, frost and snow.
Cited in Mitchell, p. 227; Inwards, p. 190.

1665. When grouse sit on dykes on the moor, rain only is expected.
Cited in Inwards, p. 190.

grove

1666. Whispering grove tells of a storm to come.

Cited in Humphreys 2, p. 62; Whitman, p. 35; Alstad, p. 42; Sloane 2, p. 117.

guinea

1667. Guinea fowls (hens) squawk more than usual before rain. Var.: Guinea hens squawk more than usual before a rain.
Cited in Inwards, p. 189; Hand, p. 23; Freier, p. 52.

1668. Guineas are unusually noisy shortly before a heavy rain.
Cited in Smith, p. 14.

1669. If guineas cry in the afternoon, there will be rain. Var.: Guineas crying in the afternoon signify rain.
Cited in Hyatt, p. 25. Recorded in Bryant (Nebr.).

1670. If when snow is on the ground, the guinea hens cry, hallow, and caw, and the turkey moves around, there will surely be a thaw.
Cited in Lee, p. 65.

1671. If while standing on a post a guinea endlessly cries, "poor trash," a rain is foretold.
Cited in Hyatt, p. 25.

1672. The calling of guinea fowls or pea fowls foretell rain.
Cited in Lee, p. 65.

1673. To have guineas cry endlessly is a presage of rain.
Cited in Hyatt, p. 25.

guitar

1674. Guitar strings shorten before rain.
Cited in Dunwoody, p. 114; Garriott, p. 22; Hand, p. 22; Freier, p. 38.

gull

1675. Gulls will soar to lofty heights, and circling round utter shrill cries before a storm. Var.: Gulls will soar aloft and circling around, utter shrill cries before a storm.
Cited in Dunwoody, p. 37; Garriott, p. 23; Hand, p. 23; Freier, p. 52.

1676. The gull comes against the rain. Comment: The gull is always seen facing against a tempest.
Cited in Wilson, p. 341.

H

hail

1677. Hail, after long continued rain, indicates a clearing up. Var.: If hail appears after a long course of rain, it is a sign of clearing up.
Cited in Mitchell, p. 229; Inwards, p. 158; Wurtele, p. 294.

1678. Hail brings frost.
Recorded in Bryant (Wis.).

1679. Hail brings frost in the tail. Var.: Hail brings frost with its tail.
Cited in Denham, p. 2; Swainson, p. 209; Marvin, p. 208; Sloane 2, p. 114.

1680. It can hail in summer and not be followed by frost.
Recorded in Bryant (Wis.).

hailstorm

1681. A hailstorm by day denotes a frost at night.
Cited in Sloane 2, p. 114.

hair

1682. Human red hair curls and kinks at the approach of a storm, and re-straightens after the storm.
Cited in Garriott, p. 22; Hand, p. 22.

1683. If your hair curls expect rain.
Recorded in Bryant (Nebr.).

1684. Ladies' hair, like seaweed, scents the storm; long, long before it starts to pour their locks assume a baneful form.
Cited in Lee, p. 41.

1685. Much hair on animals predicts a cold winter.
Cited in Freier, p. 81.

1686. Wash your hair, have it set, it will rain soon thereafter.
Cited in Smith, p. 7.

halo

1687. Circles, haloes, etc., round the heavenly bodies are prognostics of coming rain.
Cited in Steinmetz 2, p. 83.

1688. Haloes predict a storm at no great distance, and the open side of the halo tells the quarter from which it may be expected.
Cited in Inwards, p. 90.

1689. Near ring (halo) far rain, far ring near rain.
Cited in Page, p. 28; Wilshere, p. 21.

hare

1690. Hares take to the open country before a snow storm.
Cited in Mitchell, p. 228; Swainson, p. 232; Dunwoody, p. 31; Inwards, p. 185.

1691. When hares leave the fells for shelter in the lowlands, snow is on the way. Var.: When hares seek shelter in the lowlands, snow is on the way.
Cited in Inwards, p. 185; Lee, p. 66.

haw

1692. Many haws, many snaws. Comment: Haws are blossoms on hawthorn trees, snaws are snows.
Cited in Mitchell, p. 226; Wilson, p. 509.

1693. Mony haws, mony snaws; mony sloes, mony cold toes. Comment: Sloes are wild plums, the fruit of the blackthorn tree.
Cited in Denham, p. 16; Mitchell, p. 226; Northall, p. 474; Inwards, p. 214.

hawk

1694. A soaring hawk indicates clear weather.
Cited in Hyatt, p. 22.

1695. If the hawk flies into a tree and searches for lice, it is a sign of rain. Var.: If a hawk flies to the top of a tall tree and searches for lice, it's a sign of rain.
Cited in Lee, p. 63; Freier, p. 52.

1696. When men-of-war hawks fly high, it is a sign of a clear sky; when they fly low, prepare for a blow.
Cited in Dunwoody, p. 37; Garriott, p. 19; Inwards, p. 196; Hand, p. 19.

hawkweed

1697. If the hawkweed closes its flowers in the afternoon, it will rain.
Cited in Lee, p. 72.

hawthorn

1698. If many hawthorn blossoms are seen, expect a severe winter.

Cited in Swainson, p. 259.

1699. It's always cold when the hawthorn blossoms.
Cited in Inwards, p. 212.

1700. When the hawthorn has too many haws we shall still have many snaws.
Cited in Inwards, p. 212.

hay

1701. A foot deep of rain will kill hay and grain; but three feet of snow will make them come mo. Comment: Mo means more.
Cited in Inwards, p. 158.

1702. Better it is to rise betimes and make hay while the sun shines, than to believe in tales and lies which idle monks and friars devise.
Cited in Dunwoody, p. 66.

1703. Make hay while the sun shines. Vars.: (a) Gather hay while the sun shines. (b) It is good to make hay while the sun shines. (c) To make hay while the sun shines. (d) You have to make hay while the sun shines.
Cited in Cheales, p. 24; Inwards, p. 213; Whiting 1, p. 202; Wilshere, p. 18; Simpson, p. 143; Whiting 2, p. 293; Mieder, p. 286.

haying season

1704. In haying season, when there is no dew, it indicates rain.
Cited in Dunwoody, p. 48; Inwards, p. 56.

haze

1705. Haze and purple sky in the west indicate fair weather. Var.: Haze and western sky purple indicate fair weather.
Cited in Dunwoody, p. 77; Inwards, p. 146.

hazy

1706. Hazy weather is thought to prognosticate frost in winter, snow in spring, fair weather in summer, and rain in autumn.
Cited in Mitchell, p. 220; Steinmetz 2, p. 275; Inwards, p. 146.

1707. When the air is hazy, so that the solar light fades gradually, and looks white, rain will most certainly follow.
Cited in Inwards, p. 82.

heat

1708. Neither heat nor cold abides always in the sky.
Cited in Wilson, p. 365.

hedgehog

1709. Observe which way the hedgehog builds her nest, to front the north or south, or east or west; for if 'tis true what common people say, the wind will blow the quite contrary way.
Cited in Inwards, p. 185; Hand, p. 39.

1710. The hedgehog commonly hath two holes or vents in his den, one towards the south and the other towards the north; and look which of them he stops, thence will great storms and winds follow.
Cited in Inwards, p. 185.

heel

1711. If your heel itches, rain is not far away.
Cited in Hyatt, p. 30.

heifer

1712. He taught us erst the heifer's tail to view when struck aloft, that showers would straight ensue.
Cited in Whitman, p. 47.

1713. Watch the heifer's tail; when stretched aloft, 'twill rain or hail.
Cited in Freier, p. 51.

hemp

1714. A lump of hemp acts as a good hydrometer and prognosticates rain when it is damp.
Cited in Garriott, p. 21; Inwards, p. 219; Hand, p. 21.

hen

1715. Broody hens in January foretell a hot dry summer.
Cited in Hyatt, p. 24.

1716. During a shower, if hens run under shelter, it will not last long, but if they stay out, it will be a long storm.
Cited in Lee, p. 64.

1717. Hens huddle together outside their houses before rain.
Cited in Page, p. 19.

1718. Hens resorting to the perch or rest covered with dust declare rain.
Cited in Inwards, p. 187.

1719. If a hen and chickens crowd into a house, it is a sign of rain.
Cited in Swainson, p. 238.

1720. If a hen goes singing to bed, it will get up with a wet head.
Cited in Hyatt, p. 24.

1721. Much cackling among hens but no eggs denotes a rain or a storm.
Cited in Hyatt, p. 24.

1722. When hens are observed to pick and pluck themselves more than usual, rain is near.
Cited in Mitchell, p. 227.

1723. When hens run about acting frightened, a windstorm is coming.
Cited in Lee, p. 64.

1724. When the hen crows, expect a storm within and without.
Cited in Dunwoody, p. 37.

heron

1725. Herons in the evening flying up and down, as if doubtful where to rest, presages some evil approaching weather. Var.: Herons flying up and down in the evening, as if not knowing where to rest, means bad weather is approaching.
Cited in Swainson, p. 240; Inwards, p. 195; Lee, p. 63.

1726. If the heron cry in the night as she flies, it presageth wind. Var.: If the heron cries in the night as it flies, it means wind.
Cited in Inwards, p. 195; Lee, p. 63.

1727. If the heron stand melancholy on the banks, it portends rain.
Cited in Inwards, p. 195.

1728. The heron forsaking the fens and soaring aloft indicates wind.
Cited in Mitchell, p. 233.

1729. The heron screeching in the mountains is a sign of storm. Var.: Heron screeching in the mountains indicates a storm.
Cited in Inwards, p. 195; Lee, p. 63.

1730. When heron fly up and down as in doubt where to rest, expect rain.
Cited in Dunwoody, p. 37; Garriott, p. 23; Hand, p. 23.

1731. When the heron flies low, the air is gross and thickening into showers. Var.: When the heron or bittern flies low, the air is gross and thickening into showers.
Cited in Swainson, p. 240; Inwards, p. 195.

hill

1732. Distant hills appear near just before rain.
Cited in Dunwoody, p. 114.

hip

1733. Many hips and haws, many frosts and snaws. Var.: Mony hips and haws, mony frosts and snaws. Comment: Hips are the ripened fruit of wild roses, haws are the blossoms of the hawthorn tree.
Cited in Denham, p. 24; Northall, p. 474; Inwards, p. 214; Sloane 2, p. 115.

hog

1734. After hogs run here and there squealing, look for a change of weather.
Cited in Hyatt, p. 28.

1735. Hogs crying and running unquietly up and down with hay or litter in their mouths, foreshadows a storm to be near at hand.
Cited in Swainson, p. 233; Garriott, p. 23; Inwards, p. 183; Hand, p. 23.

1736. Hogs in autumn constantly looking to the north signify the nearness of winter.
Cited in Hyatt, p. 28.

1737. Hogs pick and store straws, leaves, etc., before cold weather.
Cited in Dunwoody, p. 31.

1738. Hogs rubbing themselves in winter indicates an approaching thaw.
Cited in Dunwoody, p. 31; Inwards, p. 183.

1739. If hogs are slaughtered in early winter and the small part of the milt lies toward the head, the worst part of winter is to come; if the large part of the milt lies toward the head, the worst part of winter is over. Comment: Milt is the hog's spleen.
Cited in Hyatt, p. 28.

1740. If hogs in autumn pick up dried weeds and shake them, look for a stormy winter.
Cited in Hyatt, p. 28.
1741. If you slaughter hogs early in the autumn and the lungs are clear, you are warned of a light winter; if the lungs are streaked, a hard winter.
Cited in Hyatt, p. 28.
1742. Look for falling weather after hogs rush about holding cornstalks or sticks or straw in their mouths.
Cited in Hyatt, p. 28.
1743. Look for the approach of winter after hogs in autumn pull hay or straw from a stack and begin making beds.
Cited in Hyatt, p. 28.

Hollantide
1744. If ducks do slide at Hollantide, at Christmas Day they'll swim; if ducks do swim at Hollantide, at Christmas Day they'll slide.
Comment: Hollandtide refers to All Saint's Day, November 1st.
Cited in Dunwoody, p. 102; Northall, p. 454; Marvin, p. 211; Inwards, p. 66; Wilshere, p. 14.

Holy Innocents' Day
1745. A bright day on Innocents Day, a year of plenty; a dark, wet day, a year of scarcity. Comment: Holy Innocents' Day is December 28th.
Cited in Lee, p. 151.

Holy Thursday
1746. Fine on Holy Thursday, wet on Whit Monday; fine on Whit Monday, wet on Holy Thursday. Comment: Holy Thursday is Ascension Day, Whit Monday is the Monday following Whitsuntide.
Cited in Inwards, p. 71.

Holyrood Day
1747. If dry be the buck's horn on Holyrood morn, 'tis worth a vest of gold; but if wet it be seen ere Holyrood e'en, bad harvest is foretold. Var.: If dry be the buck's horn on Holyrood morn, 'tis worth a kist of gold; but if wet it be seen ere Holyrood e'en, bad

harvest is foretold. Comment: Holyrood Day occurs on the fourteenth of September. Kist means chest.
Cited in Dunwoody, p. 29; Inwards, p. 62.

honey

1748. The amount of honey in the bees' hives tells the length of the winter.
Recorded in Bryant (Wis.).

hornet

1749. Hornets build nests high before warm summers. Var.: Hornets build nests high before warm summers, and low before cold and early winters.
Cited in Dunwoody, p. 57; Inwards, p. 205.

1750. Hornets, wasps, and gnats sting more frequently against wet weather than in fair.
Cited in Steinmetz 1, p. 111.

1751. If hornets build nests close to the ground, it means heavy snow.
Recorded in Bryant (W.Va.).

1752. If hornets build nests high up in the trees, there will not be much snow.
Recorded in Bryant (W.Va.).

1753. If hornets build their nests low, a mild winter will follow; if high a hard winter.
Cited in Hyatt, p. 18.

1754. The higher the hornets build their nests, the deeper the snow. Var.: As high as the hornet's nest so will be the snow next winter.
Cited in Smith, p. 10; Freier, p. 84.

1755. When hornets build their nests near the ground, expect a cold and early winter. Var.: When hornets' nests are built close to the ground, a hard cold winter is coming.
Cited in Dunwoody, p. 57; Smith, p. 13.

horse

1756. A horse rolling on the ground foretells a change of weather or a storm.
Cited in Hyatt, p. 28.

1757. Horses and mules very lively without apparent cause indicate cold. Var.: Horses and mules, if very lively without apparent cause, indicate cold.
Cited in Dunwoody, p. 32; Inwards, p. 180.

1758. Horses are startled and nervous before a storm.
Cited in Freier, p. 49.

1759. Horses are unusually frolicsome for several days preceding a storm.
Cited in Hyatt, p. 28.

1760. Horses become fidgety just before a windstorm.
Cited in Hyatt, p. 28.

1761. Horses foretell the coming of rain by starting more than ordinary, and appearing in other respects restless and uneasy on the road. Vars.: (a) Horses, as well as other domestic animals, foretell the coming of rain by starting more than ordinary and appearing in other respects restless and uneasy. (b) Horses foretell rain when they appear restless and uneasy.
Cited in Dunwoody, p. 32; Garriott, p. 23; Inwards, p. 179; Hand, p. 23; Lee, p. 71.

1762. Horses race before a wind.
Cited in Lee, p. 71.

1763. Horses rolling over in the pasture, expect rain.
Cited in Freier, p. 67.

1764. Horses running with their backs to the wind denote a storm.
Cited in Hyatt, p. 28.

1765. Horses staying close together in a corner of the pasture or under a tree with their backs to the wind denote rain.
Cited in Hyatt, p. 28.

1766. Horses stretch out their necks and sniff up the air just before a fall of rain. Vars.: (a) After horses have stretched out their necks and sniffed the air, you may predict a rain. (b) If horses stretch out their necks and sniff the air, rain will ensue. (c) When horses stretch out their necks and sniff the air it will rain.
Cited in Steinmetz 1, p. 112; Swainson, p. 232; Inwards, p. 179; Sloane 2, p. 120; Hyatt, p. 28; Lee, p. 71; Freier, p. 67.

1767. Horses sweating in the stable is a sign of rain. Vars.: (a) If horses in the stable are sweating and switching their tails, a rain is at hand. (b) Sweating of horses in the stable is a sign of rain. (c) When horses sweat in the stable, it is a sign of rain.

Cited in Dunwoody, p. 115; Garriott, p. 21; Inwards, p. 179; Hand, p. 21; Hyatt, p. 29; Lee, p. 71.

1768. Horses' tails appear larger when rain is coming because the hair is standing erect.
Cited in Lee, p. 71.

1769. If a horse turns back his lips and grins, it will rain.
Cited in Lee, p. 71.

1770. If horses or mules gallop about playfully for some days, cold weather is approaching.
Cited in Hyatt, p. 28.

1771. If in late winter horses begin to shed their hair, spring will arrive early; if they do not shed any hair, spring will be delayed.
Cited in Hyatt, p. 29.

1772. If the hair of a horse grows long early, expect an early winter.
Cited in Dunwoody, p. 32.

1773. If young horses do rub their backs against the ground, it is a sign of great drops of rain to follow.
Cited in Inwards, p. 180.

1774. Spirited horses strut more than ordinary and are restless and uneasy before a rain.
Cited in Alstad, p. 82.

1775. The hair of horses becomes curly and rough just before a rain.
Cited in Hyatt, p. 29.

1776. To hear more clearly than usual the tread of horse hooves on the road is a warning of rain.
Cited in Hyatt, p. 29.

1777. Trust not the horse if it is frisky nor the sun if it turns its back.
Comment: If the sun looks pale and hazy.
Cited in Whitman, p. 26.

1778. When horses and cattle stretch out their necks and sniff the air it will rain.
Cited in Dunwoody, p. 32; Garriott, p. 23; Hand, p. 23.

1779. When horses are restless and paw with their hoof, you'll soon hear the patter of rain on your roof.
Cited in Freier, p. 49.

1780. When horses assemble in the corner of a field, with heads to leeward, expect rain.
Cited in Dunwoody, p. 32; Inwards, p. 179.

hot

1781. As hot as it is in the summer, that's how cold it is in the winter.
Recorded in Bryant (Tex.).

1782. In hot climates the wind sets from the sea to the land during the
day; from the land to the sea during the night.
Cited in Sloane 1, p. 150.

hound

1783. Sign, too, of rain: his outstreched feet the hound extends, and
curves his body to the ground.
Cited in Lee, p. 70.

ice

1784. A thick coating of ice on the trees in February is a sign of a large fruit crop.
Cited in Hyatt, p. 34.

1785. Expect not fair weather in winter from one night's ice.
Cited in Dunwoody, p. 91; Marvin, p. 207; Wilson, p. 240.

1786. Ice in November brings mud in December.
Cited in Marvin, p. 211.

1787. Ice remaining long on the trees in winter is an indication of a good fruit year.
Cited in Hyatt, p. 34.

1788. If ice bears before Christmas, it won't bear a goose after.
Cited in Wilshere, p. 15.

1789. If ice will bear a man before Christmas, it will not bear a mouse afterward.
Cited in Dunwoody, p. 101.

1790. If the ice crack much, expect frost to continue.
Cited in Dunwoody, p. 54.

1791. If the ice will bear a goose before Christmas, it will not bear a duck afterwards. Var.: If the ice will bear a goose before Christmas, it will bear a duck after.
Cited in Denham, p. 62; Marvin, p. 211; Inwards, p. 34.

1792. If there's ice in November that will bear a duck, there'll be nothing after but sludge and muck. Vars.: (a) Ice in November to bear a duck, the rest of the winter'll be slush and muck. (b) If there be ice in November that will bear a duck, there will be nothing thereafter but sleet and muck.
Cited in Swainson, p. 141; Dunwoody, p. 99; Marvin, p. 210; Page, p. 51.

1793. Thin ice and thick ice look the same from a distance.
Cited in Mieder, p. 323.

1794. Thin ice carries no weight.
Cited in Mieder, p. 323.

1795. Try the ice before you venture on it.
Cited in Mieder, p. 323.

1796. When the ice before Martinmas bears a duck, then look for a winter o' mire and muck. Var.: If Martinmas ice can bear a

duck, the winter will be all mire and muck. Comment: Martinmas is November 11th.
Cited in Northall, p. 455; Wright, p. 118; Wilshere, p. 14.

Indian Summer

1797. An Indian Summer is foretold by spider webs on the trees in autumn.
Cited in Hyatt, p. 19.

1798. If we don't get our Indian Summer in October or November we will get it in winter. Var.: If we do not get an Indian Summer in October or November, we will get it in the winter.
Cited in Dunwoody, p. 89; Garriott, p. 45; Whitman, p. 45; Inwards, p. 32; Hand, p. 45; Sloane 2, p. 41; Freier, p. 77.

indigo

1799. Just before rain or heavy dew the wild indigo closes or folds its leaves. Var.: The wild indigo folds its leaves before a rain.
Cited in Dunwoody, p. 68; Inwards, p. 212; Alstad, p. 93.

insect

1800. As soon as you observe insects carrying material for nests, you will know that cold weather is not far away.
Cited in Hyatt, p. 17.

1801. Insects fly low before rain.
Cited in Mitchell, p. 228.

1802. Insects flying in numbers just at evening show weather changing to rain.
Cited in Dunwoody, p. 57; Lee, p. 56.

1803. The early appearance of insects indicates an early spring and good crops. Var.: Early insects, early spring, good crops.
Cited in Sloane 2, p. 124; Freier, p. 82.

1804. The mating of insects in August is a presage of a delayed autumn.
Cited in Hyatt, p. 17.

1805. When insects are especially annoying, there is going to be a hard winter.
Cited in Smith, p. 12.

1806. When little black insects appear on the snow, expect a thaw.
Cited in Dunwoody, p. 56; Inwards, p. 208.

itch

1807. There will be a change of weather after your head has itched.
Cited in Hyatt, p. 30.
1808. When a scar from an operation itches, it means that it is going to rain.
Cited in Smith, p. 7.
1809. Your nose itching three times within an hour is an indication of rain within twenty-four hours.
Cited in Hyatt, p. 30.

J

Jack Frost

1810. Jack Frost in Janiveer nips the nose of the nascent year.
Cited in Swainson, p. 19; Inwards, p. 37.

jack rabbit

1811. When you see jack rabbits active and running around in the
daytime, it will rain soon.
Cited in Reeder, no pp.

jackdaw

1812. Jackdaws are unusually clamorous before rain.
Cited in Dunwoody, p. 37.

Janiveer

1813. Janiveer freezes the pot by the fire. Var.: January freezes the pot
by the fire.
Cited in Denham, p. 23; Swainson, p. 19; Northall, p. 430;
Inwards, p. 37; Wilson, p. 410.
1814. Janiveer sow oats, get golden groats.
Cited in Cheales, p. 21.
1815. The blackest month in all the year is the month of Janiveer.
Cited in Denham, p. 26; Swainson, p. 19; Northall, p. 432;
Wright, p. 10; Inwards, p. 36.
1816. Who in Janiveer sows oats gets gold and groats, who sows in
May gets little that way.
Cited in Wilshere, p. 7.

January

1817. A black January means a fat graveyard.
Recorded in Bryant (Utah).
1818. A favorable January brings us a good year. Var.: A favorable
January, a good year.
Cited in Dunwoody, p. 93; Garriott, p. 44; Marvin, p. 216;
Whitman, p. 41; Inwards, p. 36; Hand, p. 44.
1819. A summerish January indicates a winterish spring. Var.: A
summerish January, a winterish spring.
Cited in Dunwoody, p. 94; Inwards, p. 36.
1820. A warm January, a cold May.

Cited in Garriott, p. 44; Whitman, p. 42; Inwards, p. 37; Hand, p. 44; Freier, p. 82.

1821. A wet January, a wet spring. Var.: Wet January, wet spring.
Cited in Dunwoody, p. 94; Inwards, p. 36; Wilshere, p. 6.

1822. Always expect a thaw in January.
Cited in Dunwoody, p. 94; Whitman, p. 41; Inwards, p. 36; Hand, p. 44; Freier, p. 81.

1823. As the weather is on the first three days of January, so will it be during the three months of winter.
Cited in Hyatt, p. 13.

1824. As the weather is on the first three days of January, so will it be the entire year.
Cited in Hyatt, p. 13.

1825. Dry January, plenty of wine.
Cited in Dunwoody, p. 93; Inwards, p. 36.

1826. Fog in January brings a wet spring.
Cited in Dunwoody, p. 93; Inwards, p. 36.

1827. Have rivers much water in January, then the autumn will forsake them. But are they small in January, then brings the autumn surely much wine.
Cited in Dunwoody, p. 93.

1828. If birds begin to whistle in January, frosts to come.
Cited in Inwards, p. 36.

1829. If grain grows in January, there will be a year of great need.
Cited in Dunwoody, p. 83; Inwards, p. 36.

1830. If January calends be summerly gay, 'twill be winterly weather till the calends of May. Vars.: (a) If January calends be summerly gay, it will be winterly weather till the calends of May. (b) If the calends of January be smiling and gay, you'll have wintry weather till the calends of May. Comment: Calends refers to the first day of the new moon and the first day of the month in the ancient Roman calendar.
Cited in Denham, p. 26; Northall, p. 431; Wright, p. 9; Inwards, p. 36; Wilson, p. 410; Wilshere, p. 7.

1831. If the grass grows green in January, it will grow the worse for it all year. Vars.: (a) If grass grows in January, it will grow badly all the year. (b) If the grass grows in January, it grows the worse for all the year. (c) The grass that grows in Janiveer grows no more all the year.

Cited in Denham, p. 24; Mitchell, p. 230; Steinmetz 1, p. 137; Swainson, p. 20; Cheales, p. 20; Dunwoody, p. 93; Northall, p. 432; Garriott, p. 44; Wright, p. 10; Inwards, p. 36; Hand, p. 44; Page, p. 43; Wilshere, p. 6.

1832. If there be no snow before January, there will be the more in March and April.
Cited in Whitman, p. 41; Inwards, p. 37.

1833. If you see grass in January, lock your grain in the granary.
Cited in Swainson, p. 20; Northall, p. 432; Humphreys 1, p. 430; Humphreys 2, p. 10; Whitman, p. 41; Inwards, p. 36; Lee, p. 135; Wilshere, p. 7.

1834. In January much rain and little snow is bad for mountains, valleys, and trees.
Cited in Dunwoody, p. 93; Inwards, p. 36.

1835. In January should sun appear, March and April pay full dear.
Var.: In January if the sun appear, March and April pay full dear.
Cited in Swainson, p. 24; Cheales, p. 20; Dunwoody, p. 93; Northall, p. 432; Brunt, p. 67; Inwards, p. 36.

1836. It's January if a bushel of snow falls through a keyhole.
Recorded in Bryant (Ohio).

1837. January and February fill or empty the granary. Vars.: (a) January and February, do fill or empty the granary. (b) January or February, do fill or empty the granary.
Cited in Swainson, p. 23; Humphreys 1, p. 430; Humphreys 2, p. 10; Whitman, p. 41.

1838. January blossoms fill no man's cellar.
Cited in Humphreys 1, p. 430; Humphreys 2, p. 10; Whitman, p. 41; Brunt, p. 67; Lee, p. 134.

1839. January commits the fault, and May bears the blame.
Cited in Brunt, p. 67; Inwards, p. 37.

1840. January fill dyke, February black and white.
Cited in Dunwoody, p. 93.

1841. January never lies dead in a dyke gutter.
Cited in Denham, p. 23.

1842. January warm, the Lord have mercy. Comment: Because a premature growth of vegetation is likely to suffer severe damage from spring frosts.

Cited in Dunwoody, p. 93; Garriott, p. 44; Humphreys 1, p. 430; Humphreys 2, p. 10; Whitman, p. 41; Brunt, p. 67; Inwards, p. 36; Hand, p. 44; Wilshere, p. 6; Freier, p. 82; Whiting 2, p. 342. Recorded in Bryant (Wis.).

1843.   January wet no wine you get. Vars.: (a) If January wet, remains empty the barrel. (b) Is January wet?-the barrel remains empty. Cited in Dunwoody, p. 93; Humphreys 1, p. 430; Humphreys 2, p. 10; Whitman, p. 41; Inwards, p. 36; Lee, p. 134.

1844.   Much rain in January, no blessing to the fruit. Cited in Dunwoody, p. 93; Inwards, p. 36.

1845.   Never shut out the January sun. Cited in Mieder, p. 337.

1846.   On one of the first three days of January a rain betokens a wet February. Cited in Hyatt, p. 10.

1847.   The first three days of January rule the coming three months. Var.: The first three days of January foretell the type of weather in January, February, and March. Cited in Dunwoody, p. 100; Marvin, p. 216; Inwards, p. 38; Smith, p. 3.

1848.   The first twelve days of January rule the next twelve months; the first day equals January, the second February, etc. Cited in Smith, p. 3.

1849.   The last twelve days of January rule the weather of the whole year. Cited in Dunwoody, p. 100; Inwards, p. 36.

1850.   The weather during the first twelve days of January will control the weather for the whole year. Cited in Hyatt, p. 13.

1851.   When oak trees bend with snow in January, good crops may be expected. Cited in Inwards, p. 37.

1852.   Who in January sows oats, gets gold and groats; who sows in May, gets little that way. Cited in Denham, p. 26; Swainson, p. 24; Northall, p. 432; Wilson, p. 410.

January 1

1853. A rain on January 1st forecasts seven rainy New Year's Days in succession.
Cited in Hyatt, p. 10.

1854. January 1st, morning red, foul weather, and great need.
Cited in Dunwoody, p. 93.

1855. Whatever the weather is like on the first day of the year, the rest of the year will be characteristic.
Cited in Smith, p. 3.

January 2

1856. January 2nd shows the nature and state of September. Var.: January 2nd, as the weather is this day so it will be in September.
Cited in Swainson, p. 28; Dunwoody, p. 93.

January 4

1857. January 4th does show the nature and state of November.
Cited in Swainson, p. 28.

January 8

1858. January 8th before noon declared the nature of June, and after noon the nature of May.
Cited in Swainson, p. 28.

January 10

1859. January 10th shows the nature of October before noon, and after noon the nature of September.
Cited in Swainson, p. 29.

January 11

1860. January 11th before noon, this day declares the nature of December, and after noon the nature of November.
Cited in Swainson, p. 29.

January 12

1861. If the sun shines on January 12th, it foreshows much wind.
Cited in Swainson, p. 30.

1862. January 12th does foreshow the nature and condition of the whole year, and does confirm the eleven days going before.
Cited in Swainson, p. 29.

January 14
1863. January 14th will either be the coldest or wettest day of the year.
Cited in Inwards, p. 38; Wurtele, p. 296.

January 22
1864. If the sun shine on 22nd January, there shall be much wind.
Cited in Inwards, p. 38.

jellyfish
1865. When many jellyfish appear in the sea, a period of storms will follow.
Cited in Lee, p. 60.

July
1866. A shower of rain in July, when the corn begins to fill, is worth a plough of oxen, and all that belongs there till. Var.: A shower in July, when the corn begins to fill, is worth a yoke of oxen, and all belongs there till.
Cited in Mitchell, p. 230; Cheales, p. 22; Northall, p. 441; Wright, p. 89; Inwards, p. 57; Wilson, p. 730.
1867. As July, so next January. Var.: As July, so the next January.
Cited in Dunwoody, p. 98; Garriott, p. 45; Inwards, p. 58; Hand, p. 45.
1868. In July shear your rye.
Cited in Northall, p. 442.
1869. In July, some reap rye; in August, if one will not, the other must.
Cited in Denham, p. 52.
1870. July, God send thee calm and fair, that happy harvests we may see; with quiet time and healthsome air, and man to God may thankful be.
Cited in Dunwoody, p. 98; Garriott, p. 45; Whitman, p. 43; Inwards, p. 57; Hand, p. 45.
1871. Much thunder in July injures wheat and barley.
Cited in Inwards, p. 57.

1872. Ne'er trust a July sky. Var.: Never trust a July sky.
Cited in Garriott, p. 45; Whitman, p. 43; Inwards, p. 57; Hand, p. 45.

1873. No tempest, good July, lest corn come off blue by. Var.: No tempest, good July, lest the corn look ruely.
Cited in Denham, p. 53; Swainson, p. 110; Northall, p. 441; Inwards, p. 57.

1874. The first Friday in July is always wet.
Cited in Inwards, p. 58.

1875. Whatever July and August do not boil, September cannot fry.
Cited in Dunwoody, p. 98; Garriott, p. 45; Whitman, p. 43; Inwards, p. 58; Hand, p. 45.

1876. When the months of July, August, and September are unusually hot, January will be the coldest month.
Cited in Dunwoody, p. 98; Inwards, p. 58.

July 1

1877. A rainy July 1st denotes rain, or rain off and on, for the next three weeks.
Cited in Hyatt, p. 10.

1878. If the first of July be rainy weather, 'twill rain more or less for forty days together. Var.: If the first of July be rainy weather, it will rain more or less for four weeks together.
Cited in Denham, p. 52; Swainson, p. 112; Dunwoody, p. 98; Northall, p. 441; Wright, p. 85; Whitman, p. 38; Wilson, p. 263; Wilshere, p. 11.

1879. Rain on July 1st brings seventeen rainy days that month.
Cited in Hyatt, p. 10.

July 2

1880. If it rain on the second of July, such weather shall be forty days after: day by day.
Cited in Swainson, p. 112.

July 4

1881. It always rains on the fourth of July.
Recorded in Bryant (Wash.).

1882. July 4th is always rainy, because so much ammunition and fireworks is shot off in the air.

Cited in Hyatt, p. 32.

July 10

1883.　As the weather on July 10th, so it will be for seven weeks. Var.:
If it rains on July 10th, it will rain seven weeks.
Cited in Dunwoody, p. 91; Whitman, p. 39; Inwards, p. 59.

July 25

1884.　Puffy white clouds on July 25th foretells much snow in the
coming winter.
Cited in Lee, p. 150.

July 26

1885.　If it rains on July 26th, it will rain for the following two weeks.
If it is dry, expect two weeks of dryness.
Cited in Sloane 2, p. 42.

June

1886.　A calm June puts the farmer in tune.
Cited in Northall, p. 440; Wright, p. 68.

1887.　A cold and wet June spoils mostly the whole year. Vars.: (a) A
cold and wet June spoils the rest of the year. (b) A cold wet June
practically spoils the whole year.
Cited in Dunwoody, p. 97; Garriott, p. 45; Whitman, p. 43;
Inwards, p. 56; Hand, p. 45; Hyatt, p. 34; Freier, p. 79.

1888.　A dripping June brings all things in tune. Vars.: (a) A dripping
June puts all things in tune. (b) A dripping June sets all in tune.
(c) A dry May and a dripping June brings all things in tune.
Cited in Dunwoody, p. 97; Wright, p. 68; Inwards, p. 56;
Simpson, p. 60.

1889.　A dry June, much corn; a wet June, no corn at all.
Cited in Hyatt, p. 34.

1890.　A good leak in June sets all in tune. Var.: A leak in June sets the
corn in tune.
Cited in Northall, p. 440; Inwards, p. 56; Page, p. 48; Wilshere,
p. 11.

1891.　A wet June makes a dry September.
Cited in Wright, p. 84; Inwards, p. 57.

1892. Calm weather in June, sets corn in tune.
Cited in Denham, p. 49; Dunwoody, p. 97; Garriott, p. 44;
Wilshere, p. 11; Mieder, p. 646.

1893. If June be sunny, harvest comes early.
Cited in Inwards, p. 56.

1894. If north wind blows in June, rye will be splendid at harvest time.
Var.: If north wind blows in June, good rye harvest.
Cited in Dunwoody, p. 97; Inwards, p. 56.

1895. It is a good June that brings some rain and some dry weather.
Cited in Inwards, p. 56.

1896. It never clouds up in a June night for a rain.
Cited in Dunwoody, p. 45; Whitman, p. 43; Inwards, p. 56;
Hand, p. 45.

1897. June damp and warm does not make the farmer poor. Var.: June
warm and damp does not make the farmer poor.
Cited in Dunwoody, p. 97; Garriott, p. 45; Whitman, p. 43;
Inwards, p. 56; Hand, p. 45; Wilshere, p. 11; Freier, p. 77.

1898. Wet June, dry September.
Cited in Sloane 2, p. 42.

1899. When it is the hottest in June, it will be the coldest in the next
February at corresponding day. Var.: When it is hottest in June,
it will be coldest in the corresponding days of the next February.
Cited in Dunwoody, p. 97; Inwards, p. 57.

1900. When the wind goes to the west early in June, expect wet
weather until the end of August.
Cited in Inwards, p. 56.

June 1

1901. If it is raining on June 1st, it will rain fifteen days before
clearing off.
Cited in Hyatt, p. 10.

June 8

1902. If on the eighth of June it rains, it foretells wet harvests, men
hath sain. Vars.: (a) If it raineth on the eighth of June, a wet
harvest men will see. (b) If on the eighth of June it rain, to
expect a wet harvest you may be fain. (c) To expect a wet
harvest you may be fain, if on the eighth of June it should rain.

Cited in Cheales, p. 24; Dunwoody, p. 97; Northall, p. 440; Wright, p. 67; Inwards, p. 57; Wilshere, p. 11.

June 27

1903. If it rain on June 27th, it will rain for seven weeks.
Cited in Whitman, p. 38; Inwards, p. 57.

June bug

1904. If you hear June bugs in the morning, it's going to be a hot day.
Cited in Smith, p. 11.

1905. The person who kills a June bug causes a rain.
Cited in Hyatt, p. 18.

K

katydid

1906. Katydids cry three months before frosts.
Cited in Dunwoody, p. 57.

1907. The first frost of the season occurs six weeks after we hear the first katydid. Var.: Six weeks after the first katydid, look for frost.
Cited in Dunwoody, p. 54. Recorded in Bryant (Vt.).

1908. When katydids come, spring will soon follow.
Cited in Smith, p. 11.

1909. When the katydid says "Kate," he announces ten days till a frost.
Cited in Sloane 2, p. 41.

1910. When you hear a katydid, it will be six weeks till cold weather.
Recorded in Bryant (W.Va.).

kelp

1911. A lump of kelp acts as a good hygrometer, and prognosticates rain when it becomes damp.
Cited in Mitchell, p. 230.

1912. A piece of kelp or seaweed hung up will become damp before rain.
Cited in Inwards, p. 215.

killdee

1913. If the killdees hollar today, tomorrow will be windy.
Cited in Smith, p. 12.

kingfisher

1914. The peaceful kingfishers are met together about the decks, and prophesy calm weather.
Cited in Swainson, p. 241; Inwards, p. 196; Lee, p. 65.

kite (bird)

1915. If kites fly high, fair weather is at hand. Vars.: (a) If kites fly high, fair weather is coming. (b) Kites flying unusually high are said to indicate fair weather.
Cited in Swainson, p. 241; Dunwoody, p. 37; Inwards, p. 220; Lee, p. 75.

kite (toy)

1916. If you start to fly a kite and it soars straight up into the air at once, the weather will remain clear; but if the kite flies to one side or goes into a tailspin, the weather will soon change.
Cited in Hyatt, p. 32.

knife

1917. If a knife is dropped, then a fork, and they lie crossing each other, a storm may be expected.
Cited in Hyatt, p. 31.

knot

1918. Knots get tighter before a rain.
Cited in Freier, p. 38.

L

Lady Day

1919. If the wind is in the north on Lady Day it will remain in that quarter until Midsummer Day. Comment: Lady Day is March 25th (Annunciation Day), Midsummer Day is June 24th.
Cited in Wilshere, p. 19.

1920. On Lady Day the latter the cold comes over the water. Var.: On Lady Day the latter the cold comes on the water. Comment: Latter Lady Day is April 6th.
Cited in Northall, p. 451; Inwards, p. 53.

ladybug

1921. When ladybugs swarm expect a warm.
Cited in Sloane 2, p. 42.

lamp

1922. If a lamp or a lantern has an unceasing flicker or sputter, it denotes rain in summer and snow in winter.
Cited in Hyatt, p. 31.

lamp wick

1923. Lamp wicks crackle before a rain.
Cited in Freier, p. 63.

1924. Lamp wicks crackle, candles burn dim, soot falls down, smoke descends, walls and pavements are damp, and disagreeable odors arise from ditches and gutters before rain.
Cited in Garriott, p. 22; Hand, p. 22.

land

1925. The land and the sea are heated together, but the land and the sea have their own kinds of weather.
Cited in Alstad, p. 23.

lark

1926. Field larks, congregating in flocks, indicate severe cold.
Cited in Dunwoody, p. 38; Garriott, p. 40; Inwards, p. 195; Hand, p. 40.

1927. If larks fly high and sing long, expect fine weather. Vars.: (a) Expect fine weather if larks fly high and sing long. (b) Larks, when they sing long and fly high, forebode fine weather.

Cited in Swainson, p. 241; Dunwoody, p. 20; Inwards, p. 195; Lee, p. 61; Freier, p. 21.

1928. When field larks rise before they sing at dawn, with an overcast sky, expect rain; but when they fly very high, singing as they rise, expect a fine day.
Cited in Inwards, p. 195.

1929. When ground larks soar high on a cloudy day, clearing skies will follow.
Cited in Lee, p. 46.

leaf

1930. An early falling of leaves denotes an early winter.
Cited in Hyatt, p. 16.

1931. Before rain the leaves of the lime, sycamore, plane, and poplar trees show a great deal more of their under surface when trembling in the wind.
Cited in Garriott, p. 25; Inwards, p. 212; Hand, p. 25.

1932. Early falling leaves indicate an early fall.
Cited in Dunwoody, p. 82.

1933. If in autumn the tops of trees are bare but leaves hang on the sides, they prophesy a mild winter; if they have fallen from the sides but remain on the tops, a severe winter.
Cited in Hyatt, p. 16.

1934. If in the fall of the leaves in October many of them wither and hang on the boughs, it betokens a frosty winter and much snow.
Cited in Dunwoody, p. 66.

1935. If on the trees the leaves still hold, the coming winter will be cold. Var.: If on the trees the leaves long hold, the coming winter will be cold.
Cited in Whitman, p. 49; Inwards, p. 209; Wilshere, p. 18.

1936. If the falling leaves remain under the trees and are not blown away by the wind, expect a fruitful year to follow.
Cited in Dunwoody, p. 66.

1937. If the leaves are slow to fall, expect a cold winter.
Cited in Dunwoody, p. 66.

1938. If the leaves wither on their branches in late autumn, instead of beginning to fall in October as normal, an extra cold winter is due.
Cited in Page, p. 50.

1939. If tree leaves curl up to form cups, rain will soon fill the cups.
Cited in Hyatt, p. 16.

1940. If tree leaves turn up on Monday, a rain will turn up before Sunday.
Cited in Hyatt, p. 16.

1941. If tree leaves with silvery or whitish undersides turn upside down or inside out, rain is on its way. Comment: The cottonwood, elm, maple, oak and willow are ordinarily named.
Cited in Hyatt, p. 16.

1942. Leaves and straws playing in the air when no breeze is felt, the down of plants flying about, and feathers floating and playing on the water, show that winds are at hand.
Cited in Inwards, p. 209.

1943. Leaves still hanging on the branches in late October and early November foretell much snow that winter.
Cited in Hyatt, p. 16.

1944. Leaves turned up so as to show the underside indicate rain. Vars.: (a) When leaves show their backs, it will rain. (b) When the leaves on trees turn up their under sides, it will rain. (c) When the leaves show their under sides, be very sure that rain betides. (d) When trees show the underside of their leaves, giving a much lighter appearance than usual, it is a sure sign of wet weather.
Cited in Dunwoody, p. 82; Humphreys 2, p. 80; Whitman, p. 49; Boughton, p. 124; Inwards, p. 209; Sloane 2, p. 42; Lee, p. 53; Page, p. 11.

1945. Stand at the foot of a tree after it has leafed in the spring: if you can see the sky through the leaves, a pleasant summer may be predicted; if you cannot see the sky, a hot and dry summer.
Cited in Hyatt, p. 16.

1946. The curling up of tree leaves before they fall in autumn is a sign of an open winter.
Cited in Hyatt, p. 16.

1947. The silver maple shows the lining of its leaf before a storm.
Cited in Dunwoody, p. 68; Inwards, p. 211; Lee, p. 71.

1948. Tree leaves turning yellow in August warn you of an early autumn.
Cited in Hyatt, p. 16.

1949. When dry leaves rattle on the trees, expect snow.
Cited in Dunwoody, p. 75; Inwards, p. 209.

1950. When leaves fall early, fall and winter will be mild; when leaves fall late, winter will be severe.
Cited in Freier, p. 78.

1951. When leaves of trees are thick, expect a cold winter. Var.: It's a sign of a severe winter when leaves are thick on trees.
Cited in Dunwoody, p. 584; Smith, p. 13.

1952. When the leaves of the sugar maple tree are turned upside down expect rain. Vars.: (a) Silver maple leaves turn over before a rain. (b) When leaves on a maple tree turn upside down it's a sign of rain.
Cited in Dunwoody, p. 82; Garriott, p. 25; Inwards, p. 211; Hand, p. 25; Smith, p. 7; Freier, p. 57.

1953. When the leaves of trees curl, with the wind from the south, it indicates rain. Var.: When the leaves of trees curl during a south wind, it will rain.
Cited in Garriott, p. 25; Inwards, p. 209; Hand, p. 25; Lee, p. 71.

1954. When the live oaks turn the pale side of their leaves up, it will rain.
Cited in Reeder, no pp.

1955. When wind blows leaves bottom side up it will rain.
Cited in Smith, p. 7.

leaf crops

1956. Leaf and grain crops are planted in the light of the moon.
Cited in Freier, p. 83.

leap year

1957. A leap year is never a good sheep year. Var.: Leap year was ne'er a good sheep year.
Cited in Denham, p. 17; Northall, p. 492; Whitman, p. 46.

1958. In leap year the weather always changes on Friday.
Cited in Dunwoody, p. 89.

leech

1959. A leech in a bottle is considered a good weather prophet, in dry, calm weather the leech will remain at the bottom; before rain or snow it will cling to the side of the bottle at the top of the water, and before wind it will be found in active motion.
Cited in Mitchell, p. 228.

1960. A leech placed in a jar of water will remain at the bottom until rain is approaching, when it will rise to the surface, and if thunder is to follow will frequently crawl out of the water.
Cited in Dunwoody, p. 72.

1961. Leeches kept in glasses are observed to move about frequently before rain. Var.: Leeches kept in glass jars move about more frequently just before rain.
Cited in Steinmetz 2, p. 290; Dunwoody, p. 72.

Lent

1962. Dry Lent, fertile year. Var.: A dry Lent means a fertile year.
Cited in Dunwoody, p. 102; Inwards, p. 70; Lee, p. 151; Page, p. 37.

1963. Never come Lent, never come winter. Comment: Never come Lent means a late Lent.
Cited in Inwards, p. 70.

light

1964. Refractions of light of any remarkable kind frequently forbode rain.
Cited in Hand, p. 22.

lighten

1965. If in a clear and starry night it lighten in the southeast, it foretelleth great store of wind and rain to come from those parts. Comment: Lighten means to shine with or like lightning.
Cited in Inwards, p. 168.

1966. If it lightens in the north during the day, expect an immediate rain; if at night, a downpour within twenty-four hours.
Cited in Hyatt, p. 6.

1967. To see it lighten in the south is an indication of a drought.
Cited in Hyatt, p. 6.

1968. When it lightens only from the northwest, look for rain the next day.
Cited in Swainson, p. 217; Inwards, p. 169.

lightning

1969. An old dog draws lightning.
Cited in Smith, p. 3.

1970. As a protection against lightning, during a storm a spade may be thrown out into the yard.
Cited in Hyatt, p. 33.

1971. Don't be afraid of the lightning you can see, because you will not see the lightning that kills you.
Recorded in Bryant (N.Y.).

1972. Dry weather always accompanies lightning in the east.
Cited in Hyatt, p. 6.

1973. During the day, lightning in the northwest betokens rain at once or that night; after sunset, rain before morning.
Cited in Hyatt, p. 6.

1974. Expect more lightning in summer and autumn than in spring or winter.
Cited in Steinmetz 1, p. 140.

1975. February lightning forecasts a frost sufficient to kill on the corresponding date in May.
Cited in Hyatt, p. 6.

1976. Forked lightning at night, the next day clear and bright.
Cited in Dunwoody, p. 80.

1977. If a man sits on a fence and curses, he will be struck by lightning.
Cited in Hyatt, p. 33.

1978. If a tree is struck by lightning, the wood will not do to burn in the fireplace, as it will bring bad luck.
Recorded in Bryant (Miss.).

1979. If lightning strikes a tree near a spring and splinters from the tree fall into the water, these splinters can be placed in the spring and this will cause it to rain.
Recorded in Bryant (Miss.).

1980. If there be lightning without thunder after a clear day, there will be a continuance of fair weather.
Cited in Dunwoody, p. 80.

1981. If there be sheet lightning with a clear sky on spring, summer, and autumn evenings, expect heavy rains.
Cited in Dunwoody, p. 82.
1982. If you cover a mirror when a storm is raging, the lightning will not strike you.
Recorded in Bryant (N.Y.).
1983. Lightning brings heat.
Cited in Dunwoody, p. 80; Inwards, p. 167.
1984. Lightning follows a draft.
Cited in Smith, p. 3.
1985. Lightning from the north presages winds.
Cited in Swainson, p. 217.
1986. Lightning from the south or west brings both wind and rain from these parts. Var.: If from the south or the west it lightens, expect both wind and rain from these parts.
Cited in Swainson, p. 217; Inwards, p. 169.
1987. Lightning in December means a cold spell.
Cited in Hyatt, p. 6.
1988. Lightning in March means unseasonable weather all year.
Cited in Hyatt, p. 6.
1989. Lightning in spring indicates a good fruit year.
Cited in Dunwoody, p. 81; Inwards, p. 30.
1990. Lightning in summer indicates good healthy weather.
Cited in Dunwoody, p. 82; Marvin, p. 216.
1991. Lightning in the north in summer is a sign of heat.
Cited in Dunwoody, p. 81.
1992. Lightning in the north means an immediate rain.
Cited in Arora, p. 6.
1993. Lightning in the north will be followed by rain in twenty-four hours. Var.: Lightning in the north, rain before twenty-four hours.
Cited in Dunwoody, p. 81; Inwards, p. 169. Recorded in Bryant (Miss.).
1994. Lightning in the south, brings little else but drought.
Cited in Sloane 2, p. 43.
1995. Lightning in the south low on the horizon indicates dry weather.
Cited in Dunwoody, p. 81; Inwards, p. 169.
1996. Lightning in winter makes prudent navigators reef their sails.
Cited in Inwards, p. 167.

1997. Lightning is attracted to mirrors.
Cited in Sloane 2, p. 43.

1998. Lightning never strikes in the same place thrice.
Cited in Smith, p. 2.

1999. Lightning never strikes twice in the same place. Vars.: (a) Lightning never strikes the same place twice. (b) Lightning never strikes twice in the same place except when it forgets where it struck last. (c) Lightning seldom strikes twice in the same place. Cited in Inwards, p. 171; Taylor 2, p. 221; Smith, p. 3; Simpson, p. 133; Dundes, p. 94; Whiting 2, p. 374; Arora, p. 5; Mieder, p. 377. Recorded in Bryant (N.Mex.).

2000. Lightning sours milk.
Cited in Sloane 2, p. 43; Smith, p. 3.

2001. Lightning under north star will bring rain in three days.
Cited in Dunwoody, p. 81; Inwards, p. 169.

2002. Lightning without thunder after a clear day, there will be a continuance of fair weather.
Cited in Inwards, p. 168.

2003. Northeast lightning is considered by some an omen of rain within twenty-four hours; by others an omen of dry weather.
Cited in Hyatt, p. 6.

2004. Place a shoe under each bed post and you will not be struck by lightning.
Cited in Smith, p. 3.

2005. Red lightning foretells a dry spell. Comment: You are seeing lightning from a storm that is passing you by.
Cited in Sloane 2, p. 43.

2006. Sheet lightning, without thunder, during night, having a whitish color, announces unsettled weather.
Cited in Mitchell, p. 223; Inwards, p. 168.

2007. The person who counts ten between a flash of lightning and a peal of thunder will never be hit.
Cited in Hyatt, p. 33.

2008. The safest place to be during a storm is near a spot once struck by lightning; it never strikes the same thing twice.
Cited in Hyatt, p. 32.

2009. There is lightning lightly before thunder.
Cited in Wilson, p. 464.

2010. Three consecutive nights of lightning in the north will bring rainy weather.
Cited in Hyatt, p. 6.

2011. 'Tis the tops of mountains that the lightning strikes, and the thunderbolts always fall on the tallest buildings and trees.
Recorded in Bryant (Can. [Ont.]).

2012. Unless lightning is from the north, there will be no rain.
Cited in Smith, p. 8.

2013. Wear shoes to bed during a lightning storm and you will not be struck.
Cited in Smith, p. 2.

2014. When the flashes of lightning appear very pale, it argues the air to be full of waterish meteors; and if red and fiery, inclining to winds and tempests.
Cited in Dunwoody, p. 81; Inwards, p. 170.

2015. Where lightning strikes, go dig your well.
Cited in Sloane 2, p. 44.

2016. Yaller gal, yaller gal, flashin' through the night, summer storms will pass you by unless the lightnin's white. Comment: This refers to "heat lightning," which is simply lightning from a storm that is passing you by, or has passed you by, seen far to the south, north, or east. Seen through dusty storm air, it appears red. White lightning is usually that which is seen through clearer air and from a storm which is in a western quadrant and on its way towards you.
Cited in Sloane 1, p. 107; Sloane 2, p. 43; Lee, p. 86.

2017. You will not be struck by lightning, if during a storm you wear your belt twisted.
Cited in Hyatt, p. 32.

2018. Your house can be protected against lightning, if you throw your scissors out into the yard during a storm.
Cited in Hyatt, p. 33.

lightning bug

2019. A great many lightning bugs in June foretells a hot summer.
Cited in Hyatt, p. 18.

2020. If lightning bugs fly high, there will be dry weather; if low, wet weather.
Cited in Hyatt, p. 18.

lizard

2021. When lizards chirrup, it is a sure indication of rain.
Cited in Dunwoody, p. 72.

lobster

2022. When lobsters heighten their holes about the surface of the ground, it is a sign of approaching rain.
Cited in Dunwoody, p. 50.

locust

2023. A locust singing after sunset is forecasting hot weather for the next day.
Cited in Hyatt, p. 18.

2024. After you have heard locusts, it will be six weeks till frost.
Cited in Hyatt, p. 18.

2025. Locusts sing when the air is hot and dry.
Cited in Freier, p. 28.

2026. Noisy locusts are a warning of a dry spell.
Cited in Hyatt, p. 18.

2027. The locust sings when it is to be hot and clear.
Cited in Lee, p. 57.

2028. Three months from the time you hear the first locusts, the first frost will come.
Recorded in Bryant (S.C.).

2029. When locusts are heard, dry weather will follow, and frost will occur in six weeks.
Cited in Dunwoody, p. 57; Inwards, p. 73; Sloane 2, p. 125; Lee, p. 57.

2030. When the locust yells, it's going to get extremely hot.
Cited in Smith, p. 11.

log

2031. An easy splitting log indicates rain.
Cited in Dunwoody, p. 82.

Lookout Mountain

2032. When Lookout Mountain has its cap on, it will rain in six hours.
Comment: Lookout Mountain is a mountain range in Georgia, Tennessee, and Alabama.

Cited in Dunwoody, p. 45; Garriott, p. 13; Inwards, p. 144; Hand, p. 13; Sloane 2, p. 14; Lee, p. 93; Freier, p. 44.

loon

2033. Hunters say that the direction in which the loon flies in the morning will be the direction of the wind next day.
Cited in Dunwoody, p. 38.

2034. When loons fly out to sea, it will be fine weather; when they fly toward land, it indicates bad weather.
Cited in Lee, p. 61.

Lord

2035. The Lord sends sun at one time, rain at another.
Cited in Mieder, p. 385.

2036. The Lord tempers the wind to the shorn lamb.
Cited in Mieder, p. 385.

lunation

2037. As the fourth and fifth day's weather, so's that lunation altogether.
Cited in Whitman, p. 68.

macaw

2038. The feathers of the blue macaw turn a greenish hue before rain.
Cited in Inwards, p. 189.

mackerel

2039. The mackerel's cry is never long dry.
Cited in Northall, p. 473.

magpie

2040. For anglers in spring it is always unlucky to see single magpies;
but two may always be regarded as a favourable omen. Comment: The reason is, that in cold and stormy weather one magpie
alone leaves the nest in search of food, while the other remains
with the eggs or young ones.
Cited in Inwards, p. 193.

2041. In spring time, when magpies fly abroad singly, the weather
either is or will soon be stormy; but when both birds are seen
together, the weather must be mild or will soon become so.
Cited in Steinmetz 1, p. 110.

2042. Magpies, flying three or four together and uttering harsh cries,
predict windy weather.
Cited in Dunwoody, p. 38; Inwards, p. 193.

March

2043. A bushel of March dust is a thing worth the ransom of a king.
Vars.: (a) A bushel of March dust is worth a king's ransom. (b)
A bushel of March dust on the leaves is worth a king's ransom.
Cited in Denham, p. 36; Steinmetz 1, p. 137; Dunwoody, p. 36;
Northall, p. 434; Taylor 1, p. 114; Inwards, p. 45; Wilson, p.
511; Lee, p. 135; Whiting 2, p. 82.

2044. A damp, rotten March gives pain to farmers.
Cited in Dunwoody, p. 95; Inwards, p. 46.

2045. A dry March, a wet April, a dry May, and a wet June, is
commonly said to bring all things in tune.
Cited in Northall, p. 477; Inwards, p. 47.

2046. A dry March never begs its bread. Vars.: (a) A dry and cold
March never begs bread. (b) A dry cold March never begs bread.
Comment: The dry cold winds of March prepare the soil for
seeds, which germinate and produce fruit in the autumn.

Cited in Denham, p. 31; Steinmetz 1, p. 137; Swainson, p. 55; Dunwoody, p. 95; Garriott, p. 44; Marvin, p. 207; Whitman, p. 42; Inwards, p. 45; Hand, p. 44.

2047. A dry March, wet April, and cool May fill barn, cellar, and bring much hay.
Cited in Dunwoody, p. 95; Marvin, p. 212; Inwards, p. 48.

2048. A frosty winter and a dusty March, and a rain about April, another about Lammas time when the corn begins to fill, is well worth a plough of gold, and all her pins theretill. Comment: Lammas Day is August 1st.
Cited in Swainson, p. 59.

2049. A March sun sticks like a lock of wool.
Cited in Swainson, p. 54; Inwards, p. 46.

2050. A March wisher is never a good fisher. Comment: March when blustering and stormy is not a good month for fishing.
Cited in Denham, p. 31; Dunwoody, p. 95; Northall, p. 437; Marvin, p. 205; Inwards, p. 46.

2051. A peck of March dust and a shower in May makes the corn green and the fields gay. Var.: A peck of March dust and a shower in May makes the corn green and the meadow gay.
Cited in Swainson, p. 60; Cheales, p. 21; Dunwoody, p. 95; Northall, p. 434; Inwards, p. 45; Wilson, p. 511.

2052. A peck of March dust is worth a king's ransom. Var.: A peck of March dust is worth an earl's ransom.
Cited in Denham, p. 31; Swainson, p. 55; Cheales, p. 21; Taylor 1, p. 114; Boughton, p. 125; Inwards, p. 45; Wilson, p. 511; Wilshere, p. 9; Simpson, p. 176.

2053. A warm March will bring a cold April.
Cited in Mieder, p. 406.

2054. A wet March makes a sad autumn. Var.: A wet March, a sad autumn.
Cited in Steinmetz 2, p. 193; Inwards, p. 48; Lee, p. 135.

2055. A wet March makes a sad harvest.
Cited in Swainson, p. 55; Inwards, p. 46.

2056. A windy March and a showery April make a beautiful May.
Cited in Denham, p. 31; Dunwoody, p. 95; Inwards, p. 47; Wilshere, p. 9; Freier, p. 81.

2057. A windy March and a sunny April make a beautiful May.
Cited in Wilshere, p. 8.

2058. A windy March foretells a fine May.
Cited in Wilshere, p. 9.

2059. As it rains in March, so it rains in June.
Cited in Dunwoody, p. 94; Inwards, p. 48; Hand, p. 44.

2060. As many mistises in March, so many frostises in May.
Cited in Inwards, p. 48.

2061. As much dew in March, so much fog rises in August.
Cited in Inwards, p. 48.

2062. As much fog in March, so much rain in summer.
Cited in Inwards, p. 48.

2063. Better slaughter in the country than March should come in mild.
Cited in Inwards, p. 45.

2064. Better to be bitten by a snake than to feel the sun in March.
Cited in Inwards, p. 46.

2065. Dust in March brings grass and foliage.
Cited in Dunwoody, p. 95; Inwards, p. 45; Hand, p. 42.

2066. Every pint of March dust brings a peck of September corn and
a pound of October cotton.
Cited in Mieder, p. 171. Recorded in Bryant (S.C.).

2067. For every night it doesn't freeze in March, it will freeze in May.
Cited in Boughton, p. 125.

2068. For wheat a peck of dust in March is worth a king's ransom; or
wet and soddy, the land must go to oats and corn.
Cited in Dunwoody, p. 68.

2069. He who freely lops in March will get his lap full of fruit.
Comment: Lop means to cut off branches from a tree or vine.
Cited in Swainson, p. 57.

2070. If March borrows from April three days, and they be ill, April
borrows of March again three days of wind and rain. Var.:
March borrows of April three days, and they are ill; April
borrows of March again three days of wind and rain.
Cited in Inwards, p. 50; Wilshere, p. 9.

2071. If March comes in a raring, she'll go out a tearing! Comment:
To go tearing refers to behaving boisterously.
Cited in Hyatt, p. 13.

2072. In beginning or in end March its gifts will send.
Cited in Dunwoody, p. 94; Inwards, p. 45.

2073. In March the birds begin to search; in April the corn begins to
fill; in May the birds begin to lay.

Cited in Wilson, p. 511.

2074. In March the cuckoo starts; in April a'tune his bill; in May, a'sing all day; in June, a'change his tune; in July, away a'fly; in August away a'must; in September, you'll allers remember; in October, 'ull never get over.
Cited in Wilson, p. 511.

2075. March birds are best.
Cited in Swainson, p. 58.

2076. March borroweth from April, three days and they were ill: the first was frost, the second was snaw; and the third as cauld, as ever't blaw. Vars.: (a) March borrowed from April three days, and they were ill. (b) March borrowed from April three days, and they were ill; the first of them is wun and weet, the second it is snaw and sleet, the third of them is peel-a-bane, and freezes the wee bird's neb to the stane. (c) March borrowed from April three days, and they were ill; the one was sleet, the other was snow, the third was the worst that e'er did blow.
Cited in Denham, p. 37; Swainson, p. 65; Northall, p. 437; Inwards, p. 49; Wilson, p. 511.

2077. March borrows its last three days from April. Comment: The weather at the end of March and the beginning of April is usually similar.
Cited in Page, p. 45.

2078. March borrows of April three days, and they are ill; April returns them back again, three days, and they are rain.
Cited in Swainson, p. 65; Northall, p. 436.

2079. March comes in like a lamb and goes out like a lion. Var.: If March comes in like a lamb, it will go out like a lion.
Cited in Hand, p. 42; Hyatt, p. 13.

2080. March comes in like a lion and goes out like a lamb. Vars.: (a) A March that comes in like a lion will go out as quiet as a newborn lamb. (b) If March comes in like a lion, it will go out like a lamb. (c) If March comes in like a roaring lion it will go out like a lamb. (d) March, black ram, comes in like a lion and goes out like a lamb. (e) March comes in like a wolf and goes out like a lamb. (f) March, hackham, comes in like a lion and goes out like a lamb. Comment: Black ram is an obscure expression that could refer to the constellation Aires; black ram is sometimes spelled "backham," "balkham," or "hackham."

Cited in Denham, p. 31; Swainson, p. 57; Cheales, p. 21; Dunwoody, p. 95; Northall, p. 435; Garriott, p. 44; Taylor 1, p. 111; Whitman, p. 42; Inwards, p. 47; Taylor 2, p. 238; Sloane 2, p. 44; Smith, p. 8; Wilson, p. 511; Lee, p. 8; Page, p. 45; Whiting 1, p. 282; Simpson, p. 147; Freier, p. 79; Whiting 2, p. 403; Arora, p. 16; Mieder, p. 406. Recorded in Bryant (Ark., Fla., Miss., N.C., S.C., Utah. Can. [Ont.]).

2081. March comes in with an adder's head and goes out with a peacock's tail. Vars.: (a) If March comes in with adder's head, it goes out with peacock's tail. (b) March comes in with adder's head, but goes out like a peacock's tail. (c) March comes in with adder's heads and goes out with peacocks' tails.
Cited in Denham, p. 32; Swainson, p. 57; Dunwoody, p. 95; Inwards, p. 47.

2082. March damp and warm, will do the farmer much harm.
Cited in Dunwoody, p. 95; Garriott, p. 44; Marvin, p. 207; Humphreys 2, p. 10; Whitman, p. 42; Inwards, p. 46; Hand, p. 44.

2083. March does from April gain three days and they're in rain, returned by April in 's bad kind, three days and they're in wind.
Cited in Northall, p. 436; Wright, p. 31.

2084. March dry, good rye, March wet, good wheat.
Cited in Northall, p. 434; Marvin, p. 212; Inwards, p. 45.

2085. March dust and March win' bleaches as well as simmer's sin.
Var.: March dust and March wun' bleach as well as summer's sun.
Cited in Swainson, p. 56; Inwards, p. 45.

2086. March dust and May sun, makes corn white and maids dun.
Var.: March wind and May sun makes clothes white and maids dun. Comment: Dun means dingy.
Cited in Denham, p. 39; Northall, p. 434; Wilson, p. 511.

2087. March dust on an apple leaf, brings all kinds of fruit to grief.
Var.: March dust on an apple leaf brings all kinds of grief.
Cited in Northall, p. 436; Inwards, p. 45.

2088. March dust to be sold, worth a ransom of gold.
Cited in Swainson, p. 55.

2089. March flings, April fleyes. Comment: Flings means kicks; fleyes means warms.
Cited in Swainson, p. 59; Inwards, p. 47.

2090. March flowers bring summer showers.
Cited in Freier, p. 77.

2091. March flowers make no summer bowers.
Cited in Steinmetz 1, p. 137; Dunwoody, p. 67; Northall, p. 435; Garriott, p. 44; Marvin, p. 207; Whitman, p. 42; Inwards, p. 45; Hand, p. 44; Lee, p. 135.

2092. March grass never did good.
Cited in Denham, p. 31; Swainson, p. 55; Dunwoody, p. 95; Marvin, p. 207; Taylor 1, p. 115; Inwards, p. 44; Wilson, p. 511.

2093. March grows, never dows. Comment: Dows means thrives.
Cited in Inwards, p. 47.

2094. March in January, January in March. Var.: March in Janiveer, Janiveer in March, I fear.
Cited in Denham, p. 23; Swainson, p. 23; Dunwoody, p. 95; Northall, p. 435; Garriott, p. 44; Wright, p. 11; Whitman, p. 42; Inwards, p. 37; Hand, p. 42; Wilson, p. 511; Wurtele, p. 298; Lee, p. 134; Page, p. 43; Wilshere, p. 8.

2095. March many weathers. Vars.: (a) In March, many weathers. (b) March many weathers rained and blowed, but March grass never did good.
Cited in Denham, p. 32; Swainson, p. 55; Northall, p. 435; Garriott, p. 42; Whitman, p. 42; Inwards, p. 44; Taylor 2, p. 238; Hand, p. 42; Wilson, p. 511; Page, p. 45.

2096. March many weathers; April and May the key of the whole year.
Cited in Wilshere, p. 9.

2097. March said to Averil, I see three hoggs on yonder hill; an' if ye'll lend me dayis three, I'll find a way to gar them dee. The first o' them was wind an' weet; the second o' them was snaw an' sleet; the third o' them was sic' a freeze, it froze the birds' nebs to the trees. When the three days were past and gane, the silly hoggs cam' hirplin hame. Comment: Hoggs are year-old sheep. This is the oldest English, Scottish version of the proverb about borrowing days.
Cited in Inwards, p. 50.

2098. March sun lets snow stand on a stone.
Cited in Swainson, p. 54; Inwards, p. 46.

2099. March water and May sun makes clothes clear and maidens dun.
Comment: Dun means dingy.

Cited in Inwards, p. 48.

2100. March water is worse than a stain in cloth.
Cited in Inwards, p. 46.

2101. March wet and windy, makes the barn full and findy. Var.:
March wet and windy makes the barn full and finnie. Comment:
Finnie means the feel of the grain as indicating quality.
Cited in Dunwoody, p. 95; Inwards, p. 46.

2102. March whisker was never a good fisher. Comment: In Fuller's
Gnomologia, 1732, "wisher" is written as whisker. A blustering
March is unfavorable to the angler, although good for the farmer.
Cited in Wilson, p. 511.

2103. March will search, April will try; May will tell you whether you
live or die. Vars.: (a) March search, April try, May will prove
if you live or die. (b) March will search, April will try, May
shew if you live or die.
Cited in Cheales, p. 23; Dunwoody, p. 95; Northall, p. 435;
Wright, p. 31; Wilshere, p. 9.

2104. March wind and May sun, makes clothes clear and maidens dun.
Comment: Dun means dingy.
Cited in Denham, p. 39; Swainson, p. 60; Dunwoody, p. 35;
Northall, p. 430; Wright, p. 31; Inwards, p. 47.

2105. March wind, kindles the ether, and blooms the whin. Vars.: (a)
March wind kindles the adder and bloom the thorn. (b) March
wind wakes the adder and blooms the whin. Comment: Kindles
means wakes, ether means adder, whin means grass.
Cited in Denham, p. 39; Swainson, p. 56; Northall, p. 435;
Inwards, p. 46; Wilson, p. 512.

2106. March winds and April rains bestow great blessings in May.
Cited in Hyatt, p. 34.

2107. March winds and April showers bring forth May flowers. Vars.:
(a) March winds and April showers bring May flowers. (b)
March winds and April showers usher in May flowers. (c) March
winds, April showers, bring forth May flowers. (d) March winds
bring April showers.
Cited in Denham, p. 36; Cheales, p. 21; Northall, p. 435;
Garriott, p. 44; Whitman, p. 42; Inwards, p. 48; Hand, p. 44;
Page, p. 47; Wilshere, p. 8; Freier, p. 79; Mieder, p. 406.

2108. March windy and April rainy; makes May the pleasantest month
of any.

Cited in Whiting 1, p. 283.

2109. March yeans the lammie and buds the thorn, and blows through the flint of an ox's horn. Comment: Yeans the lammie means to bring forth young sheep.
Cited in Inwards, p. 44.

2110. Rain in March, poor harvest.
Cited in Dunwoody, p. 95.

2111. Snow in March is bad for fruit and grapevine.
Cited in Dunwoody, p. 95; Inwards, p. 46; Hand, p. 42.

2112. So many fogs in March you see, so many frosts in May will be.
Cited in Inwards, p. 48.

2113. So many frosts in March, so many in May.
Cited in Denham, p. 32; Inwards, p. 48.

2114. The March sun causes dust, and the wind blows it about.
Cited in Wilson, p. 511.

2115. The sun in March raises but does not melt. Var.: The March sun raises, but dissolves not. Comment: Raises means heats, dissolves means melts.
Cited in Swainson, p. 54; Inwards, p. 46; Wilson, p. 511.

2116. Thunder and lightning in March denote much fruit and plenty of grain.
Cited in Hyatt, p. 34.

2117. Thunder in March, it brings sorrow.
Cited in Inwards, p. 46.

2118. Whatever March does not want April brings along.
Cited in Dunwoody, p. 95; Inwards, p. 52.

2119. When March is like April, April will be like March. Var.: When March has April weather, April will have March weather.
Cited in Dunwoody, p. 95; Garriott, p. 44; Hand, p. 44.

2120. When March kills, April flays.
Cited in Inwards, p. 47.

2121. Winds in March and rains in April promise great blessings in May.
Cited in Dunwoody, p. 94; Garriott, p. 44; Whitman, p. 42; Hand, p. 44.

2122. Windy March and rainy April make a beautiful May.
Cited in Boughton, p. 125.

March 1

2123. On the first of March the crows begin to search. Var.: On the first of March the crows begin to search, by the first of April they are sitting still, by the first of May they are a' flown away; croupin' greedy back again, wi' October wind and rain. Comment: Crows are supposed to begin pairing on this day.
Cited in Denham, p. 39; Swainson, p. 61; Wright, p. 24; Wilson, p. 262; Simpson, p. 82.

2124. Snow on March 1st can be followed by snow anytime during the next thirty days.
Cited in Hyatt, p. 11.

March 6

2125. If it freezes on March 6th, it will freeze on the fortieth day thereafter, April 15th.
Cited in Hyatt, p. 11.

March 10

2126. A freeze on March 10th may be followed by freezing weather anytime during the next forty days.
Cited in Hyatt, p. 11.

2127. If it does not freeze on the tenth of March, a fertile year may be expected.
Cited in Dunwoody, p. 95; Wright, p. 27; Inwards, p. 49.

2128. On March 10th the direction of the wind will remain unchanged for forty days.
Cited in Hyatt, p. 9.

2129. Rain on March 10th means rain for forty days.
Cited in Hyatt, p. 10.

March 21

2130. March 21st spring begins.
Cited in Wright, p. 33.

2131. Where the wind is at twelve o'clock on the twenty-first of March, there she'll bide for three months afterward.
Cited in Wright, p. 33; Inwards, p. 69.

marigold

2132. If the marigold does not open its petals by seven in the morning, it will rain or thunder that day. Vars.: (a) If the marigold does not open by seven o'clock in the morning, you may expect rain that day. (b) If the marigold should open at six or seven in the morning and not close until four in the afternoon, we may reckon on settled weather. (c) If the small Cape marigold should open at six or seven in the morning, and not close till four in the afternoon, we may reckon on settled weather.
Cited in Swainson, p. 259; Dunwoody, p. 66; Inwards, p. 215; Freier, p. 29.

marsh

2133. Marshes give off an eerie light before a rain.
Cited in Freier, p. 24.

martin

2134. Martins fly low before and during rainy weather. Var.: Martins fly low before and during rain.
Cited in Dunwoody, p. 38; Garriott, p. 19; Inwards, p. 195; Hand, p. 23; Lee, p. 64.

2135. No killing frost after martins. Var.: No killing frost after martins are seen.
Cited in Dunwoody, p. 38; Garriott, p. 40; Inwards, p. 195; Hand, p. 40; Freier, p. 77.

2136. When martins appear, winter has broken. Var.: When martins appear, winter is broken.
Cited in Dunwoody, p. 38; Garriott, p. 40; Inwards, p. 195; Hand, p. 40.

May

2137. A cold May and windy makes a fat barn and a findy. Vars.: (a) A cold and a windy May, a full barn will find ye. (b) A cold and windy May will fill the barn. (c) A cold May and wind makes full barns and a findy. (d) A cold May and windy a full barn will be finely. (e) A cold May is kindly and fills the barn finely. Comment: A findy is a plentiful harvest.

Cited in Denham, p. 4; Steinmetz 1, p. 137; Steinmetz 2, p. 193; Swainson, p. 89; Dunwoody, p. 97; Northall, p. 438; Inwards, p. 54; Wilson, p. 132; Page, p. 47.

2138. A cold May brings many things.
Cited in Dunwoody, p. 97.

2139. A cold May enriches no one.
Cited in Dunwoody, p. 96.

2140. A cold May gives full barns and empty churchyards.
Cited in Inwards, p. 54.

2141. A cold May, plenty of corn and hay. Var.: A cold May is good for corn and hay.
Cited in Cheales, p. 22; Inwards, p. 54.

2142. A cold wet May will fill your barns with grain and hay. Vars.: (a) A cold wet May brings a barn full of hay. (b) A cold wet May fills the barn full of hay.
Cited in Boughton, p. 123; Smith, p. 8. Recorded in Bryant (Wis.).

2143. A cool May brings good wheat.
Recorded in Bryant (N.C.).

2144. A dry May and a dripping June brings all things into tune. Vars.: (a) A dry May and dripping June brings everything in tune. (b) A dry May and a dripping June keeps everything in tune.
Cited in Northall, p. 439; Inwards, p. 55; Wilson, p. 207; Page, p. 48; Wilshere, p. 10.

2145. A dry May and a leaking June, bring all things into tune. Var.: A leaky May and a dry June puts all in tune.
Cited in Swainson, p. 93; Wilshere, p. 10.

2146. A dry May and a leaking June make the farmer whistle a merry tune.
Cited in Dunwoody, p. 96; Boughton, p. 123; Inwards, p. 55.

2147. A dry May is followed by a wet June.
Cited in Dunwoody, p. 96; Garriott, p. 44; Whitman, p. 43; Hand, p. 44; Hyatt, p. 13.

2148. A dry May portends a wholesome summer.
Cited in Wilshere, p. 10.

2149. A farmer can rely upon a cool dark May giving him full crops.
Cited in Hyatt, p. 34.

2150. A hot May makes a fat churchyard. Var.: A hot May makes a crowded cemetery.

Cited in Denham, p. 43; Swainson, p. 89; Dunwoody, p. 96; Northall, p. 438; Garriott, p. 44; Whitman, p. 43; Inwards, p. 53; Hand, p. 44; Wilson, p. 388; Page, p. 47; Wilshere, p. 10. Recorded in Bryant (Ark., N.Y.).

2151. A leaking May and a warm June bring on the harvest very soon.
Cited in Inwards, p. 55.

2152. A leaky May and a dry June keep the poor man's head abune.
Comment: Abune means above.
Cited in Inwards, p. 55.

2153. A May cold is a thirty-day cold.
Cited in Wilson, p. 519.

2154. A May flood and windy makes a full barn and a findy. Comment: A findy is a plentiful harvest.
Cited in Steinmetz 1, p. 137.

2155. A May flood never did good. Var.: A May flood never did any good. Comment: Flood means a wet May.
Cited in Denham, p. 43; Steinmetz 1, p. 137; Steinmetz 2, p. 193; Swainson, p. 91; Dunwoody, p. 96; Northall, p. 431; Inwards, p. 53; Wilson, p. 519; Wilshere, p. 10; Mieder, p. 216.

2156. A May wet was never kind yet.
Cited in Inwards, p. 53.

2157. A misty May and a hot June bring cheap meal and harvest soon.
Cited in Inwards, p. 55.

2158. A normal wet and cool May brings a wet June.
Cited in Dunwoody, p. 96.

2159. A rainy May brings plenty of corn and hay.
Cited in Inwards, p. 53.

2160. A red gay May, best in any year; February full of snow is to the ground most dear; a whistling March, that makes the ploughman blithe; a moisty April, that fits him for the scythe.
Cited in Inwards, p. 55.

2161. A snowstorm in May is worth a wagon-load of hay.
Cited in Inwards, p. 54.

2162. A wet May and a winnie, makes a fou stacky and a finnie. Comment: Fou stacky means a full stack; finnie means the feel of the grain as indicating quality.
Cited in Swainson, p. 91; Inwards, p. 54.

2163. A wet May makes a big load of hay. Vars.: (a) A wet May brings a good load of hay. (b) Cold, wet May, barn full of hay. Cited in Inwards, p. 53; Sloane 2, p. 45; Page, p. 47; Wilshere, p. 10.

2164. A wet May makes a lang-tailed hay. Cited in Inwards, p. 54.

2165. A wet May will fill a byre full of hay. Cited in Denham, p. 44; Northall, p. 439; Inwards, p. 54.

2166. A windy May makes a fair year. Cited in Swainson, p. 91.

2167. An abnormal warm May brings a wet June. Cited in Dunwoody, p. 96.

2168. Be it weal or be it woe, beans should blow before May go. Cited in Swainson, p. 91.

2169. Be sure of hay till the end of May. Cited in Swainson, p. 91.

2170. Blossoms in May are not good. Cited in Swainson, p. 90; Inwards, p. 53.

2171. Cold May enriches no one. Cited in Inwards, p. 54.

2172. Cool and evening dew in May brings wine and much hay. Cited in Dunwoody, p. 96; Inwards, p. 54.

2173. Dry May brings nothing. Var.: Dry May brings nothing gay. Cited in Dunwoody, p. 96; Garriott, p. 44; Whitman, p. 43; Inwards, p. 53; Hand, p. 44.

2174. Dry May, wet June. Cited in Sloane 2, p. 45; Smith, p. 8.

2175. For a warm May, the parsons pray. Comment: Meaning more burial fees. Cited in Inwards, p. 53.

2176. For an east wind in May 'tis your duty to pray. Cited in Inwards, p. 54.

2177. Haddocks are good when dipped in May flood. Vars.: (a) The haddocks are good when dipped in May flood. (b) Three dips in May flood make the fish in the sea good. Cited in Swainson, p. 90; Inwards, p. 53.

2178. He that would live for aye must eat sage in May. Comment: Aye means eternity. Cited in Swainson, p. 92.

2179. He who mows in May will have neither fruit nor hay.
Cited in Swainson, p. 92.

2180. He who sows oats in May, gets little that way.
Cited in Denham, p. 47; Swainson, p. 92.

2181. If May be cold and wet, September will be warm and dry, and vice versa.
Cited in Inwards, p. 55.

2182. If May will be a gardener, he will not fill the granaries.
Cited in Dunwoody, p. 96; Inwards, p. 53.

2183. It is unlucky to go on the water the first Monday in May.
Cited in Wright, p. 53.

2184. Look at your corn in May, and you'll come sorrowing away; look at it again in June, and you'll come singing another tune.
Cited in Dunwoody, p. 96.

2185. Many thunderstorms in May, and the farmer sings "hey! hey!"
Cited in Dunwoody, p. 96; Inwards, p. 54.

2186. May and June are twin sisters.
Cited in Denham, p. 49; Swainson, p. 92.

2187. May bees don't fly in November.
Cited in Whiting 1, p. 285.

2188. May, come she early, or come she late, she'll make the cow to quake. Var.: Come it early or come it late, in May comes the cow quake. Comment: Cow quake is a kind of spring grass; the cold winds that prevail in May cause the grass to tremble.
Cited in Denham, p. 48; Swainson, p. 92; Dunwoody, p. 97; Northall, p. 440.

2189. May damp and cool fills the barns and wine vats.
Cited in Dunwoody, p. 96; Garriott, p. 44; Whitman, p. 43; Inwards, p. 53; Hand, p. 44; Freier, p. 78.

2190. May is a broom of a bright-faced angel that sweeps the sky clean for the coming summer.
Cited in Sloane 1, p. 79.

2191. May makes or mars the wheat.
Cited in Wilson, p. 519.

2192. May never goes out without a wheatear. Comment: A wheatear is a spike of wheat.
Cited in Wilson, p. 519.

2193. May rain kills lice.
Cited in Denham, p. 43.

2194. Mist in May and heat in June makes the harvest right soon. Vars.: (a) Mist in May, and heat in June makes the harvest come right soon. (b) Mist in May, and heat in June will bring the harvest right soon. (c) Mists in May, heat in June make the harvest come right soon.
Cited in Dunwoody, p. 97; Northall, p. 439; Inwards, p. 55; Wilshere, p. 10.

2195. Rain in May makes bread for the whole year.
Cited in Wilshere, p. 10.

2196. Rain in the beginning of May is injurious to wine. Var.: Rain in the beginning of May injures wine.
Cited in Dunwoody, p. 96; Inwards, p. 53.

2197. Shear your sheep in May and you'll shear them away.
Cited in Swainson, p. 92.

2198. The first flood in May takes the smolts away. Comment: Smolt is fair weather.
Cited in Northall, p. 439.

2199. The month of May seeks warmth to exchange for bread.
Cited in Inwards, p. 54.

2200. The more thunder in May, the less in August and September.
Cited in Inwards, p. 54.

2201. Water in May is bread all the year.
Cited in Swainson, p. 90.

2202. Wet May, dry July.
Cited in Dunwoody, p. 96; Garriott, p. 44; Whitman, p. 43; Hand, p. 44; Freier, p. 79.

2203. Wet May, dry June.
Cited in Smith, p. 8.

2204. Wet May makes short corn and long hay; dry May makes long corn and short hay.
Cited in Whiting 1, p. 285.

2205. Who sows in May gets little that way.
Cited in Northall, p. 432.

May 1

2206. After a rain on May 1st, you may predict twenty rainy days for the month.
Cited in Hyatt, p. 10.

2207. Hoar frost on May 1st indicates a good harvest.
Cited in Dunwoody, p. 96; Inwards, p. 55; Hand, p. 42.
2208. Rain on May 1st foretells extremely dry weather.
Cited in Smith, p. 8.
2209. Rain on May 1st ruins blackberries.
Cited in Hyatt, p. 34.
2210. The fair maid, who the first of May, goes to the fields at break of day, and washes in dew from the hawthorn tree, will ever after handsome be. Var.: The maid who would be fair to view, let her lave her face with early dew on May day in the morning.
Cited in Northall, p. 492; Wright, p. 52; Inwards, p. 54.

May 8
2211. If on the eighth of May it rain, it foretells a wet harvest.
Cited in Garriott, p. 42; Whitman, p. 38; Hand, p. 42.

meadow lark
2212. A meadow lark singing before sunrise means rain that day.
Cited in Hyatt, p. 22.

metal
2213. If metal plates and dishes sweat it is a sign of bad weather.
Cited in Garriott, p. 21; Hand, p. 21; Freier, p. 34.

meteor
2214. After an unusual fall of meteors, dry weather is expected.
Cited in Inwards, p. 100.
2215. If many meteors in summer, expect thunder.
Cited in Inwards, p. 100.
2216. If meteors shoot towards the north, expect a north wind next day.
Cited in Dunwoody, p. 74; Inwards, p. 100.
2217. Many meteors in summer means much snow in winter.
Recorded in Bryant (W.Va.).
2218. Many meteors presage much snow next winter.
Cited in Dunwoody, p. 74; Inwards, p. 100.
2219. Meteors in greater numbers than usual signify unpleasant weather.
Cited in Hyatt, p. 3.

mice

2220. If mice run about more than usual, wet weather may be expected. Var.: An exceptional scampering about by mice will be followed by rain.
Cited in Swainson, p. 232; Inwards, p. 184; Hyatt, p. 29.

2221. Mice will run and frolic before a storm.
Cited in Freier, p. 50.

Midsummer Eve

2222. If it rains on Midsummer Eve, all the filberts will be spoiled.
Cited in Denham, p. 52; Swainson, p. 106; Inwards, p. 57.

milk

2223. Milk cream makes most freely with a north wind.
Cited in Dunwoody, p. 85.

2224. Milk or cream souring in the night means a thunderstorm next day.
Cited in Hyatt, p. 31.

milkweed

2225. Milkweed closing at night indicates rain. Var.: Milkweed closes at night before a rainy spell.
Cited in Dunwoody, p. 66; Garriott, p. 25; Hand, p. 25; Sloane 2, p. 45.

2226. The autumnal air filled with cotton from milkweed pods indicates no snow on Christmas.
Cited in Hyatt, p. 15.

2227. When the milkweed closes its pod, expect rain.
Cited in Freier, p. 39.

Milky Way

2228. One expects a continuation of excellent weather after the Milky Way has glittered with unusual brilliance.
Cited in Hyatt, p. 4.

2229. The direction in which the Milkmaid's Path (Milky Way) points will be the course of the wind on the following day.
Cited in Hyatt, p. 4.

2230. The edge of the Milky Way, which is the brightest, indicates the direction from which the approaching storm will come.

Cited in Dunwoody, p. 73; Inwards, p. 100.

miner

2231. Underground miners can smell rain coming.
Cited in Freier, p. 35.

mirage

2232. A mirage is followed by a rain.
Cited in Dunwoody, p. 52; Inwards, p. 150.

2233. Sailors say that mirages indicate a storm and lower their sails.
Cited in Freier, p. 65.

mist

2234. A general mist before the sun rises, near the full moon, fair weather.
Cited in Mitchell, p. 232.

2235. A misty morning may have a fine day. Vars.: (a) A cloudy morning may turn to a fair day. (b) A foul morn may turn into a fine day. (c) A foul morn may turn to a fine day. (d) A foul morn turns into a fine day.
Cited in Marvin, p. 203; Inwards, p. 74; Wilson, p. 283; Mieder, p. 419. Recorded in Bryant (Wis.).

2236. A moorn hag mist, is worth gold in a kist. Comment: Hag is mist with small rain, kist is a chest.
Cited in Northall, p. 461.

2237. A Scotch mist will wet an Englishman to the skin. Var.: A Scottish mist will wet an Englishman to the skin.
Cited in Denham, p. 11; Wilson, p. 705.

2238. An old moon's mist never died of thirst.
Recorded in Bryant (N.Y.).

2239. Black mist indicates coming rain.
Cited in Swainson, p. 200; Inwards, p. 145.

2240. Gray mists at dawn, the day will be warm.
Cited in Page, p. 5; Wilshere, p. 3.

2241. If mist rise to the hill tops and there stay, expect rain shortly.
Cited in Inwards, p. 140.

2242. If mists rise in low ground and soon vanish, expect fair weather.
Cited in Mitchell, p. 232; Dunwoody, p. 20; Inwards, p. 145.

2243. If mists rise to the hill tops, rain in a day or two.
Cited in Mitchell, p. 232; Freier, p. 66.

2244. If the mists ascend, fair weather; if they descend, foul.
Cited in Steinmetz 2, p. 275.

2245. If there be a general mist before sunrise near the full of the moon, the weather will be fine for some days.
Cited in Dunwoody, p. 61.

2246. In the evenings of autumn and spring vapor (mist) arising from a river is regarded as sure proof of coming frost. Var.: In the evenings of autumn and spring, vapor arising from a river is regarded as a sure indication of coming frost.
Cited in Mitchell, p. 220; Steinmetz 2, p. 275; Dunwoody, p. 21; Inwards, p. 145.

2247. Mist in autumn is a sign of rain.
Cited in Marvin, p. 212.

2248. Mist in May and heat in June make the harvest right soon.
Cited in Denham, p. 47; Swainson, p. 93.

2249. Mist in spring is a sign of snow.
Cited in Marvin, p. 212.

2250. Mist in spring is worse than poison.
Cited in Marvin, p. 212.

2251. Mist in summer is a sign of heat.
Cited in Marvin, p. 212.

2252. Mist in the valley, good man go on your journey.
Cited in Whitman, p. 34.

2253. Mist in winter is a sign of snow.
Cited in Marvin, p. 212.

2254. Mist rising from the pond, fair weather tomorrow.
Cited in Freier, p. 36.

2255. Mists dispersing on the plain, scatter away the clouds and rain; but when they rise to the mountain tops, they'll soon descend in copious drops.
Cited in Humphreys 2, p. 49; Whitman, p. 35; Inwards, p. 145; Alstad, p. 125.

2256. Mists, if they rise on low ground and soon vanish, fair weather.
Cited in Freier, p. 65.

2257. So many mists in March you see, so many frosts in May will be.
Vars.: (a) Mist in March bring rain, or in May, frosts again. (b) Mists in March are followed by frosts in May.

Cited in Denham, p. 36; Swainson, p. 60; Cheales, p. 24; Dunwoody, p. 52; Northall, p. 436; Hyatt, p. 12; Wilshere, p. 9; Simpson, p. 152.

2258. The mists of the valley walk up the hill, and run back as rain to power the mill.
Cited in Alstad, p. 27.

2259. Thin, white, fleecy, broken mist slowly ascending the sides of a mountain whose top is uncovered, predicts a fair day.
Cited in Mitchell, p. 220; Steinmetz 2, p. 275; Dunwoody, p. 20; Inwards, p. 140.

2260. Three foggy or misty mornings indicate rain.
Cited in Dunwoody, p. 53; Hyatt, p. 12.

2261. When the mist comes from the hill, then good weather it doth spill; when the mist comes from the sea, then good weather it will be.
Cited in Denham, p. 18; Swainson, p. 198; Northall, p. 461; Inwards, p. 140; Wilson, p. 536; Freier, p. 66.

2262. When the mist creeps up the hill, fisher out and try your skill, when the mist begins to nod, fisher then put up your rod. Vars.: (a) Whem the mist creeps up the hill, fisher, out and try your skill; when the mist begins to nod, fisher, then put by your rod. (b) When the mist creeps up the hill, fisher out and try your skill; when the mist begins to nod, fisher, then put past your rod.
Cited in Mitchell, p. 220; Steinmetz 2, p. 272; Swainson, p. 198; Dunwoody, p. 20; Humphreys 2, p. 49; Whitman, p. 34; Inwards, p. 140; Alstad, p. 125; Lee, p. 29. Recorded in Bryant (Mich., Wis.).

2263. When the mist is on the hill, then good weather it doth spoil; when the mist takes to the sea, then good weather it will be.
Cited in Dunwoody, p. 54.

2264. Where there are high hills, and the mist which hangs over the lower lands draws towards the hills in the morning, and rolls up to the top, it will be fair; but if the mist hangs upon the hills, and drags along the woods there will be rain.
Cited in Inwards, p. 145.

2265. White mist in winter indicates frost.
Cited in Steinmetz 2, p. 275; Swainson, p. 200; Dunwoody, p. 21; Inwards, p. 145.

mist (harr)

2266. A northern harr, brings fine weather from far. Comment: Harr is a mist with small rain.
Cited in Northall, p. 461; Inwards, p. 145.

mole

2267. A mole coming to the top of the ground in winter discloses an early spring.
Cited in Hyatt, p. 29.

2268. If a long dry spell is likely, moles dig holes as usual, but they build no hills on the surface, and their presence can be detected simply by the slight outline of their tunnels.
Cited in Page, p. 15.

2269. If a mole burrows deeply and therefore casts up a high mound, an early autumn is forecasted.
Cited in Hyatt, p. 29.

2270. If moles have only a few worm basins, we will have a mild winter.
Cited in Freier, p. 81.

2271. If the mole dig his hole two feet and a half deep, expect a very severe winter; if two feet deep, not so severe; if one foot deep, a mild winter.
Cited in Dunwoody, p. 32.

2272. If the mound of a mole-run is higher than usual, the animal is burrowing deeper to escape dry weather.
Cited in Hyatt, p. 29.

2273. Moles raise more hillocks than ususal a day or two before rain, and it is a sign of thaw when, after a long frost, they begin to work again.
Cited in Mitchell, p. 228.

2274. The sand mole makes a mournful noise just before frost.
Cited in Dunwoody, p. 33.

2275. When moles throw up more earth than usual, expect rain. Vars.: (a) If moles throw up more earth than usual, rain is indicated. (b) When the moles throw up the earth, rain follows soon.
Cited in Steinmetz 1, p. 112; Dunwoody, p. 32; Inwards, p. 184; Sloane 2, p. 121; Lee, p. 66.

2276. When the mole throws up fresh earth during a frost, it will thaw in less than forty-eight hours. Var.: When moles throw up earth during a frost, it will thaw within forty-eight hours.
Cited in Inwards, p. 184; Lee, p. 66.

month

2277. The chill is on from near and far, in all the months that have an R.
Cited in Sloane 1, p. 141.

2278. The month that comes in good will go out bad. Var.: The month that comes in good, goes out bad.
Cited in Dunwoody, p. 93; Garriott, p. 44; Whitman, p. 41; Inwards, p. 72; Hand, p. 44; Lee, p. 145; Freier, p. 81.

moon

2279. A dry moon is far north and soon seen.
Cited in Dunwoody, p. 59.

2280. A fog and a small moon bring an easterly wind soon.
Cited in Wright, p. 81.

2281. A growing moon and a flowing tide are lucky times to marry in.
Recorded in Bryant (Miss.).

2282. A moon slung low in the south during February will introduce thirty days of agreeable weather.
Cited in Hyatt, p. 3.

2283. A south moon indicates bad weather.
Cited in Dunwoody, p. 63.

2284. A watery moon brings rain soon.
Cited in Alstad, p. 165.

2285. Clear moon, frost soon.
Cited in Mitchell, p. 223; Steinmetz 2, p. 284; Swainson, p. 188; Dunwoody, p. 21; Humphreys 1, p. 439; Humphreys 2, p. 41; Whitman, p. 27; Walton, no pp.; Inwards, p. 88; Alstad, p. 120; Sloane 2, p. 46; Smith, p. 3; Wurtele, p. 293; Lee, p. 124; Wilshere, p. 135; Freier, p. 36. Recorded in Bryant (Tex., Utah).

2286. If a quarter of the moon sets so it could spill if it had anything in it, it will rain.
Cited in Smith, p. 6.

2287. If the first moon of August is hanging on the point, it is a forecaster of a wet August.
Recorded in Bryant (N.Y., S.C.).

2288. If the moon be fair throughout and rainy at the close, the fair weather will probably return on the fourth or fifth day.
Cited in Dunwoody, p. 60; Inwards, p. 91.

2289. If the moon does not have a ring and yet several nearby stars are grouped about it in an irregular circle, you may prepare for rain.
Cited in Hyatt, p. 3.

2290. If the moon holds water, dry weather can be expected.
Recorded in Bryant (N.Y., S.C.).

2291. If the moon is dish-shaped so that it will hold snow, it portends a hard winter.
Recorded in Bryant (Nebr.).

2292. If the moon is in a cup shape, it is said it can hold water and no rain will fall.
Recorded in Bryant (Oreg.).

2293. If the moon is in the south sky, it is a sign of warm weather.
Cited in Smith, p. 11.

2294. If the moon is rainy throughout, it will be clear at the change, and perhaps the rain will return a few days after.
Cited in Dunwoody, p. 59; Inwards, p. 94.

2295. If the moon is seen between the scud and broken clouds during a gale, it is expected to scuff away the bad weather. Var.: The moon seen between broken clouds during a gale will scuff away bad weather.
Cited in Dunwoody, p. 60; Inwards, p. 91. Recorded in Bryant.

2296. If the moon rises clear, expect fair weather.
Cited in Freier, p. 35.

2297. If the moon show a silver shield, be not afraid to reap your field; but if she rises haloed round, soon we'll tread on deluged ground. Vars.: (a) If the moon show a silver shield, be not afraid to reap your field; but if she rises halved round, soon we'll tread on deluged ground. (b) If the moon shows a silver shield, don't be afraid to reap your field, but if she rises haloed round, soon you'll walk on flooded ground.
Cited in Swainson, p. 188; Dunwoody, p. 60; Northall, p. 463; Garriott, p. 27; Humphreys 2, p. 41; Whitman, p. 26; Inwards,

p. 89; Hand, p. 27; Sloane 2, p. 126; Wilson, p. 542; Wilshere, p. 20; Freier, p. 55.

2298. If you are unable to hang a dipper on the edge of the moon, a wet month may be predicted; if you are able, a dry month.
Cited in Hyatt, p. 3.

2299. If you see a moon through glass, it is trouble while it lasts.
Recorded in Bryant (Can. [P.E.I.]).

2300. In the decay of the moon, a cloudy morning bodes a fair afternoon.
Cited in Denham, p. 8.

2301. The further the moon is to the south the greater the drought; the further west the greater the flood, and the further northwest the greater the cold.
Cited in Dunwoody, p. 59.

2302. The moon grows fat on clouds. Comment: The moon is above the horizon through the early evening when the cloud disappearance in question occurs, only during the first and second quarters, or while it is nightly growing fuller and rounder.
Cited in Marvin, p. 216; Humphreys 2, p. 42; Whitman, p. 28; Inwards, p. 97.

2303. The moon loses its outline ten hours before rain.
Cited in Sloane 2, p. 48.

2304. The moon on its side means rain, for all the water spills out.
Recorded in Bryant (Kans.).

2305. The moon swallows the wind.
Cited in Inwards, p. 88.

2306. Tipped moon wet; cupped moon dry.
Cited in Freier, p. 56.

2307. Watch for cold weather when the moon is in the north, warm weather when the moon is in the south.
Cited in Hyatt, p. 3.

2308. When luna (moon) lowers, then April showers.
Cited in Denham, p. 43; Northall, p. 438.

2309. When the moon has a white look, and when her outline is not very clear, rain or snow is looked for.
Cited in Mitchell, p. 223; Steinmetz 2, p. 284; Inwards, p. 88.

2310. When the moon is darkest near the horizon, expect rain.
Cited in Dunwoody, p. 63; Garriott, p. 27; Hand, p. 27; Freier, p. 35.

2311. When the moon is tilted, it will pour.
Cited in Reeder, no pp.

2312. When the moon is turned up on end, it is a rain sign.
Cited in Smith, p. 6.

2313. When the moon is visible in the daytime, the days are relatively cool.
Cited in Dunwoody, p. 59.

2314. When the moon is wet, no rain you get; when the moon is dry, a rain is nigh.
Cited in Hyatt, p. 2.

2315. When the moon lies on its points the moisture is all out so it will not rain.
Cited in Smith, p. 6.

2316. When the moon runs high, expect cool or cold weather.
Cited in Dunwoody, p. 59.

2317. When the moon runs low, expect warm weather.
Cited in Dunwoody, p. 64.

2318. When the moon's a boat, the earth's afloat.
Recorded in Bryant (Ohio).

2319. When the moon's outline is not clear, rain is to be expected.
Cited in Sloane 2, p. 126.

moon (back)

2320. If the moon is on its back in the third quarter, it is a sign of rain.
Cited in Inwards, p. 93; Wurtele, p. 299.

2321. If the moon lies on its back, it is a dry moon. Var.: It is sure to be a dry moon if it lies on its back, so that you can hang your hat on its horns.
Cited in Inwards, p. 93; Hand, p. 37. Recorded in Bryant (Miss.).

2322. If the moon turns on its back in the third quarter, it is a sign of rain.
Cited in Dunwoody, p. 63.

2323. The bonny moon is on her back, mend your shoon and sort your thack. Comment: Shoon means shoes, thack means thatch.
Cited in Swainson, p. 182; Whitman, p. 26; Inwards, p. 92.

2324. When the moon is on his back, the hunter hangs up his horn.
Recorded in Bryant (N.M.).

2325. When the moon lies on her back then the sou'west wind will crack; when she rises up and nods, the northeasters dry the sod.
Cited in Whitman, p. 26; Inwards, p. 93; Hand, p. 37.

2326. When the moon lies on her back, she sucks the wet into her lap.
Var.: If the moon lies on her back, she sucks the wet into her lap.
Cited in Garriott, p. 37; Whitman, p. 26; Hand, p. 37; Sloane 2, p. 46; Freier, p. 56.

2327. When the moon lies on its back its a sure sign of rain.
Cited in Smith, p. 6.

moon (blue)

2328. A moon with a blue cast is a sign of rain.
Cited in Hyatt, p. 2.

moon (burr)

2329. Far burr, near rain, near burr, far rain. Var.: Near burr, far rain. Comment: Burr is the cloud or dark circle about the moon.
Cited in Northall, p. 463; Humphreys 1, p. 438; Marvin, p. 210; Humphreys 2, p. 38; Whitman, p. 67; Inwards, p. 89; Wilson, p. 556.

moon (changing)

2330. A moon changing in the morning is the beginning of unsettled weather; in the afternoon, settled weather.
Cited in Hyatt, p. 2.

2331. A moon that has changed during the night will commence a wet season; the nearer the change to midnight, the sooner the rain.
Cited in Hyatt, p. 2.

2332. A Saturday change in the moon is enough, as it is always followed by a severe storm.
Cited in Dunwoody, p. 63.

2333. A Saturday's change and a Sunday's full moon, once in seven years is once too soon. Var.: Saturday's change and Sunday's prime, once is enough in seven years' time.
Cited in Swainson, p. 192; Northall, p. 462; Inwards, p. 94.

2334. A Saturday's change and a Sunday's full comes too soon whene'er it wull.
Cited in Inwards, p. 94.

2335. A Saturday's change (moon) brings the boat to the door, but a Sunday's change brings it up on mid-floor.
Cited in Swainson, p. 192.

2336. Five changes of the moon in one month denotes cool weather in summer and cold in winter.
Cited in Dunwoody, p. 60.

2337. Five phases of the moon in any month reverses the weather for thirty days.
Cited in Hyatt, p. 2.

2338. If the moon change on Sunday, there will be a flood before the month is out.
Cited in Swainson, p. 191; Marvin, p. 215; Inwards, p. 94.

2339. If the moon changes in the morning, it indicates warm weather; if in the evening, cold weather.
Cited in Dunwoody, p. 64.

2340. If the moon changes with the wind in the east, the weather during that moon will be foul. Var.: If the moon changes while the wind is in the east, disagreeable weather will follow.
Cited in Dunwoody, p. 59; Inwards, p. 89; Hyatt, p. 2.

2341. If the new moon embraces the old moon, stormy weather is foreboded. Var.: If the old moon embraces the new moon, stormy weather is foreboded.
Cited in Steinmetz 2, p. 284; Dunwoody, p. 16.

2342. Moon changing in the morning indicates warm weather; in the evening, cold weather.
Cited in Inwards, p. 94.

2343. Phases of the moon occurring in the evening, expect fair weather.
Cited in Dunwoody, p. 60.

2344. Saturday change (moon), and Sunday full, is always wet, and always wull.
Cited in Wright, p. 81.

2345. Saturday new, and Sunday full never was fine, and never wull. Var.: Saturday's new, and Sunday's full was never fine, and never wull.
Cited in Cheales, p. 20; Marvin, p. 215.

2346. Saturday's change and Sunday's full (moon), never brought good and never will. Var.: Saturday change and Sunday full, never did good and never will.

Cited in Swainson, p. 192; Cheales, p. 20; Whitman, p. 28; Inwards, p. 94; Wilshere, p. 20.

2347. Saturday's moon and Sunday's prime, once is enough in seven years' time.
Cited in Marvin, p. 215; Inwards, p. 94.

2348. Saturday's moon, Sunday seen; the foulest weather there ever hath been.
Cited in Wright, p. 81; Inwards, p. 570.

2349. The moon and the weather may change together; but change of the moon does not change the weather: if we'd no moon at all and that may seem strange, we still should have weather that's subject to change.
Cited in Northall, p. 463; Garriott, p. 27; Humphreys 1, p. 439; Wright, p. 82; Humphreys 2, p. 41; Whitman, p. 26; Hand, p. 27; Wilshere, p. 20; Freier, p. 56.

2350. The three days of the change of the moon from the way to the wane we get no rain.
Cited in Dunwoody, p. 64.

2351. To see the old moon in the arms of the new one, is reckoned a sign of fine weather; and so is the turning up of the horns of the new moon. Var.: The old moon seen in the new moon's arms is a sign of fair weather.
Cited in Swainson, p. 182; Dunwoody, p. 61; Inwards, p. 96. Recorded in Bryant (Mich.).

2352. Weather prophets prophesy twelve months ahead by observing the first change of the moon in January: if it occurs during the day, a wet year; if during the night, a dry year.
Cited in Hyatt, p. 2.

2353. When phases (changes) of the moon occur in the morning, expect rain. Var.: When changes of the moon occur in the morning, expect rain.
Cited in Dunwoody, p. 63; Inwards, p. 94.

moon (crescent)

2354. If in the spring a crescent moon hangs like a cradle (rests on its back), the summer will be dry. Var.: If in the spring a crescent moon hangs crossways, (stand on one or both horns), a wet spring is signified.
Cited in Hyatt, p. 3.

2355. If the crescent moon stands upright with a north wind blowing, west winds usually follow, and the month will continue stormy to the end.
Cited in Inwards, p. 92.

2356. The crescent moon spills or holds the rain for the coming month.
Cited in Sloane 2, p. 47.

2357. When the crescent moon can hold water it will be dry. Var.: If the crescent of the moon holds water, we will have a dry spell.
Cited in Whitman, p. 27; Freier, p. 56.

moon (dark)

2358. Root crops are planted in the dark of the moon.
Cited in Freier, p. 83.

moon (Friday)

2359. A Friday's moon, come when it will, comes too soon.
Cited in Dunwoody, p. 63; Northall, p. 463.

2360. A Friday's moon is a month too soon.
Cited in Inwards, p. 94.

moon (full)

2361. A few days after full or new moon, changes of weather from good to bad or bad to good are thought more probable than at other times.
Cited in Mitchell, p. 223; Inwards, p. 90.

2362. A full moon in August brings frost.
Cited in Dunwoody, p. 59.

2363. Any year with thirteen moons that are full will make a crib full of corn and a barn full of hay.
Cited in Smith, p. 4.

2364. Fish bite best as the moon grows full.
Cited in Whitman, p. 28.

2365. Full moon in October without frost, no frost till full moon in November.
Cited in Dunwoody, p. 61; Inwards, p. 64; Sloane 2, p. 49; Wilshere, p. 20.

2366. If the full moon rise pale, expect rain.
Cited in Dunwoody, p. 61; Garriott, p. 27; Inwards, p. 91; Hand, p. 27.

2367. If the full moon rises clear, expect fine weather.
Cited in Dunwoody, p. 60; Garriott, p. 27; Hand, p. 27.
2368. If the full moon rises red, expect wind.
Cited in Dunwoody, p. 62; Marvin, p. 216; Inwards, p. 97.
2369. If there be a general mist before sunrise near the full of the moon, the weather will be fine for some days.
Cited in Inwards, p. 97.
2370. In western Kansas it is said that when the moon is near full it never storms.
Cited in Dunwoody, p. 60; Inwards, p. 97.
2371. Look at a full moon and you'll die.
Recorded in Bryant (Oreg.).
2372. Near full moon a misty sunrise bodes fair weather and cloudless skies.
Cited in Marvin, p. 216; Whitman, p. 28; Inwards, p. 97.
2373. Plant cotton, corn, watermelon, beans, squash, and peas just before the full of the moon.
Recorded in Bryant (Miss.).
2374. The full moon brings fair weather.
Cited in Marvin, p. 216; Inwards, p. 97. Recorded in Bryant (Okla., Tex., Wis.).
2375. The full moon eats clouds. Comment: When the moon is full the evaporation of clouds is easily observed, thus the full moon appears to eat the clouds.
Cited in Swainson, p. 185; Dunwoody, p. 60; Marvin, p. 216; Humphreys 2, p. 42; Whitman, p. 28; Inwards, p. 97.
2376. The nearer to twelve in the afternoon, the drier the moon, the nearer to twelve in the forenoon, the wetter the moon.
Cited in Inwards, p. 95.
2377. Two full moons in a calendar month bring on a flood. Vars.: (a) Two full moons within the same month give us rain. (b) When there are two full moons in one month, there are sure to be large floods.
Cited in Dunwoody, p. 60; Inwards, p. 97; Hyatt, p. 2.

moon (half)

2378. If there is a half moon, it will probably rain.
Cited in Smith, p. 6.

moon (halo)

2379. A circle (halo) around the moon, it's going to rain soon. Vars.: (a) A circle around the moon means stormy weather. (b) Circle around the moon, rain soon.
Cited in Hyatt, p. 3. Recorded in Bryant (Miss., Nebr.).

2380. A circle (halo) of the moon never filled a pond.
Cited in Freier, p. 47.

2381. A far brugh, a near storm. Comment: Brugh means halo.
Cited in Marvin, p. 210; Inwards, p. 89.

2382. A halo about the moon foretells rain next day say some, within three days say others; the time frequently being determined by the size of the halo: the smaller the halo, the sooner the rain.
Cited in Hyatt, p. 3.

2383. A large ring (halo) around the moon and low clouds indicate rain in twenty-four hours; a small ring and high clouds rain in several days.
Cited in Dunwoody, p. 61; Garriott, p. 27; Hand, p. 27.

2384. A lunar (moon) halo indicates rain, and the larger the halo the sooner the rain may be expected.
Cited in Garriott, p. 27; Marvin, p. 210; Hand, p. 27.

2385. A lunar (moon) halo indicates rain, and the number of stars enclosed, the number of days of rain. Var.: If there is a circle around the moon and there are stars within the circle, it will rain in the number of days there are in the circle.
Cited in Dunwoody, p. 60. Recorded in Bryant (Ill., Miss., Nebr.).

2386. A ring around the moon is a sign of rain. Vars.: (a) Ring around the moon, brings a storm soon. (b) Ring around the moon, rain soon.
Cited in Whitman, p. 27; Hyatt, p. 3; Smith, p. 6; Arora, p. 6.

2387. A ring around the moon means snow. Var.: A ring around the moon is a sign of snow.
Cited in Smith, p. 9. Recorded in Bryant (Wash.).

2388. As many rings as the moon has, so many will be the days until rain.
Cited in Hyatt, p. 3.

2389. Double circles around the moon portray very severe weather.
Cited in Sloane 2, p. 126.

2390. During cold weather a lunar halo discloses warmer weather; during warm weather, colder weather.
Cited in Hyatt, p. 3.

2391. If the moon rises haloed round, soon we'll tread on deluged ground.
Cited in Northall, p. 463.

2392. If there are two moon-rings, it will snow within twenty-four hours.
Cited in Hyatt, p. 3.

2393. Large halo 'round the moon, heavy rains very soon.
Cited in Lee, p. 125.

2394. Moon in a circle indicates storm, and number of stars in circle the number of days before storm. Vars.: (a) A ring around the moon foretells rain, the number of stars within the ring denote the number of days that will pass before rain comes. (b) A ring around the moon means bad weather; the number of stars in the ring tells how many days off. (c) If a moon-ring has stars, the number of stars will enumerate the foul days approaching. (d) Ring around the moon means a change in the weather, stars in the ring means that it is many days off. (e) When there is a circle around the moon, count the number of stars in it. The number tells the number of days before a major change in the weather.
Cited in Dunwoody, p. 63; Inwards, p. 98; Smith, p. 3; Wurtele, p. 294; Reeder, no pp. Recorded in Bryant (Ind., Ohio).

2395. Moondogs foretell puppy weather (mildly unsettled). Comment: Moondogs are luminous appearances seen in connection with lunar halos.
Recorded in Bryant (Mich.).

2396. Ring around the moon, ice cold by noon.
Recorded in Bryant (Wash.).

2397. Ring around the moon means a storm the next day.
Recorded in Bryant (Ind., Ohio).

2398. Ring around the moon, rain before noon.
Recorded in Bryant (Ind., Ohio).

2399. The bigger the ring, the nearer the wet. Comment: This proverb refers to a ring around the moon.
Cited in Humphreys 1, p. 438; Humphreys 2, p. 38; Whitman, p. 27; Sloane 1, p. 12; Wurtele, p. 294; Lee, p. 85.

2400. The circle of the moon never filled a pond; the circle of the sun wets a shepherd.
Cited in Swainson, p. 187; Garriott, p. 27; Whitman, p. 67; Hand, p. 27.

2401. The larger the halo about the moon the nearer the rain clouds and the sooner the rain may be expected.
Cited in Dunwoody, p. 60; Garriott, p. 21; Hand, p. 21. Recorded in Bryant (Mich.).

2402. The moon with a circle brings water in her beak.
Cited in Swainson, p. 186; Dunwoody, p. 60; Garriott, p. 27; Marvin, p. 210; Humphreys 2, p. 38; Whitman, p. 27; Inwards, p. 89; Hand, p. 27; Wurtele, p. 295; Lee, p. 84; Wilshere, p. 20; Freier, p. 47. Recorded in Bryant (Wis.).

2403. The open side of the halo tells the quarter from which the wind or rain may be expected.
Cited in Inwards, p. 89.

2404. When round the moon there is a brugh, the weather will be cold and rough. Comment: A brugh is a halo.
Cited in Denham, p. 17; Northall, p. 463; Marvin, p. 210; Inwards, p. 89; Freier, p. 47.

2405. When the ring round the moon is far, rain is soon. When the ring round the moon is near, rain is far away.
Cited in Wilshere, p. 20.

2406. Whenever there is a red ring around the moon it will rain the next day.
Cited in Smith, p. 6.

moon (horns)

2407. A moon standing on its horns will within three days begin a wet spell lasting the whole month.
Cited in Hyatt, p. 3.

2408. Horns of the moon obscure, rain.
Cited in Mitchell, p. 232.

2409. If the ends (horns) of the moon extend downward, water is pouring over the lip of the dipper (a wet month); if upward, the dipper retains the water (a dry month). Vars.: (a) If the corners of the moon are turned down, it will be a wet month. (b) If the points of the moon curve downward, water is running out of the apron (a wet month); if upward, the apron withholds the water (a dry month). (c) If the tips of the moon tilt downward, much water will flow under the bridges (a wet month); if upward the water now flowing beneath the bridges will be reduced to a mere trickle (a dry month). Comment: But in each case, the opposite interpretation is also held.

Cited in Hyatt, p. 3; Smith, p. 6.

2410. If the horns of the moon are hidden on the third or fourth day, a rain is imminent.

Cited in Hyatt, p. 2.

2411. If the horns on the new moon are sharper, fair weather.

Cited in Freier, p. 55.

2412. If the upturned horns of a moon lying on its back lean toward the northwest, you can look for a chilly month with rain.

Cited in Hyatt, p. 3.

2413. In winter, when the moon's horns are sharp and well defined, frost is expected.

Cited in Mitchell, p. 223; Steinmetz 2, p. 284; Dunwoody, p. 21; Inwards, p. 92.

2414. Sharp horns do threaten windy weather. Comment: When the crescent moon's ends, or "horns," are clear and sharply defined, it means that high altitude air is unusually clear as the result of high speed winds aloft. As winds begin aloft and lower to earth, "sharp horns" on the moon would predict a windy day.

Cited in Humphreys 1, p. 439; Humphreys 2, p. 42; Whitman, p. 68; Sloane 2, p. 46.

2415. When the horns of the moon are sharp, it indicates dry weather. Vars.: (a) If the horns of the moon are sharp and pointed, clear weather, maybe frost; if the points are dull, expect rain. (b) If the horns of the new moon are sharp, expect dry weather; if blunted rain.

Cited in Dunwoody, p. 59; Freier, p. 55. Recorded in Bryant (Mich.).

2416. When you can hang your oilskins on the horns of the moon, (that is when they point upward) you'll not need them.

Recorded in Bryant (Mich.).

moon (Michaelmas)

2417. The Michaelmas moon rises aye alike soon. Comment: The moon, rising more northerly, rises earlier. Michaelmas Day is September 29th.
Cited in Wilson, p. 529.

2418. The Michaelmas moon, rises nine nights alike soon.
Cited in Denham, p. 57; Northall, p. 463.

moon (mist)

2419. A moon veiled by vapor is a foreshadow of different weather within the next twenty-four hours.
Cited in Hyatt, p. 2.

2420. An old moon in a mist, is worth gold in a kist; but a new moon's mist will never lack thirst. Vars.: (a) An old moon mist ne'er died of thirst. (b) As safe as treasure in a kist, is the day in an old moon's mist. (c) Auld moon mist ne'er died of thirst. (d) The new moon's mist is better than gold in a kist. Comment: A kist is a chest.
Cited in Denham, p. 15; Swainson, p. 184; Northall, p. 461; Inwards, p. 97; Freier, p. 63.

2421. If mist's in the new moon, rain in the old; if mist's in the old moon, rain in the new. Var.: Mists in the old moon, rain in the new; rain in the old moon, mists in the new.
Cited in Wright, p. 81; Inwards, p. 92; Freier, p. 55.

moon (new)

2422. A hundred hours after the new moon regulates the weather for the month.
Cited in Inwards, p. 95.

2423. A moon becoming new on Monday notifes you of fine weather. Comment: According to others, this is a notice of rain for forty days.
Cited in Hyatt, p. 2.

2424. A new moon and a windy night sweeps cobwebs out of the drab house of February.
Cited in Sloane 1, p. 41.

2425. A new moon lying on her back means wet weather.
Cited in Mitchell, p. 224.

2426. A new moon soon seen is long thought of.
Cited in Swainson, p. 182.

2427. A new moon with a north wind will hold under the full.
Cited in Dunwoody, p. 61.

2428. A new moon with sharp horns, threatens windy weather.
Cited in Denham, p. 4; Inwards, p. 92.

2429. If it rains the first day of a new moon, there will be extreme precipitation. If it rains the second day, it will be less moisture; and if it rains the third day, it will be very little.
Cited in Smith, p. 6.

2430. If the first new moon in January lies on its side or back (horns upward), predict a wet year; if on its belly (horns downward), a dry year.
Cited in Hyatt, p. 3.

2431. If the new moon appears with the points of the crescent turned up, the month will be dry. If the points are turned down, it will be wet.
Cited in Dunwoody, p. 61; Garriott, p. 36; Inwards, p. 91; Hand, p. 36; Wurtele, p. 299.

2432. If the new moon holds the old moon in her lap, fair weather.
Cited in Freier, p. 55.

2433. If the new moon is far north, it will be cold for two weeks, but if far south, it will be warm.
Cited in Dunwoody, p. 61; Inwards, p. 92.

2434. If the new moon is on its face, wet weather ahead. Comment: New moon on its face refers to a crescent moon facing down.
Recorded in Bryant (Can. [Ont.]).

2435. If the new or full moon changes in fair or warm part of the day, it indicates a warm moon, and if it changes in the cool part of the day, it indicates that the weather will be cool during the moon.
Cited in Dunwoody, p. 59.

2436. If you see a new moon through the brush, it will bring you a month of trouble; if you see it in the clear, it will bring good luck.
Recorded in Bryant (Miss., N.Mex.).

2437. Never look at the new moon over your left shoulder, it is bad luck.
Recorded in Bryant (Kans.).

2438. New moon far in north in summer, cool weather; in winter, cold.
Cited in Dunwoody, p. 59.

2439. New moon far in the south indicates dry weather for a month.
Var.: The new moon appearing far to the south foretells dry weather.
Cited in Dunwoody, p. 59; Inwards, p. 92. Recorded in Bryant (Mich.).

2440. New moon on its back indicates wind; standing on its point indicates rain in summer and snow in winter.
Cited in Dunwoody, p. 61.

2441. On the fourth day of the new moon, if bright, with sharp horns, no winds nor rain till the month be finished.
Cited in Mitchell, p. 232.

2442. The fifth day of the new moon indicates the general character of the weather until the full of the moon.
Cited in Dunwoody, p. 60.

2443. The nearer to a new moon on Christmas, the harder will be the rest of the winter.
Cited in Sloane 2, p. 34.

2444. The new moon appearing far to the north indicates increasing cold.
Recorded in Bryant (Mich.).

2445. The new moon lying on her back or being ill-made is a prognostic of wet weather.
Cited in Inwards, p. 93.

2446. The new moon standing up lets the water out.
Recorded in Bryant (Mich.).

2447. When the new moon comes in at midnight, or within thirty minutes before or after, the following month will be fine.
Cited in Wright, p. 81; Inwards, p. 95.

2448. When the new moon lies on her back, she sucks the wet into her lap.
Cited in Northall, p. 464; Inwards, p. 93.

moon (old)

2449. Threatening clouds, without rain, in old moon, indicate drought.
Cited in Inwards, p. 97.

moon (pale)

2450. A dim or pale moon indicates rain, a red moon indicates wind. Vars.: (a) A pale moon doth rain, the red moon doth blow, but the white moon doth neither rain nor blow. (b) If the moon is pale and dim, it is a sign of rain. (c) Pale moon doth rain, red moon doth blow, white moon neither rain nor snow.
  Cited in Dunwoody, p. 62; Northall, p. 463; Humphreys 2, p. 25; Whitman, p. 27; Inwards, p. 88; Sloane 2, p. 46; Hyatt, p. 2; Smith, p. 6; Wilson, p. 608; Lee, p. 124; Page, p. 28; Wilshere, p. 20; Freier, p. 34. Recorded in Bryant (Mich.).

2451. If the moon rises pale, expect rain; if it rises clear, expect fair.
  Cited in Sloane 2, p. 46.

moon (red)

2452. A red-tinged girdle about the moon betokens rain in summer and snow in winter.
  Cited in Hyatt, p. 2.

2453. If moon red be, of water speaks she.
  Cited in Hyatt, p. 2.

2454. If on her cheeks you have seen the maiden's blush, the ruddy moon foreshows the winds will rush.
  Cited in Dunwoody, p. 63.

2455. If the moon rises large and red among clouds, expect rain within twelve hours.
  Recorded in Bryant (Mich.).

2456. When the moon is red, wind. Var.: If the moon is red, it will be windy.
  Cited in Mitchell, p. 232; Smith, p. 12.

2457. When the moon rises red and appears large, with clouds, expect rain in twelve hours.
  Cited in Dunwoody, p. 62; Garriott, p. 27; Inwards, p. 91; Hand, p. 27; Freier, p. 35.

moon (Saturday)

2458. A Saturday's moon, come when it will, it comes too soon. Vars.: (a) A Saturday moon, if it comes once in seven years, comes once too soon. (b) A Saturday's moon always comes too soon. (b) A Saturday's moon in March is enough for seven years. (c)

If Saturday's moon comes once in seven years, it comes too soon.
Cited in Denham, p. 9; Swainson, p. 192; Cheales, p. 20; Dunwoody, p. 59; Northall, p. 461; Wright, p. 82; Marvin, p. 215; Inwards, p. 94; Wilson, p. 699.

2459. If the moon on a Saturday be new or full, there always was rain, and there always wull.
Cited in Inwards, p. 95.

moon (wane)

2460. In the wane of the moon, a cloudy morning bodes a fair afternoon. Vars.: (a) In the decay of the moon, a cloudy morning bodes a fair afternoon. (b) In the old moon, a cloudy morning means a fair afternoon. (c) In the old of the moon, a cloudy morning means a fair afternoon. (d) In the wane of the moon, a cloudy morning is a sign of a fair afternoon. (e) In the waning of the moon, a cloudy morning bodes a fair afternoon.
Cited in Swainson, p. 185; Dunwoody, p. 59; Northall, p. 463; Marvin, p. 211; Inwards, p. 91; Freier, p. 66. Recorded in Bryant (Mich., Wis.).

2461. Sow peas and beans in the wane of the moon, who soweth them sooner soweth them too soon.
Cited in Swainson, p. 188; Whitman, p. 28. Recorded in Bryant (Wis.).

moonlight

2462. Cut stove wood on moonlight nights and it will burn good.
Recorded in Bryant (Miss.).

2463. Moonlight dries no mittens. Var.: Moonlight does not dry mittens.
Cited in Whiting 2, p. 423. Recorded in Bryant (N.C.).

moonlit

2464. Moonlit nights have the heaviest frosts.
Cited in Humphreys 1, p. 439; Humphreys 2, p. 41; Whitman, p. 67; Sloane 2, p. 46. Recorded in Bryant (Mich.).

moonrise

2465. An uncloudy moonrise reveals clear weather for the next twenty-four hours.
Cited in Hyatt, p. 2.

moonset

2466. If a Friday moonset is bright, rain will come before Tuesday.
Cited in Hyatt, p. 2.

morning

2467. A bright morning is followed by a dark day. Vars.: (a) It is often observed, a bright morning is succeeded by a dark rainy day. (b) The pleasantest morning is sometimes followed by the darkest night.
Cited in Whiting 1, p. 296.

2468. A foul morn may turn to a fine day. Vars.: (a) A cloudy morning may turn to a fair day. (b) A foul morn may turn into a fine day. (c) A foul morn turns into a fine day.
Cited in Marvin, p. 203; Wilson, p. 283; Mieder, p. 419.
Recorded in Bryant (Wis.).

2469. A gaudy morning bodes a wet afternoon. Comment: Gaudy means overly bright.
Cited in Wilson, p. 297.

2470. A wet morning may turn to a dry afternoon.
Cited in Wilshere, p. 6.

2471. Morn dry, rain nigh; morn wet, no rain yet.
Cited in Sloane 2, p. 36.

2472. Morning red, foul weather and great need.
Cited in Inwards, p. 37.

2473. When the morn is dry, the rain is nigh. When the morn is wet, no rain you get.
Cited in Humphreys 2, p. 59.

moss

2474. A hard winter always follows the appearance of moss on the south side of trees in autumn.
Cited in Hyatt, p. 16.

2475. Moss dry, sunny sky; moss wet, rain you'll get.
Cited in Sloane 2, p. 48.

2476. When the mountain moss is soft and limpid, expect rain. Var.:
When mountain moss is dry and brittle expect clear weather.
Cited in Dunwoody, p. 67; Garriott, p. 21; Hand, p. 21; Freier,
p. 37.

mulberry
2477. When the mulberry has shown green leaf, there will be no more
frost. Var.: When the mulberry buds and puts forth leaves, fear
no more frosts.
Cited in Inwards, p. 211; Sloane 2, p. 131.

mule
2478. A mule rolling on the ground at mid-day foretells a storm before
night.
Cited in Hyatt, p. 28.
2479. If in summer a mule refuses to eat or drink, and stands looking
over the barnyard or pasture gate with his head toward the house,
a drought may be expected.
Cited in Hyatt, p. 28.

mullet
2480. Mullet run south on the approach of cold northerly wind and
rain.
Cited in Dunwoody, p. 50; Inwards, p. 199.

mushroom
2481. Mushrooms and toadstools are numerous before rain. Var.:
Mushrooms and toadstools are plentiful before rain.
Cited in Dunwoody, p. 67; Garriott, p. 25; Hand, p. 25; Freier,
p. 41.
2482. Mushrooms in November disclose a light winter.
Cited in Hyatt, p. 15.
2483. When mushrooms spring up during the night, expect rain. Vars.:
(a) A lot of mushrooms popping up overnight warns you of rain.
(b) The sudden growth of mushrooms presageth rain.
Cited in Dunwoody, p. 67; Inwards, p. 216; Sloane 2, p. 112;
Hyatt, p. 15.

2484. When the moon is at the full, mushrooms you may freely pull;
but when the moon is on the wane, wait ere you think to pluck
again.
Cited in Northall, p. 481; Wright, p. 82.

muskrat

2485. A muskrat building away from the water denotes a flood.
Cited in Hyatt, p. 29.
2486. As soon as the muskrat starts to build its house, cold weather is
on its way.
Cited in Hyatt, p. 29.
2487. Built by the muskrat in autumn, a small house reveals a mild
winter and a large one a bitter winter.
Cited in Hyatt, p. 29.
2488. If a muskrat nest is deep in the ground, expect a harsh winter; if
near the surface of the ground, an open winter.
Cited in Hyatt, p. 29.
2489. If muskrats build houses in shallow water, it betokens a warm
winter; if in deep water, a cold winter.
Cited in Hyatt, p. 29.
2490. It is a sure sign of a lot of deep snows for winter, if the muskrats
build their houses high.
Cited in Hyatt, p. 29.
2491. The muskrats build their houses twenty inches higher and very
much warmer in early and long winters than in short ones.
Cited in Dunwoody, p. 32.
2492. When muskrats build larger houses in deeper water, it's going to
be a cold winter.
Cited in Freier, p. 84.
2493. You will never see a muskrat cutting corn stalks and carrying
them underground, unless a hard winter is coming.
Cited in Hyatt, p. 29.

nervous temperament

2494. Persons of a nervous temperament have a sense of dread or a depression of spirits preceding a fall of rain.
Cited in Inwards, p. 218.

nettle

2495. Dead nettles in abundance late in the year are a sign of a mild winter.
Cited in Dunwoody, p. 65; Inwards, p. 212; Sloane 2, p. 131.

New Year

2496. Till New Year, sweat; till May, no heat.
Cited in Swainson, p. 28.

New Year's Day

2497. A southern wind on New Year's Day will return every three days all winter.
Cited in Hyatt, p. 8.

2498. As the direction of the wind is on New Year's Day, so will it be mostly all year.
Cited in Hyatt, p. 8.

2499. At New Year's Day a cock's stride, at Candlemas an hour wide.
Comment: Candlemas Day is February 2nd. The text refers to the lengthening of days.
Cited in Swainson, p. 28; Northall, p. 444.

2500. At New Year's tide, the days lengthen a cock's stride.
Cited in Denham, p. 23; Swainson, p. 27.

2501. For the next three months the wind will not deviate from its path on New Year's Day.
Cited in Hyatt, p. 8.

2502. If the weather on New Year's Day is so mild that a turtledove coos, you may expect a good harvest that year.
Cited in Hyatt, p. 34.

2503. If the wind before sunrise on New Year's Day is from a certain point, during the next two months you will not find it out of that point for more than forty-eight hours.
Cited in Hyatt, p. 8.

2504. If the wind comes from the south on New Year's Day, it will come from the south every day during January.

Cited in Hyatt, p. 8.

2505. If the wind is blowing from the northwest on New Year's morning, for forty days it will continue from that direction.
Cited in Hyatt, p. 9.

2506. If water drips from the eaves of a house on New Year's Day, an excellent crop year is indicated.
Cited in Hyatt, p. 34.

2507. In whatever direction the wind blows on New Year's morning, it will blow every twenty-four hours, or will not shift for more than twenty-four hours, during the next forty days.
Cited in Hyatt, p. 8.

2508. New Year's Day occurring on Sunday presages a dry summer.
Cited in Hyatt, p. 13.

2509. On New Year's Day a southerly wind begins forty days of clement weather.
Cited in Hyatt, p. 9.

2510. The course of the wind on New Year's Day will be retraveled every forty-eight hours, for a few minutes at least, throughout the following forty-eight days.
Cited in Hyatt, p. 8.

2511. The quarter of the wind at five o'clock in the morning on New Year's Day will be its direction three times that year.
Cited in Hyatt, p. 8.

2512. The trend for the year's weather is determined by the day on which the New Year falls.
Cited in Freier, p. 87.

2513. The weather on New Year's Day rules the weather of the three following months.
Cited in Hyatt, p. 13.

2514. Wet weather on New Year's Day is an omen of a rainy January.
Cited in Hyatt, p. 10.

2515. Wind in the south on New Year's Day means a dry summer; wind in the north, a wet summer.
Cited in Hyatt, p. 9.

New Year's Eve

2516. If New Year's Eve night, wind blow south, it betokenth warmth and growth; if west, much milk, and fish in the sea; if north, much cold, and storms there will be; if east, the trees will bear

much fruit, if northeast, flee it man and beast. Var.: If on New Year's Eve the wind blows south, it betokeneth warmth and growth; if west, much milk and fish in the sea, if north, cold and storms there will be; if west, the trees will bear much fruit, if northeast, then flee it, man and beast.
Cited in Denham, p. 23; Swainson, p. 167; Dunwoody, p. 86; Northall, p. 457; Wright, p. 129; Inwards, p. 69; Wilshere, p. 16.

2517. If the weather is fair on New Year's Eve night, it reveals the coming year as one of abundance.
Cited in Hyatt, p. 909.

night

2518. A blustering night, a fair day.
Cited in Inwards, p. 74; Wilson, p. 71; Mieder, p. 430.

2519. If the night is hot, a day will usually be fair.
Recorded in Bryant (Wis.).

2520. Night brings out the stars.
Cited in Mieder, p. 430.

2521. The darker the night, the brighter the candle.
Cited in Mieder, p. 430.

2522. The longest night will have an end.
Cited in Wilson, p. 482.

North Star

2523. When the North Star starts to twinkle, it will probably bring rain.
Comment: The North Star is Polaris.
Cited in Freier, p. 54.

November

2524. A cold November, a warm Christmas.
Cited in Wilshere, p. 14.

2525. A cold November signifies a mild winter. Var.: A cold November a warm winter.
Cited in Page, p. 51; Wilshere, p. 14.

2526. As November, so the following March. Var.: November weather is duplicated during March.
Cited in Dunwoody, p. 99; Garriott, p. 45; Whitman, p. 44; Inwards, p. 65; Hand, p. 45; Hyatt, p. 14.

2527. As the wind is in the month of November, so will it be in the month of December.
Cited in Dunwoody, p. 86.

2528. Ice in November brings mud in December.
Cited in Whitman, p. 44; Inwards, p. 65; Wilshere, p. 15.
Recorded in Bryant (Wis.).

2529. If on the first three days in November the wind travels from the south, the winter will be mild.
Cited in Hyatt, p. 9.

2530. If the ice in November will bear a duck, there'll be nothing after but sludge and muck. Vars.: (a) If there be ice in November to bear a duck, the rest of the winter'll be slush and muck. (b) If there is ice in November that will bear a duck, there will be nothing thereafter but sleet and muck.
Cited in Northall, p. 443; Inwards, p. 65; Sloane 1, p. 153; Hyatt, p. 11; Wilshere, p. 15.

2531. November take flail, let ships no more sail.
Cited in Denham, p. 61; Swainson, p. 141; Cheales, p. 23; Dunwoody, p. 99; Northall, p. 443; Inwards, p. 62; Sloane 2, p. 95; Wilson, p. 582.

2532. When in November the water rises, it will show itself the whole winter.
Cited in Dunwoody, p. 92; Inwards, p. 65.

November 1

2533. A wet and cloudy November 1st is an indication of a wet winter.
Cited in Hyatt, p. 10.

2534. If the beech acorn is wet on November 1st, so will the winter be wet.
Cited in Lee, p. 151.

2535. On the first of November, if the weather hold clear, an end of wheat sowing do make for the year.
Cited in Swainson, p. 142; Northall, p. 443; Wilson, p. 263.

November 11

2536. A fair November 11th which is also cold and dry, forecasts an open winter.
Cited in Hyatt, p. 13.

2537. Bad weather on November 11th forecasts an open winter.

Cited in Hyatt, p. 13.

2538. If geese on November 11th walk over ice, they will walk in mud at Christmas.
Cited in Hyatt, p. 11.

2539. If trees are not stripped of leaves before November 11th, a raw winter is betokened.
Cited in Hyatt, p. 16.

November 21

2540. As November 21st, so the winter. Var.: As the weather is on November 21st, so will it be all winter.
Cited in Dunwoody, p. 92; Inwards, p. 67; Hyatt, p. 13.

November 25

2541. The weather of November 25th will be the weather of February.
Cited in Hyatt, p. 14.

nut

2542. Nuts with a thick covering denote a hard winter. Var.: Unusually thick nutshells predict a severe winter.
Cited in Dunwoody, p. 67; Freier, p. 77.

2543. The approaching winter will be severe if: nuts have thick shells, trees have a heavy coat of moss on their trunks, and animals have thick coats of fur.
Recorded in Bryant (N.Y.).

2544. When there are plenty of nuts, expect a hot dry harvest.
Cited in Inwards, p. 212.

nuthatch

2545. The first nuthatch seen is a sign that spring is here for good.
Cited in Smith, p. 11.

oak

2546. Beware of an oak, it draws the stroke; avoid an ash, it counts the flash; creep under the thorn, it can save you from harm. Comment: Advice on where to shelter from lightning during a thunderstorm.
Cited in Simpson, p. 167.

2547. If the oak bear many acorns it foreshows a long and hard winter.
Cited in Swainson, p. 259; Sloane 2, p. 131.

2548. If the oak is out in leaf before the ash, it will be a dry summer; if the ash first, wet. Vars.: (a) If the oak before the ash comes out, there has been or there will be drought. (b) If the oak is out before the ash, then you'll only get a splash; but if the ash precedes the oak, then you may expect a soak. (c) If the oak is out before the ash, 'twill be a summer of wet and splash, but if the ash is before the oak, 'twill be a summer of fire and smoke. (d) If the oak's before the ash, then you'll only get a splash. (e) If the oak's before the ash, then you'll only get a splash; if the ash's before the oak, then you may expect a soak. (f) Oak before ash have a splash, ash before oak have a soak. (g) Oak, smoke; ash, splash. (h) When buds the oak before the ash, you'll only get a summer splash. (i) When the oak is before the ash, the summer will be dry and mash. (j) When the oak is out before the ash, then you ought to see it rain; when the ash is out before the oak, then you may expect a drought.
Cited in Mitchell, p. 226; Dunwoody, p. 93; Swainson, p. 260; Cheales, p. 24; Northall, p. 474; Marvin, p. 211; Whitman, p. 49; Boughton, p. 124; Inwards, p. 210; Wilson, p. 584; Page, p. 11; Wilshere, p. 21; Simpson, p. 167; Freier, p. 77. Recorded in Bryant (N.C., S.C.).

2549. When oak trees bend in January, good crops may be expected.
Cited in Dunwoody, p. 116.

2550. When oak trees hang full, expect a severe winter with much snow.
Cited in Sloane 2, p. 131.

2551. When the oak comes out before the ash, there will be fine weather in harvest; but when the ash comes out before the oak, the harvest will be wet. Var.: Oak before ash means a fine

harvest; ash before oak, harvest wet.
Cited in Inwards, p. 210; Wilshere, p. 21.

2552. When the oak puts on his gosling gray, 'tis time to sow barley, night and day. Var.: When the oak puts on its gosling gray, it's time to sow barley night and day.
Cited in Denham, p. 46; Swainson, p. 12; Dunwoody, p. 89; Northall, p. 479; Freier, p. 82.

2553. When the oaks are in the gray, then, farmers, plant away.
Cited in Taylor 2, p. 268.

oak ball

2554. If in the autumn you open an oak ball (oak tree gall) and find a worm, it denotes a warm winter because the worm is naked; but if the oak ball contains a fly instead of the larval worm, it denotes a cold winter because the fly is covered with hair.
Cited in Hyatt, p. 16; Freier, p. 81.

oak leaf

2555. Don't plant your corn until the oak leaf is as big as a squirrel's ear.
Cited in Freier, p. 81.

2556. Go fishing when oak leaves get as large as a hog's ears.
Recorded in Bryant (N.Y., S.C.).

oat

2557. He who sows his oats in May will get little that way.
Cited in Freier, p. 82.

2558. I looked at my oats in May, and came sorrowing away, I went again in June, and came away singing a thankful tune.
Cited in Northall, p. 491.

2559. Warm and wet for oats and corn, cool and dry for wheat and rye.
Recorded in Bryant (Ill., N.Y.).

October

2560. A rainy October, a windy January.
Cited in Hyatt, p. 10.

2561. A warm October, a cold February.
Cited in Wilshere, p. 14.

2562. As the weather in October, so will it be the next March.
Cited in Dunwoody, p. 99; Garriott, p. 45; Inwards, p. 64; Hand, p. 45; Freier, p. 83.

2563. For every fog in October a snow in the winter, heavy or light according as the fog is heavy or light.
Cited in Inwards, p. 64; Wilshere, p. 14.

2564. Good October, a good blast, to blow the hogs acorn and mast.
Vars.: (a) A good October and a good blast, so blow the hog acorn and the mast. (b) Grant October, a good blast, to blow the hog acorn and mast!
Cited in Denham, p. 59; Cheales, p. 23; Dunwoody, p. 99; Northall, p. 442; Inwards, p. 64.

2565. Ice in October to bear a duck, rest of winter is as wet as muck.
Cited in Wilshere, p. 14.

2566. If in October and November there be snow and frost, then January and February are likely to be open and mild. Var.: If there be snow in October and November, then January and February are likely to be open and mild.
Cited in Mitchell, p. 233; Steinmetz 1, p. 136.

2567. If in the fall of the leaves in October many of them wither on the bough and hang there, it betokens a frosty winter and much snow.
Cited in Inwards, p. 64.

2568. If October brings much frost and wind, then are January and February mild. Var.: If October brings heavy frosts and winds, then will January and February be mild.
Cited in Dunwoody, p. 92; Whitman, p. 49; Inwards, p. 64; Hand, p. 45; Freier, p. 83.

2569. If the end of October and beginning of November be warm and rainy, then January and February are likely to be frosty and cold, except after a very dry summer. Var.: If the latter end of October and the beginning of November are warm and rainy, then January and February are likely to be open and mild.
Cited in Mitchell, p. 233; Steinmetz 1, p. 136.

2570. If there is snow and frost in October, January will be mild.
Cited in Wilshere, p. 14.

2571. Much rain in October, much wind in December.
Cited in Dunwoody, p. 99; Garriott, p. 45; Whitman, p. 44;

Inwards, p. 64; Hand, p. 45; Hyatt, p. 10; Wilshere, p. 14; Freier, p. 83.

2572. October and November cold indicate that the following January and February are mild.
Cited in Inwards, p. 64.

2573. October has twenty-one fair days. Var.: October always has twenty-one fine days. Comment: This probably refers to Indian summer.
Cited in Sloane 2, p. 49; Wilshere, p. 14.

2574. October warm with certain speed, makes February cold indeed.
Recorded in Bryant (N.Y.).

2575. Plenty of rain in October and November on the North Pacific coast indicates a mild winter; little rain in these months will be followed by a severe winter.
Cited in Dunwoody, p. 69; Inwards, p. 64.

2576. There are always nineteen fine days in October.
Cited in Inwards, p. 64.

2577. Warm October, cold February.
Cited in Dunwoody, p. 99; Garriott, p. 45; Whitman, p. 44; Inwards, p. 64; Hand, p. 45.

2578. When birds and badgers are fat in October, expect a cold winter.
Cited in Inwards, p. 64.

2579. When it freezes and snows in October, January will bring mild weather; but if it is thundering and heat lightning, the winter will resemble April in temper.
Cited in Dunwoody, p. 99; Inwards, p. 64.

old man

2580. The burial of an old man causes rain.
Cited in Wurtele, p. 300.

onion

2581. Halve an onion, sprinkle water on one half, and plant the two halves near each other: if the watered half comes up first, the weather that summer will be rainy.
Cited in Hyatt, p. 15.

2582. Onion's skin very thin, mild winter's coming in: onion's skin thick and rough, coming winter cold and rough.

Cited in Swainson, p. 260; Dunwoody, p. 67; Northall, p. 475; Wright, p. 103; Whitman, p. 50; Inwards, p. 214; Sloane 2, p. 49; Lee, p. 134; Freier, p. 80.

2583. Thick skins on onions means a hard winter is coming. Var.: When a hard winter is promised, onions grow thick skins.
Cited in Smith, p. 14; Page, p. 10; Wilshere, p. 23.

2584. Thin onion skins in autumn signify a mild winter; thick onion skins, a cold winter.
Cited in Hyatt, p. 15.

2585. When the onion's skin is thin and delicate, expect a mild winter; but when the bulb is covered by a thick coat, it is held to foreshow a severe season.
Cited in Inwards, p. 214; Sloane 2, p. 132.

owl

2586. A hoot owl at night is a sign of snow.
Cited in Smith, p. 9.

2587. A screeching owl indicates cold or storm. Vars.: (a) A screeching owl indicates a storm. (b) The screeching of an owl indicates cold or storm.
Cited in Dunwoody, p. 38; Marvin, p. 216; Inwards, p. 194; Sloane 2, p. 130.

2588. An owl crying in the day in the late fall is a sign of a snowstorm.
Cited in Hyatt, p. 22.

2589. If a number of owls hoot at the same time in daylight, a change of weather will follow.
Cited in Hyatt, p. 22.

2590. If an owl hoots about two o'clock in the afternoon, rain may be expected.
Cited in Hyatt, p. 22.

2591. If an owl hoots anytime during the day; it will storm within twenty-four to forty-eight hours.
Cited in Hyatt, p. 22.

2592. If an owl hoots in daytime while sitting on a fence, rain is in the air.
Cited in Hyatt, p. 22.

2593. If an owl hoots in the morning, colder weather is coming.
Cited in Hyatt, p. 22.

2594. If an owl hoots in the morning part of the day, look for rain before three days.
Recorded in Bryant (Miss.).

2595. If an owl hoots just at dusk, rain is signified.
Cited in Hyatt, p. 22.

2596. If during the day an owl hoots among trees on high ground or back in the hills, it warns you of dry weather; if down in the timber along a branch or creek, wet weather.
Cited in Hyatt, p. 22.

2597. If owls scream during bad weather, there will be a change.
Cited in Inwards, p. 194.

2598. If owls scream in foul weather, it will change to fair.
Cited in Dunwoody, p. 38.

2599. If you can hear the far-off hooting of an owl, you may look for rain.
Cited in Hyatt, p. 22.

2600. If you hear an owl hoot in the winter, bad weather is ahead.
Cited in Smith, p. 1.

2601. Owls hooting indicate rain. Vars.: (a) The hooting of the owl brings rain. (b) The hooting of the owl fortells rain. (c) When the owl sings, we shall have rain.
Cited in Dunwoody, p. 38; Lee, p. 64; Freier, p. 22. Recorded in Bryant (Wis.).

2602. Screech owls are most noisy just before rain.
Cited in Inwards, p. 194.

2603. The cry of the screech owl becomes harsher before rain.
Cited in Page, p. 19.

2604. The crying of an owl in fair weather indicates storm.
Cited in Marvin, p. 216.

2605. The crying of an owl in storm indicates fair weather.
Cited in Marvin, p. 216.

2606. The screaming of an owl in bad weather indicates change of weather.
Cited in Marvin, p. 216.

2607. When owls whoop much at night, expect fair weather. Var.: The hooting of an owl at night indicates fair weather.
Cited in Swainson, p. 243; Dunwoody, p. 20; Marvin, p. 216.

oxen

2608. If oxen be seen to lie along upon the left side, it is a token of fair weather.
Cited in Inwards, p. 181.

2609. If oxen lick their front hooves, it is a sign of rain.
Cited in Sloane 2, p. 120.

2610. If oxen lie on their right side, look towards the south, and lick their hoofs; it is a sign of rain.
Cited in Steinmetz 1, p. 111.

2611. If oxen turn up their nose and sniff the air, a sign of rain.
Cited in Sloane 2, p. 120.

2612. If oxen turn up their nostrils and sniff the air, or if they lick their forefeet or lie on their right side, it will rain.
Cited in Dunwoody, p. 14; Inwards, p. 182.

2613. When oxen do lick themselves against the hair, it betokens rain to follow shortly after. Var.: If oxen lick their hair the wrong way, a sign of rain.
Cited in Wright, p. 79; Inwards, p. 181; Sloane 2, p. 120.

oxeye

2614. The great white oxeye closes before rain.
Cited in Inwards, p. 216; Sloane 2, p. 112.

pain

2615. Headache, toothache, pain in corns, rheumatism, neuralgic pains, etc, are felt by some people before change from dry to wet or mild to cold.
Cited in Mitchell, p. 228.

2616. You can foretell a rain by the joints or bones of your body becoming stiff or paining.
Cited in Hyatt, p. 30.

palm

2617. A piece of blessed palm burned in the stove on Palm Sunday protects your house all year against lightning.
Cited in Hyatt, p. 33.

2618. Protect your house during a storm by burning palm branches blessed on Palm Sunday.
Cited in Hyatt, p. 33.

Palm Sunday

2619. A bad Palm Sunday denotes a year of failing crops. Var.: A bad Palm Sunday will lead to poor crops.
Cited in Lee, p. 152; Freier, p. 91.

2620. A wet Palm Sunday means a sunny Easter.
Cited in Hyatt, p. 10.

2621. A wet Palm Sunday will be followed by seven weeks of rain.
Cited in Hyatt, p. 10.

2622. From whatever quarter the wind blows on Palm Sunday, it will continue to blow for the greater part of the coming summer.
Cited in Wright, p. 35; Inwards, p. 70.

2623. If the weather is not clear on Palm Sunday, it means a bad year.
Cited in Dunwoody, p. 103; Inwards, p. 70.

2624. When there is fair weather on Palm Sunday, there will be fair weather on Easter.
Cited in Boughton, p. 124.

paper

2625. Paper and straw fly about before rain.
Cited in Steinmetz 1, p. 112.

parrot

2626. A chattering parrot is an indication of rain.
       Cited in Hyatt, p. 22.

2627. Parrots and canaries dress their feathers and are wakeful the
       evening before a storm Var.: Parrots and canaries excessively
       dress their feathers before a rain.
       Cited in Dunwoody, p. 38; Inwards, p. 189; Alstad, p. 81.

2628. Parrots whistling indicate rain. Var.: When parrots whistle,
       expect rain.
       Cited in Garriott, p. 23; Inwards, p. 189; Hand, p. 23; Sloane
       2, p. 129; Freier, p. 22.

parsley

2629. Parsley sown on Good Friday bears a heavier crop than that
       sown on any other day.
       Cited in Wright, p. 38.

partridge

2630. Partridges drum only in fall when a mild and open winter
       follows.
       Cited in Dunwoody, p. 33.

2631. Partridges perching high in a tree indicate that rain is coming.
       Cited in Freier, p. 53.

2632. Spruce partridges (Canada grouse) feeding heavily indicate bad
       weather.
       Cited in Lee, p. 63.

2633. The severity of the winter can be determined by how far down
       the feathers grow on a partridge's leg.
       Cited in Freier, p. 81.

2634. White partridges (willow grouse) perching high in trees mean a
       snowstorm is coming.
       Cited in Lee, p. 63.

Pastor Sunday

2635. If it rains on Pastor Sunday, it will rain every Sunday until
       Pentecost. Var.: Rain on Pastor Sunday means rain on every
       Sunday until Pentecost. Comment: Pastor Sunday is the second
       Sunday after Easter.
       Cited in Whitman, p. 38; Inwards, p. 71; Lee, p. 152.

pavement

2636. If pavements appear rusty, rain will follow. Var.: If pavements ook rusty, rain may be expected soon.
Cited in Inwards, p. 220; Lee, p. 74; Freier, p. 25.

peacock

2637. A peacock always struts just before rain.
Cited in Hyatt, p. 25.

2638. If peacocks cry in the night, there is rain to fall. Vars.: (a) If peacocks cry in the night, there is rain to follow. (b) The squalling of the peacock by night often foretells a rainy day.
Cited in Swainson, p. 243; Dunwoody, p. 39; Inwards, p. 189; Sloane 2, p. 130.

2639. If peacocks run about in confusion while crying, rain may be forecasted.
Cited in Hyatt, p. 25.

2640. If the peacock cries when he goes to roost, and, indeed much at any time, it is a sign of rain.
Cited in Dunwoody, p. 38.

2641. Late in the winter an extraordinary amount of crying by peacocks shows that cold weather has ended.
Cited in Hyatt, p. 25.

2642. When peacocks and guinea fowls scream and turkeys gobble, expect rain.
Cited in Dunwoody, p. 39.

2643. When the peacock loudly bawls, soon we'll have both rain and squalls. Var.: When the peacock loudly bawls, soon you'll have both rain and squalls.
Cited in Swainson, p. 243; Dunwoody, p. 14; Garriott, p. 23; Whitman, p. 48; Inwards, p. 189; Hand, p. 23; Wilson, p. 616; Lee, p. 64; Freier, p. 52.

2644. When the peacock's distant voice you hear, are you in want of rain? Rejoice, 'tis almost here. Var.: The cry of a peacock portends rain.
Cited in Dunwoody, p. 38; Inwards, p. 189; Hyatt, p. 25.

peafowl

2645. Peafowl utter loud cries before a storm, and select a low perch.
Cited in Dunwoody, p. 39; Inwards, p. 189.

2646. When the peafowl scream, it is a sign of rain.
Recorded in Bryant (Miss.).

Pentecost
2647. Rain at Pentecost forebodes evil. Var.: Rain on Pentecost forebodes evil.
Cited in Dunwoody, p. 102; Inwards, p. 72; Lee, p. 152.

persimmon
2648. The number of persimmons there are in bloom will indicate the number of snows there will be.
Cited in Smith, p. 10.
2649. When the persimmon tree is full of fruit, a hard winter is ahead. Var.: Expect a hard winter if there is a large crop of persimmons.
Cited in Smith, p. 13; Wurtele, p. 299.
2650. When the persimmons are sparse on the tree, a mild winter is on the way.
Cited in Smith, p. 13.

petrel
2651. In coastal areas the sighting of stormy petrel is a sign of bad weather.
Cited in Page, p. 20.
2652. The stormy petrel presages bad weather, and gives sailors notice of the approach of a tempest, by collecting under the sterns of the ships. Vars.: (a) Petrels gathering under the stern of a ship indicate bad weather. (b) Petrels gathering under the stern of a ship indicate foul weather.
Cited in Swainson, p. 243; Dunwoody, p. 39; Inwards, p. 197; Sloane 2, p. 130.

pheasant
2653. If pheasants go up to roost late in the evening, and are up with the sun the next morning, a good day is promised, but if they are early to perch, or late down to feed, the weather will break up.
Cited in Page, p. 18.

phoebe

2654. In winter the call of a phoebe informs you of warmer weather.
Cited in Hyatt, p. 22.

2655. The call of a pewee (phoebe) foretells the nearness of a storm.
Cited in Hyatt, p. 22.

picnic

2656. Never plan a picnic until the very day you want it; in this way
you can keep rain away.
Cited in Hyatt, p. 32.

pig

2657. Before a cold snap, pigs scurry around and make warm nests
with straw or hay.
Recorded in Bryant (Kans.).

2658. Don't rain eve'y time de pig squeal.
Cited in Mieder, p. 498.

2659. If the forward end of a pig's melt is thicker than the other end,
the first part of winter will be the colder. If the latter end is
thicker, the last part of winter will be the colder. Comment: Melt
is the spleen of a slaughtered animal.
Cited in Dunwoody, p. 32.

2660. If the pig wallows in a puddle before the first of May, the
summer will be cold.
Cited in Freier, p. 84.

2661. It will rain when pigs scratch themselves on a post.
Cited in Freier, p. 49.

2662. Pig's tails straighten when rain is near.
Cited in Inwards, p. 183.

2663. Pigs crying and running up and down with hay or litter in their
mouths, foreshow a storm to be near at hand.
Cited in Dunwoody, p. 14.

2664. Pigs uneasy, grunting and huddling together, indicate cold.
Cited in Dunwoody, p. 32.

2665. When in winter pigs rub against the side of their pen, it is a sure
sign of a thaw.
Cited in Dunwoody, p. 32.

2666. When pigs busy themselves gathering leaves and straw to make
a bed (in fall), expect a cold winter.

Cited in Dunwoody, p. 32.

2667. When pigs carry hay or straw in their mouths, there will be rain or wind.
Cited in Lee, p. 69.

2668. When pigs carry sticks, the clouds will play tricks; when they lie in the mud, no fears of a flood. Var.: Pigs carry sticks and straw before rain.
Cited in Inwards, p. 183; Lee, p. 69; Freier, p. 49.

2669. When pigs carry straw to their sties, bad weather may be expected.
Cited in Inwards, p. 183; Wurtele, p. 300.

2670. When pigs carry straws to their sty, a windstorm may be expected. Vars.: (a) Pigs can see wind. (b) Pigs see the wind. Comment: Impending gales may cause pigs to throw straw about with their snouts.
Cited in Dunwoody, p. 86; Wilson, p. 625; Wilshere, p. 22.

2671. When pigs go about with sticks in their mouths, expect a "norther" in Texas.
Cited in Dunwoody, p. 32.

2672. When pigs squeal in winter, there will be a blizzard.
Cited in Lee, p. 69.

pig's bladder

2673. Pig's bladder, when streched, fine; when flaccid, wet.
Cited in Inwards, p. 220.

pigeon

2674. If pigeons fly high, it foretells fair weather; if low, falling weather.
Cited in Hyatt, p. 25.

2675. If pigeons return home slowly, the weather will be wet.
Cited in Swainson, p. 243; Inwards, p. 189.

2676. It is a sign of rain when pigeons return slowly to the dove houses before the ususal time of day. Vars.: (a) If pigeons fly home slowly, it will rain. (b) If pigeons return home slowly, expect rain. (c) Pigeons return home early before rain. (d) Pigeons return home unusually early before rain. (e) Pigeons returning slowly to their loft are an indication of rain.

Cited in Dunwoody, p. 39; Garriott, p. 24; Hand, p. 24; Sloane 2, p. 130; Hyatt, p. 25; Lee, p. 63.

2677. Pigeons stay close to their quarters before rain.
Cited in Freier, p. 188.

2678. Pigeons wash before rain.
Cited in Inwards, p. 189.

2679. There will be a change of weather after pigeons become fitful and coo without ceasing.
Cited in Hyatt, p. 26.

pike

2680. When pike lie on the bed of a stream quietly, expect rain or wind.
Cited in Dunwoody, p. 50.

pill bug

2681. When you see pill bugs moving, it will rain soon. Comment: Pill bug refers to a number of related land crustaceans with a flat body which has the ability to roll itself into a ball.
Cited in Reeder, no pp.

pimpernel

2682. The scarlet pimpernel shut up their flowers before approaching rain. Vars.: (a) Closed is the pinkeyed pimpernel before rain. (b) Oh, pimpernel, whose brilliant flower closes against the approaching shower, warning the swain to sheltering bower from humid air secure. (c) The pimpernel shuts itself up extremely close against rainy weather. (d) When the corona, the Scotch pimpernel, contracts, expect rain.
Cited in Mitchell, p. 226; Steinmetz 1, p. 112; Dunwoody, p. 67; Inwards, p. 215; Lee, p. 52; Freier, p. 39.

2683. When the pimpernel closes in the daytime, it is a sign of rain. Vars.: (a) The closing of the scarlet pimpernel's flowers in the daytime, betokens rain and foul weather; if they be spread abroad, fair weather. (b) When the pink-eyed pimpernel closes in the daytime, it is a sign of rain.
Cited in Swainson, p. 260; Garriott, p. 25; Inwards, p. 215; Hand, p. 25.

2684. When the pimpernel is seen in the morning with its little red flowers widely extended, we may generally expect a fine day; when the petals are closed, rain will soon follow. Var.: The scarlet pimpernel closes before rain and opens wide in fair weather.
Cited in Dunwoody, p. 67; Sloane 2, p. 109.

pine cone

2685. Cones (pine) open for good, dry weather and close for bad.
Cited in Wilshere, p. 23.

2686. Pine cones hanging up in the house where they freely may enjoy the air, will close themselves against wet and cold weather, and open against hot and dry times. Vars.: (a) Pine cones hung up in the house close themselves against wet weather and open when dry. (b) Pine cones hung up in the house will close themselves against wet and cold weather, and open against hot and dry times.
Cited in Swainson, p. 259; Inwards, p. 211; Alstad, p. 51.

2687. Pine cones will close in wet weather and open in dry weather.
Cited in Sloane 2, p. 112.

pipe

2688. Expect rain if one or more of the following things happen while you are smoking: your pipe smelling stronger than usual, wheezing, becoming hot and sticky, and drawing badly. Var.: When pipes smell stronger, it's going to rain.
Cited in Hyatt, p. 31; Freier, p. 29.

2689. Pipes for smoking tobacco become indicative of the state of the air. When the scent is longer retained than usual and seems denser and more powerful, it often forebodes rain and wind. Var.: When the odor of pipes is longer retained than usual, and seems denser and more powerful, it often forebodes rain and wind.
Cited in Dunwoody, p. 116; Inwards, p. 220.

2690. Tobacco (pipe) becomes moist preceding rain. Var.: Tobacco gets moist before a rain.
Cited in Garriott, p. 21; Hand, p. 21; Freier, p. 35.

2691. When a foul pipe is fouler yet, be sure a rain we soon will get.
Cited in Alstad, p. 108.

2692. When the scent of pipes for smoking tobacco is retained longer than usual and seems denser and more powerful it often forebodes a storm. Var.: When the scent of a pipe is longer retained than usual and seems denser and more powerful it often forbodes a storm.
Cited in Garriott, p. 22; Hand, p. 22.

pipe (water)

2693. If water pipes start to sweat, a rain is betokened.
Cited in Hyatt, p. 31.

2694. When water pipes sweat, the weather is going to change.
Recorded in Bryant (Kans.).

piss ant

2695. If piss ants approach your door early in the week, rain will arrive before Sunday.
Cited in Hyatt, p. 17.

pitcher plant

2696. The pitcher plant opens its mouth before rain. Var.: The pitcher plant opens wider before a rain.
Cited in Dunwoody, p. 67; Garriott, p. 25; Hand, p. 25; Lee, p. 73; Freier, p. 39.

plover

2697. Before wet or rough weather, the plover makes a noise with its wings like the bleat of a lamb.
Cited in Inwards, p. 192.

2698. When plovers fly high and then low, making their plaintive cry, expect fine weather.
Cited in Inwards, p. 192.

pond

2699. Ponds often turn over during a storm.
Cited in Sloane 2, p. 50.

pondweed

2700. Pondweed sinks before a storm.
Cited in Lee, p. 72.

2701.   Pondweed sinks before rain.
Cited in Inwards, p. 216; Sloane 2, p. 112.

pony

2702.   Capering and scampering of wild ponies is a sign of rain, as it is
also when they leave their moorland lairs and come down in
droves to low ground.
Cited in Inwards, p. 180.

poppy

2703.   Poppies a sanguine mantle spread, for the blood of the dragon St.
Margaret shed. Comment: St. Margaret's Day is July 20th.
Cited in Swainson, p. 263.

porpoise

2704.   Porpoises are said to swim in the direction from which the wind
is coming.
Cited in Dunwoody, p. 51; Inwards, p. 198.

2705.   Porpoises in harbor indicate coming storm. Var.: Porpoises in a
harbor, expect a storm.
Cited in Dunwoody, p. 49; Inwards, p. 198; Sloane 2, p. 122.

2706.   Porpoises, or Sea Hogs, when observed to sport, and chase one
another about ships, expect then some stormy weather. Vars.: (a)
Porpoises, when they sport about ships and chase one another as
if in play, and indeed their being numerous on the surface of the
sea at any time, is rather a stormy sign. (b) Porpoises, when they
sport about ships and chase one another as if in play, it is a
stormy sign. (c) When porpoise sport and play, there will be a
storm.
Cited in Swainson, p. 249; Dunwoody, p. 50; Garriott, p. 24;
Hand, p. 24; Freier, p. 23.

2707.   Porpoises run into bays and around islands before a storm.
Cited in Dunwoody, p. 51; Inwards, p. 198.

2708.   When porpoises and whales spout about ships at sea, storm may
be expected. Var.: When porpoises and whales spout about ships
at sea, storms are coming.
Cited in Dunwoody, p. 51; Inwards, p. 198; Lee, p. 60.

2709.   When porpoises swim to windward, foul weather will ensue
within twelve hours.

Cited in Swainson, p. 249; Inwards, p. 198.

potato

2710. Plant the potatoes when the moon is dark, and to this line you always hark. Var.: Potatoes should be planted in the dark of the moon to insure a good crop; if planted in the light of the moon, they will be all top.
Cited in Garriott, p. 37. Recorded in Bryant (Nebr.).

prairie chicken

2711. Prairie chickens coming into the creeks and timber indicate cold weather.
Cited in Dunwoody, p. 39; Inwards, p. 190.

2712. When the prairie chicken fly at night to a distance from their usual water, and utter discordant cries during their flight, expect rain.
Cited in Inwards, p. 190.

2713. When the prairie chicken sits on the ground with all its feathers ruffled, expect cold weather.
Cited in Dunwoody, p. 39; Inwards, p. 190.

prairie dog

2714. Prairie dogs bank up their holes with grass and dirt before a storm; if they are playful, it is a sign of fair weather. Vars.: (a) Prairie dogs barricade their holes with grass and dirt before a storm. (b) When prairie dogs are playful, it is a sign of fair weather.
Cited in Dunwoody, p. 33; Lee, p. 67.

ptarmigan

2715. The frequently repeated cry of the ptarmigan low down on the mountains during frost and snow, indicates more snow and continued cold.
Cited in Mitchell, p. 227; Swainson, p. 243; Inwards, p. 190.

Purification Day

2716. When on the Purification sun hath shined, the greater part of winter comes behind. Comment: February 2nd is the feast of the Purification of the Blessed Virgin.

Cited in Swainson, p. 43; Whitman, p. 18; Inwards, p. 42; Freier, p. 74.

pursley

2717. The blooming of pursley (purslane) is an omen of rain.
Cited in Hyatt, p. 15.

Q

quail

2718. If a flock of quail crosses your path, a rain is three days away.
Cited in Hyatt, p. 23.

2719. If quail are hatched as late as September, it foreshows a late winter with mild weather.
Cited in Hyatt, p. 23.

2720. If quail are numerous in the autumn, an open winter may be predicted; if scarce, a hard winter.
Cited in Hyatt, p. 23.

2721. If you hear quail whistle, it is going to rain.
Cited in Hyatt, p. 22.

2722. Quail whistling before two o'clock in the afternoon signify rain.
Cited in Hyatt, p. 22.

2723. Quails are more abundant during an easterly wind.
Cited in Dunwoody, p. 39; Inwards, p. 190.

2724. When quails are heard in the evening, fair weather is indicated for next day. Var.: When quails are heard in the evening, expect fair weather next day.
Cited in Dunwoody, p. 39; Inwards, p. 190.

rabbit

2725. If while on a hunt during the winter, you find rabbits in the open fields, warmer weather is signified; if in brush piles only, colder weather.
Cited in Hyatt, p. 29.

2726. If your dogs chase out of a brush pile a rabbit which circles round and finally returns to the same hiding place, the weather is turning colder.
Cited in Hyatt, p. 29.

2727. In cold long winters rabbits are fat in October and November; in mild and pleasant winters they are poor in those months.
Cited in Dunwoody, p. 33.

2728. Rabbits build very good nests when a hard winter is ahead.
Cited in Smith, p. 12.

2729. Rabbits leave the field and head for the woods before a rain.
Cited in Freier, p. 50.

2730. Rabbits seek the woods before a severe storm. Var.: Rabbits go to the woods before a severe storm.
Cited in Dunwoody, p. 33; Lee, p. 66.

raccoon

2731. If after the first snow you find coon or possum prints, you will not have much snow or cold weather; if you don't find any prints, look for a cold winter.
Cited in Hyatt, p. 29.

2732. The first raccoon tracks show that winter has ended.
Cited in Hyatt, p. 29.

2733. Thin raccoons in autumn indicate a mild winter and fat ones a hard winter.
Cited in Hyatt, p. 29.

rain

2734. A rain making bubbles on the ground shows the weather will soon clear.
Cited in Hyatt, p. 11.

2735. A rain starting at three o'clock in the afternoon will last until three o'clock next afternoon.
Cited in Hyatt, p. 9.

2736. A small rain lays great winds. Vars.: (a) A little rain lays down a great wind. (b) A little rain stills a great wind.
Cited in Wilson, p. 662; Whiting 2, p. 523; Mieder, p. 498.

2737. A small rain may allay a great storm.
Cited in Inwards, p. 157; Mieder, p. 498.

2738. A small rain will lay a great dust. Vars.: (a) A little rain lays a great dust. (b) A small rain lays a great dust. (c) Small rain lays great dust.
Cited in Denham, p. 5; Wilson, p. 662; Mieder, p. 498.

2739. After rain comes fair weather.
Cited in Denham, p. 3; Wilshere, p. 6.

2740. After rain comes heat.
Cited in Marvin, p. 204.

2741. After rain comes sunshine. Vars.: (a) After the rain, the sun. (b) It never rains but it shines. (c) Sunshine always follows rain.
Cited in Inwards, p. 158; Mieder, p. 498.

2742. All who travel in the rain get wet.
Cited in Mieder, p. 498.

2743. Although it rain, throw not away your watering pot.
Cited in Wright, p. 43; Inwards, p. 158; Wilson, p. 12.

2744. April rains for men, May for beasts.
Cited in Inwards, p. 52.

2745. Bright rain makes fools fain. Comment: When a rain-cloud is succeeded by a little brightness in the sky, fools rejoice and think it will soon be fair weather.
Cited in Inwards, p. 158; Wilson, p. 86.

2746. Good signs of rain doesn't help the spring crop.
Recorded in Bryant (Wis.).

2747. Heavy September rain brings drought.
Cited in Whitman, p. 44.

2748. If it rains on the first day of the month, it will rain twenty days of that month.
Recorded in Bryant (Ohio).

2749. If it turns cold before a rain, expect heavy fall of short duration.
Cited in Alstad, p. 183.

2750. If late in the fall or early in the spring it rains for several days and then the sun comes out white, there will be snow before the season ends.
Cited in Hyatt, p. 10.

2751. If rain drops hang on a fence, the rain isn't over.
Recorded in Bryant (Ohio).

2752. If the first day of the month has a rain, the month will have fifteen rainy days.
Cited in Hyatt, p. 10.

2753. If the rain comes before the wind, lower your topsails and take them in; if the wind comes before the rain, lower your topsails and hoist them again. Var.: When the rain's before the wind, you should take your topsails in; when the wind's before the rain, soon you may make sail again.
Cited in Swainson, p. 219.

2754. If the rain makes large bubbles on the ground, it is a sign the rain will be long and heavy. Var.: If the rain makes large bubbles on the ground, it is a sign of good weather coming.
Recorded in Bryant (Nebr., N.Y.).

2755. If there is much rain in winter, the spring is generally dry.
Cited in Whitman, p. 45.

2756. If there were no rain, there'd be no hay to make when the sun shines.
Cited in Mieder, p. 498.

2757. In rainy weather, if the sky is tinged with green, rain will increase; if tinged with blue, it will be showery.
Cited in Alstad, p. 152.

2758. Into each life some rain must fall. Var.: Into every life a little rain must fall.
Cited in Mieder, p. 498.

2759. It does not rain but it pours down. Vars.: (a) Every time it rains, it pours. (b) It never rains but it hails. (c) It never rains but it pours. (d) It never rains but what it pours. (e) When it rains, it pours.
Cited in Denham, p. 2; Taylor 1, p. 111; Taylor 2, p. 303; Wilson, p. 663; Whiting 1, p. 356; Wilshere, p. 6; Simpson, p. 188; Whiting 2, p. 523; Mieder, p. 498.

2760. It is always darkest just before it rains.
Cited in Taylor 2, p. 193.

2761. It is pleasant to look on the rain when one stands dry.
Cited in Mieder, p. 498.

2762. It rains by planets. Var.: It's raining by planets. Comment: A local saying of antiquity referring to rain falling on one field but not on a near or adjoining one.
Cited in Denham, p. 5; Swainson, p. 212; Wilson, p. 663; Wilshere, p. 5.

2763. Its raining pitchforks and shovels.
Cited in Wilshere, p. 5.

2764. Large drops of rain indicate that it will not rain long. Var.:Big drops mean a short rain.
Recorded in Bryant (Nebr., Ohio).

2765. Look for colder weather to follow a rain that becomes thick and heavy.
Cited in Hyatt, p. 11.

2766. Many rains, many rowans; many rowans, many yawns. Comment: Rowans are the fruit of the mountain ash, and an abundance thereof is held to denote a deficient harvest; yawns are light grains of wheat, oats, or barley.
Cited in Denham, p. 54; Northall, p. 475; Inwards, p. 214; Wilson, p. 510.

2767. Marry the rain to the wind and you have a calm.
Cited in Dunwoody, p. 70.

2768. More rain, more grass.
Cited in Mieder, p. 498.

2769. More rain, more rest; fine weather not the best. Vars.: (a) More rain, more rest. (b) Some rain, some rest. (c) Some rain, some rest; fine weather isn't always best.
Cited in Inwards, p. 155; Wilson, p. 662; Lee, p. 99; Mieder, p. 498. Recorded in Bryant (N.Y.).

2770. More rain, more rest; more water will suit the ducks best.
Cited in Northall, p. 465.

2771. Nice day if it don't rain.
Recorded in Bryant (Ohio).

2772. On June 2nd a rain signifies a poor crop of blackberries.
Cited in Hyatt, p. 34.

2773. Preparation for rain scares it away.
Cited in Hyatt, p. 32.

2774. Rain and fine weather when they get lost both come back with an east wind.
Cited in Inwards, p. 118.

2775. Rain brought by a falling barometer lasts the longest.
Cited in Alstad, p. 184.

2776. Rain bubbling on the ground warns you of showers for the next three days.
Cited in Hyatt, p. 11.

2777. Rain long foretold, long last; short notice, soon past.
Cited in Dunwoody, p. 69; Humphreys 2, p. 55; Inwards, p. 156; Sloane 2, p. 51. Recorded in Bryant (Mich.).

2778. Rain never falls while the ground is wet in dry weather.
Cited in Hyatt, p. 13.

2779. Rain never melted anyone.
Cited in Mieder, p. 498.

2780. Rain on Good Friday and Easter Day, you'll have plenty of grass, but little good hay.
Cited in Wilshere, p. 19.

2781. Rain on these three days, the first two days of the month and the last Friday of the preceding month is a portent of a wet month; however, some say this applies only when Friday happens to be the final day of the previous month.
Cited in Hyatt, p. 10.

2782. Rain one day, shine the next.
Cited in Mieder, p. 498.

2783. Rain with south or southwest thunder, squalls occur late each successive day. Var.: Rain with south or southwest thunder brings squalls on successive days.
Cited in Dunwoody, p. 69; Inwards, p. 155.

2784. Rain, rain, go away; come again another day.
Recorded in Bryant (US, Can.).

2785. Rain, rain, go to Spain; fair weather come again.
Cited in Wilson, p. 662.

2786. Rain, rain, pouring; set the bulls a-roaring.
Cited in Inwards, p. 155.

2787. Sudden rains never last long. Vars.: (a) Sudden rains never last long, but when the air grows thick by degrees, and the sun, moon, and stars shine dimmer and dimmer, then it is like to rain six hours usually. (b) The faster the rain, the quicker the hold-up.
Cited in Mitchell, p. 233; Steinmetz 1, p. 116; Inwards, p. 157.

2788. The rain comes scouth when the wind is in the south. Var.: The rain comes scouth when the wind's in the south. Comment: To rain scouth, is to rain abundantly or heavily.
Cited in Inwards, p. 120; Wilson, p. 662.

2789. The rain rains every day upon the just and unjust feller, but mostly on the just, because the unjust has the just's umbrella.
Cited in Inwards, p. 155; Lee, p. 40.

2790. The rain that rains on everybody else can't keep you dry.
Cited in Mieder, p. 498.

2791. The warmer the rain, the longer it lasts.
Cited in Alstad, p. 184.

2792. To rain cats and dogs. Var.: It's raining cats and dogs. Comment: In Norse mythology both cat and dog were attendants of Odin, the Storm God.
Cited in Wilson, p. 662; Wilshere, p. 5.

2793. When chairs and tables creak and crack, it will rain. Var.: Chairs creaking louder than usual signify rain.
Cited in Inwards, p. 218; Hyatt, p. 30.

2794. When rain causes bubbles to rise in water it falls upon, the shower will last long.
Cited in Wright, p. 79; Inwards, p. 158.

2795. When rain comes before wind, halyards, sheets, and braces mind. Vars.: (a) When rain comes before wind, halyards, sheets, and braces mind; when wind comes before rain, soon you may make sail again. (b) When the rain's before the wind, 'tis time to take the topsails in, but, when the wind's before the rain, let your topsails out again. (c) When the rain's before the wind, you should take your topsails in; when the wind's before the rain, soon you may make sail again. (d) With the rain before the wind, your topsail halyards you must mind.
Cited in Steinmetz 1, p. 80; Swainson, p. 219; Cheales, p. 27; Dunwoody, p. 70; Whiting 2, p. 523. Recorded in Bryant (Md., Mich.).

2796. When rain ends, the clearing wind is usually from the west.
Cited in Sloane 1, p. 32.

2797. When the rain comes before the winds, you may reef when it begins; but when the wind comes before the rain, you may hoist your topsails up again.
Cited in Swainson, p. 219; Northall, p. 464.

2798. Years ago steamboat men on the Mississippi River used to say a rain going upstream (south to north) would be back again (north to south) within three days.
Cited in Hyatt, p. 9.

rain (afternoon)
2799. If it rains before four, it will then rain some more.
Cited in Smith, p. 8.
2800. You can tell by two what its going to do (rain?).
Cited in Whitman, p. 29.

rain (August)
2801. If it rains in August, it rains honey and wine.
Cited in Garriott, p. 45; Freier, p. 83.

rain (east)
2802. East rain continues for three days. Vars.: (a) Rain from the east lasts three days at least. (b) Rain from the east will last three days at least.
Cited in Hyatt, p. 9; Page, p. 7; Wilshere, p. 4.
2803. If rain falls during an east wind, it will continue a full day.
Cited in Dunwoody, p. 43.
2804. Rain from east, two days at least. Var.: Rain from the east, wet two days at least.
Cited in Cheales, p. 24; Northall, p. 465; Inwards, p. 155; Wilson, p. 662.
2805. The heaviest rains begin with an easterly wind, which gradually veers round to south and west, or a little northwest, when the rain usually ceases.
Cited in Inwards, p. 118.
2806. When the rain is from the east, it is for four and twenty hours at least. Vars.: (a) Rain from the east, twenty-four hours at least. (b) When it rains with the wind in the east, it rains for twenty-four hours at least. (c) When the rain comes from the east, it will rain for twenty-four hours at least.
Cited in Swainson, p. 227; Northall, p. 465; Whitman, p. 29; Brunt, p. 72; Inwards, p. 118.

rain (Easter)

2807. If it rains on Easter, it will rain seven Sundays after. Var.: Rain on Easter means its going to rain for seven Sundays in a row.
Recorded in Bryant (Nebr., Wis.).

rain (February)

2808. Rain in February is as good as manure.
Cited in Wilshere, p. 8.

rain (Friday)

2809. If rain falls on Friday, there will be no rain before next Friday.
Cited in Hyatt, p. 10.

rain (July)

2810. A shower of rain in July, when the corn begins to fill, is worth a plough of oxen, and all belongs theretill.
Cited in Wright, p. 89.

rain (June 27)

2811. If it rains on June 27th, it will rain seven weeks.
Cited in Dunwoody, p. 91.

rain (March)

2812. As it rains in March, so it rains in June.
Cited in Garriott, p. 44.

rain (May)

2813. Rain in May makes the hay.
Cited in Mieder, p. 498. Recorded in Bryant (Wis.).

rain (midday)

2814. If rain begins about noon it will continue through the afternoon.
Cited in Marvin, p. 208.
2815. If rain does not cease before noon it will continue till evening.
Cited in Marvin, p. 208.
2816. If the rain waits until noon to visit, prepare for a long visit.
Cited in Sloane 2, p. 51.

rain (midnight)

2817. If it rain at midnight with a south wind, it will generally last above twelve hours.
Cited in Inwards, p. 157.

2818. If rain ceases after midnight it will rain the next day.
Cited in Marvin, p. 208.

2819. If rain ceases before midnight it will be clear the next day.
Cited in Marvin, p. 208.

2820. Midnight rains make drowned fens. Var.: Night rains make drowned fens. Comment: A fen is a marsh.
Cited in Swainson, p. 212.

rain (Monday)

2821. If it rains on Monday, it will rain three days that week. Vars.: (a) On Monday a rain signifies three rainy days before the end of the week. (b) Monday rain never stops until it has rained for three days. (c) Rain on Monday, rain three days of the week. (d) Rain on Monday, rain two more days that week.
Cited in Hyatt, p. 9.

2822. If it rains on Monday, rain all week. Var.: Rain on Monday; rain everyday that week.
Cited in Hyatt, p. 10. Recorded in Bryant (Ill., Ind., Miss., N.Y.).

2823. Rain on Monday, sunshine next Sunday.
Cited in Hyatt, p. 10.

2824. Rain on the first Monday of the month presages three Monday rains for the month.
Cited in Hyatt, p. 10.

rain (morning)

2825. A fall of small drizzling rain, especially in the morning, is a sure sign of wind to follow.
Cited in Inwards, p. 157.

2826. An early morning rain stops before noon.
Cited in Hyatt, p. 9.

2827. For morning rain leave not your journey.
Cited in Wright, p. 96.

2828. If a morning rain ends before noon, the following day will be clear; if it continues until afternoon, the following day will be wet.
Recorded in Bryant (Mich.).

2829. If it rains before seven, it will clear before eleven. Vars.: (a) If it rains before seven, if will cease before eleven. (b) Rain before seven, cease before eleven. (c) Rain before seven, clear before eleven. (d) Rain before seven, clear by eleven; between one and two, we'll see what 'twill do. (e) Rain before seven, clear off before eleven. (f) Rain before seven, fine before eleven. (g) Rain before seven, lift before eleven. (h) Rain before seven, quit before eleven. (i) Rain before seven, shine before eleven.
Cited in Cheales, p. 24; Dunwoody, p. 70; Northall, p. 465; Marvin, p. 203; Taylor 1, p. 118; Whitman, p. 29; Inwards, p. 156; Sloane 2, p. 115; Hyatt, p. 9; Smith, p. 8; Wilson, p. 662; Wurtele, p. 296; Page, p. 5; Wilshere, p. 4; Simpson, p. 188. Freier, p. 66; Whiting 2, p. 523; Arora, p. 8; Mieder, p. 498.

2830. If it rains between eight and nine o'clock in the morning it will rain till noon.
Cited in Marvin, p. 208.

2831. If rain begins at early morning light, 'twill end ere day at noon is bright.
Cited in Dunwoody, p. 69; Marvin, p. 203.

2832. Monday mornings' rain is like the old woman's song, it doesn't last long.
Recorded in Bryant (N.C.).

2833. Morn wet, no rain yet, morn dry, rain nigh.
Cited in Alstad, p. 117.

2834. Morning rain is like the old lady's dance; it doesn't last very long. Vars.: (a) A morning's rain is like an old woman's dance: it doesn't last long. (b) Early morning rain and an old woman's dance are soon over.
Cited in Sloane 2, p. 51; Hyatt, p. 9; Mieder, p. 498.

2835. Rain at seven, fine at eleven; rain at eight, not fine till eight.
Cited in Inwards, p. 75.

2836. Rain in the morning, clear in the afternoon.
Cited in Smith, p. 8.

2837. Rain in the morning, sailors take warning.
Cited in Hyatt, p. 9.

rain (night)

2838. If rain begins about five o'clock in the evening it will rain all night.
Cited in Marvin, p. 208.

2839. If rain begins after nine o'clock in the evening it will rain the next day.
Cited in Marvin, p. 208.

rain (September)

2840. Heavy September rains bring drought.
Cited in Garriott, p. 45.

rain (south)

2841. Rain from the south prevents the drought, but rain from the west is always best.
Cited in Dunwoody, p. 69; Inwards, p. 155.

2842. Rain which sets in with a south wind on the north Pacific coast will probably last.
Cited in Dunwoody, p. 69; Inwards, p. 155.

2843. Rain, with a southeast wind, is expected to last for some time.
Cited in Mitchell, p. 229; Inwards, p. 121.

rain (summer)

2844. A light rain in the summer means a sign of coming draught.
Cited in Smith, p. 11.

2845. Happy are the fields that receive summer rain.
Cited in Dunwoody, p. 91; Inwards, p. 32.

2846. Midsummer rain spoils hay and grain.
Cited in Swainson, p. 107.

2847. Midsummer rain spoils wine, stock, and grain.
Cited in Dunwoody, p. 91; Garriott, p. 46; Whitman, p. 45; Hand, p. 46.

2848. There can never be too much rain before midsummer.
Cited in Garriott, p. 45; Whitman, p. 45; Hand, p. 45; Freier, p. 79.

rain (Sunday)

2849. Every day of the week a shower of rain, and on Sunday twain.
Comment: Twain is an archaic word for two.

Cited in Wilson, p. 228.

2850. First Sunday in month rain, it will rain every Sunday of the month. Vars.: (a) If it is raining on the first Sunday of the month, it is going to rain every other Sunday that month. (b) If it is raining on the first Sunday of the month, it is going to rain every Sunday that month.
Cited in Dunwoody, p. 101; Hyatt, p. 10.

2851. If it rains all week it will clear off before Sunday.
Cited in Smith, p. 7.

2852. If it rains on Sunday before mass, it will rain all week.
Cited in Denham, p. 11; Dunwoody, p. 101; Whitman, p. 40; Inwards, p. 73.

2853. If it rains on Sunday, it will rain four days that week.
Recorded in Bryant (Ohio).

2854. Rain before church, rain all the week, little or much. Var.: Rain afore church, rain all week, little or much.
Cited in Marvin, p. 215; Inwards, p. 156.

2855. Rain on Sunday, shine on Monday.
Recorded in Bryant (Ind.).

2856. Sunday rain is a sign of rain for seven consecutive Sundays.
Cited in Hyatt, p. 10.

rain (sunrise)

2857. If it begins to rain an hour or two before sunrise it is likely to be fair before noon, and so continue that day; but if the rain begin an hour or two after sunrise, it is likely to rain all that day. Var.: If it rains before sunrise, expect a fair afternoon.
Cited in Mitchell, p. 233.

2858. If it rains before daybreak it will cease before eight o'clock in the morning.
Cited in Marvin, p. 208.

2859. If it rains before the sun shines it will rain the next day.
Cited in Marvin, p. 208.

2860. If rain begins an hour before daybreak it will probably rain all day.
Cited in Marvin, p. 208.

2861. If rain commences before day, it will stop before 8:00 a.m.; if it begins about noon, it will continue through the afternoon; if not till 5:00 p.m., it will rain through the night; if it clears off in

the night, it will rain the next day. Var.: If rain commences before daylight, it will hold up before 8:00 a.m.; if it begins before noon, it will continue through the afternoon; if it commences after 9:00 p.m., it will rain the next day; if it clears off in the night, it will rain the next day.
Cited in Dunwoody, p. 69.

2862. Rain a short time before sunrise will be followed at least by a fine afternoon; but rain soon after sunrise, generally by a wet day.
Cited in Inwards, p. 157.

rain (sunshine)

2863. If it rains when the sun shines, it will surely rain the next day about the same hour. Vars.: (a) At whatever time it rains while the sun is shining, it will rain at the same time next day. (b) If it rains during sunshine, it will rain the same time the next day. (c) If it rains when the sun shines, it will rain the next day. (d) If it rains while the sun is shining, it'll rain the same time tomorrow. (e) Rain in sunshine, it will rain tomorrow. (f) Rain with sunshine means rain tomorrow. (g) Sunshine and rain, rain tomorrow. (h) Sunshine and shower, rain again tomorrow. (i) Sunshiny rain will soon go again. (j) Sunshiny showers, will rain tomorrow.
Cited in Swainson, p. 213; Dunwoody, p. 70; Northall, p. 464; Inwards, p. 157; Hyatt, p. 11; Smith, p. 6. Recorded in Bryant (Ky., Miss., Ohio, Tex. Can. [Ont.]).

2864. If it rains while the sun is shining, the devil is beating his grandmother. He is laughing, and she is crying. Vars.: (a) If it rains when the sun is shining, the devil is beating his wife. (b) If it rains while the sun is shining, the devil is whipping his wife and it will rain again the next day. (c) If it rains while the sun shines, the devil is beating his wife. (d) If it rains while the sun shines, the devil is beating his wife with a frying pan. (e) If the sun shines while it is raining, the devil is beating his wife, and if at this time you stick a pin into the earth, and hold your ear down to the ground, you will hear him beating her. (f) When it rains and the sun is shining, the devil is whipping his children. (g) When it rains and the sun is shining, the devil is whipping his wife. (h) When it rains while the sun shines, the devil is beating

his wife, and if you put an old rusty nail in the ground and place your ear to it, you can hear her holler.
Cited in Inwards, p. 157; Halpert, p. 105; Smith, p. 9; Wilson, p. 663; Wurtele, p. 301; Whiting 2, p. 524. Recorded in Bryant (Miss.).

2865. If the sun shines during rain, the fairies are baking. Vars.: (a) If the sun shines during rain, a sailor is going to heaven. (b) If the sun shines during rain, God is pissing. (c) If the sun shines during rain, it's a whore's wedding day. (d) If the sun shines during rain, the witches are dancing.
Cited in Wurtele, p. 301.

2866. When it is raining and the sun is shining, look under a rock; if hair is found under it, it will be the color of the hair of the person you marry.
Cited in Smith, p. 14.

rain (west)

2867. When rain comes from the west it will not continue long. Var.: When rain comes from the west, it will not last long.
Cited in Dunwoody, p. 70; Inwards, p. 155.

rain crow

2868. If a rain crow calls late in the evening, next day will be rainy.
Cited in Hyatt, p. 22.

2869. If the rain crow does not crow, it's not going to rain.
Cited in Mieder, p. 499. Recorded in Bryant (N.Y.).

2870. The calling of a rain crow (the black-billed cuckoo) is followed by a storm.
Cited in Hyatt, p. 22.

2871. When the rain crow calls, it is a sign of rain.
Recorded in Bryant (Nebr.).

2872. When you hear a rain crow calling, it will rain within three days.
Cited in Reeder, no pp.

rain lily

2873. When the rain lily (wild onion) blooms, it will rain within three days.
Cited in Reeder, no pp.

rainbow

2874. A morning rainbow indicates rain; an evening rainbow, fair weather. Vars.: (a) A rainbow at night, fair weather in sight; a rainbow at morn, fair weather all gorn. (b) A rainbow in the evening is a sign of good weather. (c) Rainbow at night, fair weather in sight; rainbow at morn, fair weather, 'tis gorn.
Cited in Dunwoody, p. 71; Smith, p. 3; Page, p. 30; Wilshere, p. 3.

2875. A rainbow anytime during the day is a boding of rain next day.
Cited in Hyatt, p. 5.

2876. A rainbow in spring indicates fair weather for next twenty-four to forty-two hours.
Cited in Dunwoody, p. 71.

2877. A rainbow in the east will be followed by a fine morrow, in the west by a wet day.
Cited in Sloane 2, p. 119.

2878. A rainbow in the morn, put your hook in the corn; a rainbow at eve, put your head in the sheave. Var.: A rainbow at morn, put your hook in the corn; a rainbow at eve, put your head in the sheave.
Cited in Dunwoody, p. 71; Marvin, p. 205; Wilson, p. 662; Lee, p. 82.

2879. A rainbow in the morning is the shepherd's warning; a rainbow at night is the shepherd's delight. Vars.: (a) A rainbow in the morning is a sailor's warning; a rainbow at night is a sailor's delight. (b) A rainbow in the morning is a shepherd's warning; a rainbow in the evening is a shepherd's grieving. (c) Rainbow at night sailor's delight; rainbow at morning sailors take warning. (d) Rainbow at night shepherd's delight; rainbow at morning shepherd take warning. (e) Rainbow in the morning, farmer take warning. (f) The rainbow in the morning is the shepherd's warning to carry his coat on his back; the rainbow at night is the shepherd's delight, for then no coat will he lack.
Cited in Denham, p. 9; Mitchell, p. 223; Steinmetz 1, p. 118; Steinmetz 2, p. 285; Dunwoody, p. 71; Northall, p. 466; Garriott, p. 22; Marvin, p. 206; Humphreys 2, p. 33; Taylor 1, p. 112; Whitman, p. 31; , no pp.; Brunt, p. 74; Inwards, p. 160; Hand, p. 22; Hyatt, p. 5; Wilson, p. 662; Lee, p. 82;

Whiting 2, p. 524; Arora, p. 5; Mieder, p. 499. Recorded in Bryant (Ill., La., Miss., Wis.).

2880. A rainbow is a sure sign of a shower.
Cited in Smith, p. 7.

2881. A rainbow is big enough for everyone to look at.
Cited in Mieder, p. 499.

2882. A rainbow seen at a great distance indicates fair weather.
Cited in Dunwoody, p. 71; Inwards, p. 161.

2883. A rainbow that comes near a campfire, or low down on the mountain side, it is a bad sign for crops. Var.: A bow low down on the mountains is a bad sign for the crops.
Cited in Dunwoody, p. 71; Inwards, p. 161.

2884. A Saturday's rainbow, a week's rotten weather.
Cited in Wright, p. 79.

2885. A weather gall (rainbow) at morn, fine weather all gone; a rainbow towards night, fair weather in sight.
Cited in Marvin, p. 206; Inwards, p. 160; Lee, p. 82.

2886. After a long drought the rainbow is a sign of rain. Var.: The rainbow, after a long drought, is the precursor of a decided change to wet weather; and it happens also that a perfect bow, after an unsettled time, is a precursor of fair weather.
Cited in Steinmetz 1, p. 117; Dunwoody, p. 70; Inwards, p. 162.

2887. After much wet, the rainbow indicates fair weather. Var.: After much wet weather, a rainbow indicates fair weather.
Cited in Steinmetz 1, p. 117; Dunwoody, p. 70.

2888. An all red rainbow is not infrequently seen at about the time of sunset when the raindrops in a shower to eastward are fairly large.
Cited in Inwards, p. 161.

2889. If a person sees a double rainbow, it will rain three days during the following week.
Cited in Hyatt, p. 5.

2890. If a rainbow appear in fair weather, foul will follow; but if a rainbow appear in foul weather, fair will follow.
Cited in Inwards, p. 159.

2891. If a rainbow breaks up all at once, there will follow serene and settled weather.
Cited in Steinmetz 1, p. 117.

2892. If a rainbow breaks up all at once, there will follow severe and unsettled weather.
Cited in Dunwoody, p. 71.

2893. If a rainbow disappears suddenly, it indicates fair weather.
Cited in Dunwoody, p. 71.

2894. If a rainbow is broken in two or three places, expect rainy weather for two or three days.
Cited in Inwards, p. 159.

2895. If it rains during a rainbow, it will rain again the next day.
Cited in Smith, p. 7.

2896. If rainbow appears in the west, the rain will soon resume.
Cited in Hyatt, p. 5.

2897. If the green be large and bright in the rainbow, it is a sign of rain; if red be the strongest color, there will be rain and wind together. Var.: The greener the rainbow the more rain; the redder, wind.
Cited in Dunwoody, p. 70; Inwards, p. 161.

2898. If the rainbow be seen in the morning, small rain will follow; if at noon, settled and heavy rains; if at night, fair weather. Var.: If the rainbow be in the morning, rain will follow; if at noon, slight and heavy rain; if at night, fair weather.
Cited in Steinmetz 1, p. 117; Dunwoody, p. 71.

2899. If the rainbow comes at night, the rain has gone quite.
Cited in Marvin, p. 205; Inwards, p. 160.

2900. If the rainbow forms and disappears suddenly, the prismatic colors being but slightly discernible, expect fair weather next day.
Cited in Inwards, p. 162.

2901. If there be a rainbow in the eve, it will rain and leave; but if there be a rainbow in the morrow, it will neither lend nor borrow. Var.: If there's a rainbow at eve, it will rain and leave.
Cited in Denham, p. 10; Swainson, p. 195; Northall, p. 466; Marvin, p. 206; Humphreys 2, p. 33; Whitman, p. 31; Inwards, p. 160; Freier, p. 48.

2902. If two rainbows appear at one time, they presage rain to come.
Cited in Wilson, p. 663.

2903. Rainbow at noon, more rain soon. Var.: Rainbow at noon, rain soon.
Cited in Hyatt, p. 5; Lee, p. 82.

2904.   Rainbow in the east indicates that the following day will be clear;
        a rainbow in the west is usually followed by more rain the same
        day. Vars.: (a) If a rainbow appears in the east, the weather will
        be dry according to some, wet according to others. (b) Rainbow
        in the eastern sky, the morrow will be dry; rainbow in the west
        that gleams, rain falls in streams.
        Cited in Dunwoody, p. 71; Hyatt, p. 5; Lee, p. 82.

2905.   Rainbow in the morning, farmer take warning.
        Cited in Hyatt, p. 5.

2906.   Rainbow in the morning, gives you fair warning.
        Cited in Smith, p. 7.

2907.   Rainbow in the morning shows that the shower is west of us, and
        that we will probably get it; rainbow in the evening shows that
        shower is east of us and is passing off.
        Cited in Dunwoody, p. 71; Inwards, p. 159; Hand, p. 22.

2908.   Rainbow in the Sierras in evening indicates no more rain.
        Cited in Dunwoody, p. 71.

2909.   Rainbow to windward, foul falls the day; rainbow to leeward,
        damp runs away.
        Cited in Wright, p. 80; Marvin, p. 206; Humphreys 2, p. 35;
        Whitman, p. 31; Inwards, p. 159; Sloane 2, p. 119; Lee, p. 81;
        Wilshere, p. 3; Freier, p. 49.

2910.   Rainbows with the new moon, rain until the end.
        Cited in Marvin, p. 206.

2911.   Seven rainbows, eight days' rain.
        Cited in Inwards, p. 162.

2912.   The appearance of two or three rainbows indicates fair weather
        for the present, but settled and heavy rains in two or three days'
        time. Var.: The appearance of two or three rainbows indicates
        fair weather for the present, but settled and heavy rains in a few
        days.
        Cited in Steinmetz 1, p. 117; Dunwoody, p. 71.

2913.   The boding shepherd heaves a sigh, for see, a rainbow spans the
        sky. Var.: The boding shepherd heaves a sigh, for see, a rainbow
        in the sky.
        Cited in Dunwoody, p. 71; Marvin, p. 206.

2914.   The fragment of a rainbow in the north, which is called the
        "Boar's Head," makes the boatmen keep a look-out for wind.
        Cited in Mitchell, p. 224.

2915. The predominance of dark red in the iris of the rainbow shows tempestuous weather; green, rain; and if blue, that the air is clearing.
Cited in Dunwoody, p. 71.

2916. The rainbow has but a bad character: she ever commands the rain to cease.
Cited in Dunwoody, p. 70; Marvin, p. 206.

2917. The weather's taking up now for yonder's the weather gaw; how bonny is the east now! Now the colors fade away. Comment: The weather gaw is a fragment of a rainbow.
Cited in Marvin, p. 206.

2918. Three rainy days for the week are foretold by a rainbow that appears during a rain.
Cited in Hyatt, p. 11.

2919. When a perfect rainbow shows only two principal colors, which are generally red and yellow, expect fair weather for several days.
Cited in Inwards, p. 161.

2920. When a rainbow appears in the wind's eye, rain is sure to follow.
Cited in Inwards, p. 159; Sloane 2, p. 119.

2921. When a rainbow is formed in an approaching cloud, expect a shower; but when in a receding cloud, fine weather.
Cited in Inwards, p. 159.

2922. When the rainbow does not touch water, clear weather will follow. Var.: When the rainbow does not reach down to the water, clear weather will follow.
Cited in Dunwoody, p. 71; Marvin, p. 206; Inwards, p. 161.

2923. When you see a rainbow before noon, that is the sign of rain soon. Var.: You may look for rain soon, if there's a rainbow before noon.
Cited in Hyatt, p. 5.

raindrop

2924. If during a hard rain the drops are large, three successive days of rain may be expected.
Cited in Hyatt, p. 11.

2925. If raindrops are large (the size of a quarter according to some), it will soon stop raining.
Cited in Hyatt, p. 11.

2926. Large raindrops betoken dry weather.
Cited in Hyatt, p. 11.

2927. The rain does not continue long, during which the drops adhere to any of the following: bushes, clotheslines, wires, windows, and screens of doors and windows.
Cited in Hyatt, p. 11.

2928. When the raindrops cause a lot of bubbles in puddles, it will continue to rain.
Cited in Reeder, no pp.

ram

2929. When grave rams and lambkins full of play butt at each others heads in mimic fray, rain.
Cited in Whitman, p. 47.

raspberry

2930. If raspberries bloom twice in one year and some of the blossoms produce fruit, it is a sign of a very mild winter.
Cited in Hyatt, p. 15.

rat

2931. If rats are more restless than usual, rain is at hand. Var.: If rats and mice are more restless, expect rain.
Cited in Swainson, p. 233; Sloane 2, p. 121.

2932. Much noise made by rats and mice indicates rain.
Cited in Dunwoody, p. 33.

2933. Rats and mice "bawling" foretell wind.
Cited in Lee, p. 67.

2934. Rats leave a ship before a storm. Var.: Rats desert a sinking ship.
Cited in Lee, p. 67; Mieder, p. 499.

2935. Rats seek protection from the wind on the eve of a storm.
Cited in Lee, p. 67.

2936. When rats move shavings away from the windward side of a house, it is going to storm.
Cited in Lee, p. 67.

raven

2937. If ravens croak three or four times and flap their wings, fine weather is expected.
Cited in Swainson, p. 244; Inwards, p. 193.

2938. If the raven crows, expect rain.
Cited in Lee, p. 61; Freier, p. 22.

2939. Ravens, when they croak continuously, denote wind; but if the croaking is interrupted or stifled, or at longer intervals, they show rain.
Cited in Inwards, p. 193.

redbird

2940. A redbird singing early in the spring forecasts cold weather.
Cited in Hyatt, p. 23.

2941. Early spring is presaged by a redbird that sings in January.
Cited in Hyatt, p. 23.

2942. If a redbird flies high, it betokens good weather; if low, bad weather approaches; if to the left, warmer weather.
Cited in Hyatt, p. 23.

2943. If a redbird near your house calls incessantly, it is calling for rain.
Cited in Hyatt, p. 23.

2944. If at the beginning of each flight a redbird flies up, clear weather is indicated, if it flies down, wet weather.
Cited in Hyatt, p. 23.

2945. If while watching a redbird it flies away to the right, colder weather approaches; if to the left, warmer weather.
Cited in Hyatt, p. 23.

2946. If you see a redbird, it is a sign of snow.
Cited in Smith, p. 9.

2947. Redbirds seen in winter denote a cold spell or a blizzard.
Cited in Hyatt, p. 23.

2948. The call of the redbird before fair weather is "pretty-pretty."
Cited in Hyatt, p. 23.

2949. The call of the redbird before rainy weather is variously given: "squirt-squirt," "wet-wet," and "wet weather-wet weather." Vars.: (a) Watch for a rainy summer and autumn following a spring in which the redbird calls "wet year-wet year." (b) When

a redbird calls "wet-wet," it will rain.
Cited in Hyatt, p. 23; Smith, p. 4.

reptile
2950. When reptiles are a browner color than usual, expect rain.
Cited in Inwards, p. 203.

rheumatism
2951. Rheumatic pain indicates bad weather. Var.: If rheumatism acts up, bad weather is in store.
Cited in Dunwoody, p. 116; Smith, p. 1.
2952. When rheumatic people complain of more than ordinary pains in the joints, it will rain. Vars.: (a) When an elderly person with rheumatism has aching knees and shoulder joints, rain is on the way. (b) When rheumatic people complain of more than ordinary pains it will probably rain.
Cited in Dunwoody, p. 14; Garriott, p. 21; Inwards, p. 217; Hand, p. 21; Smith, p. 7; Wilshere, p. 23.

river
2953. A rising stage in the river during November denotes a high stage all winter.
Cited in Hyatt, p. 13.
2954. If the river piles driftwood on its banks in March, the river will unpile it in June.
Cited in Hyatt, p. 13.
2955. Sediment floating near the surface of a river or stream foretells an immediate rise in water.
Cited in Hyatt, p. 13.
2956. When the Mississippi River breaks up in the spring, high water will rise to the top of the ice jams.
Cited in Hyatt, p. 13.

road
2957. When the roads become suddenly dry by wind after heavy rain, expect more within twenty-four hours.
Cited in Inwards, p. 220.

robin

2958. After the robin comes in spring, he'll get snow on his back three times before it stops.
Cited in Freier, p. 84.

2959. Don't cut hay when the robin's in the bush.
Cited in Freier, p. 54.

2960. First robins indicate the approach of spring. Vars.: (a) A robin is a sign of spring. (b) Robins indicate the approach of spring. (c) Spring arrives with the first robin.
Cited in Dunwoody, p. 39; Garriott, p. 40; Inwards, p. 194; Hand, p. 40; Hyatt, p. 23; Smith, p. 11.

2961. If a robin enters a house, it is prognostic of snow or frost.
Cited in Inwards, p. 194.

2962. If a robin rests on the ground with one wing spread out, expect rain within twenty-four hours.
Cited in Hyatt, p. 23.

2963. If a robin sings on a high branch of a tree, it is a sign of fine weather; but if one sings near the ground, the weather will be wet. Var.: When the robin sings on rooftops and at the top of trees, it is a good sign, but if it gets down lower, then be prepared for rain.
Cited in Inwards, p. 194; Page, p. 18.

2964. If robins are seen near houses it is a sign of rain.
Cited in Swainson, p. 244; Marvin, p. 209; Inwards, p. 194.

2965. If robins enter the barn, heavy rains can be expected.
Cited in Lee, p. 65.

2966. If the robin puts his nest near the trunk of a tree, there will be a wet spring.
Cited in Boughton, p. 124.

2967. If the robin sings in the bush, then the weather will be coarse; but if the robin sings on the barn, then the weather will be warm.
Cited in Northall, p. 473; Marvin, p. 209; Lee, p. 47.

2968. Long and loud singing of robins in the morning denotes rain. Vars.: (a) If robins sing loud, and robins sing long, it's a sign of rain. (b) Loud and long singing of robins denotes rain.
Cited in Dunwoody, p. 39; Garriott, p. 24; Inwards, p. 194; Hand, p. 24; Freier, p. 52.

2969. One robin doesn't make a spring. Var.: It takes more than a robin to make spring.

Cited in Mieder, p. 514.

2970. Rain is on its way when robins do one of three things: hop along the ground, sing on the ground, and fly close to home.
Cited in Hyatt, p. 23.

2971. Robins grow bolder and perch against the window in advance of unusually severe weather.
Cited in Dunwoody, p. 39.

2972. Robins in the bush, rain is coming.
Cited in Freier, p. 54.

2973. Robins will perch on the topmost branches of trees and whistle when a storm is approaching. Var.: If the robin sings loudly from the topmost of trees, expect a storm.
Cited in Dunwoody, p. 39; Garriott, p. 24; Inwards, p. 194; Hand, p. 24; Freier, p. 51.

2974. When robins build their nests low, there will be a windy summer.
Cited in Boughton, p. 124.

2975. When the robin sings and chirps, it is going to become hot.
Cited in Smith, p. 11.

2976. When the robin yells "three days," rain is coming.
Cited in Boughton, p. 123.

rock

2977. Rocks sweat before rain.
Cited in Dunwoody, p. 116.

2978. When sand rocks appear dull in the distance, expect rain.
Cited in Inwards, p. 219.

rook

2979. If rooks feed busily, and hurry over the ground in one direction, and in a compact body, a storm will soon follow.
Cited in Inwards, p. 190.

2980. If rooks feed in the streets of a village, it shows that a storm is near at hand. Var.: Rooks building their nests high means good weather, but when they feed in the streets, storms are imminent.
Cited in Inwards, p. 191; Wilshere, p. 22.

2981. If rooks fly to the mountains in dry weather, rain is near.
Cited in Inwards, p. 191.

2982. If rooks go far abroad it will be fine. Var.: If rooks stay at home, or return in the middle of the day, it will rain; if they go far abroad, it will be fine.
Cited in Dunwoody, p. 20; Inwards, p. 190.

2983. In autumn and winter, if rooks, after feeding in the morning, return to the rookery, and hang about it, rain is to be expected.
Cited in Inwards, p. 190.

2984. Rooks dart and swoop through the air before rain.
Cited in Dunwoody, p. 39.

2985. Rooks will not leave their nests in the morning before a storm.
Cited in Inwards, p. 191.

2986. The low flight of rooks indicates rain.
Cited in Garriott, p. 18; Inwards, p. 190; Freier, p. 21.

2987. When birds of long flight, rooks, swallows, and others, hang about the homestead, and fly up and down low, then rain and wind may be expected.
Cited in Steinmetz 1, p. 110.

2988. When rooks congregate on the dead branches of trees, there will be rain before night; if they stand on the live branches, the day will be fine.
Cited in Inwards, p. 191.

2989. When rooks fly from their nests and fly straight, umbrellas and raincoats can be left at home all day. Should they twist and turn on leaving their nests, rough weather is approaching.
Cited in Page, p. 21.

2990. When rooks fly high, the next day will be fair.
Cited in Lee, p. 62.

2991. When rooks fly sporting high in air, it shows that windy storms are near. Var.: If rooks when flying high, dart down and wheel about in circles, wind is foreshown.
Cited in Swainson, p. 245; Northall, p. 473; Inwards, p. 190.

2992. When rooks go home to roost, if they fly high, the next day will be fair; if low, it will be wet.
Cited in Inwards, p. 190.

2993. When rooks seem to drop in their flight, as if pierced by a shot, it is considered to foretell rain. Vars.: (a) If rooks fly irregularly and high and seem to fall, expect rain. (b) When rooks seem to drop in their flight, as if pierced by a shot, it is said to foreshow

rain.
Cited in Swainson, p. 244; Dunwoody, p. 39; Inwards, p. 190.
2994. When rooks sit in rows on dikes and palings, wind is to come.
Var.: When rooks sit in rows on dikes and fences, wind is coming.
Cited in Inwards, p. 190; Lee, p. 62.

rooster

2995. A crowing rooster during rain indicates fair weather. Vars.: (a) A rain during which a rooster crows never lasts long. (b) After a rooster crows on a rainy morning, a fair afternoon can be expected.
Cited in Dunwoody, p. 39; Hyatt, p. 25.
2996. A rooster crowing anytime in the middle of the day indicates a change of weather.
Cited in Hyatt, p. 24.
2997. A rooster crowing before sunset tells you the weather tomorrow will be the same.
Cited in Hyatt, p. 25.
2998. A rooster crowing between roosting-time and midnight presages rain between seven and eight, around eight, near nine, about ten, and at midnight.
Cited in Hyatt, p. 24.
2999. If a rooster anytime during the day jumps up on a fence or gatepost and crows, a rain is indicated.
Cited in Hyatt, p. 25.
3000. If a rooster crows early in the morning while sitting upon a fence, it will rain before breakfast.
Cited in Hyatt, p. 25.
3001. If a rooster crows in the middle of the day, it will rain.
Recorded in Bryant (Ind.).
3002. If a rooster crows in the morning, it is a sailor's warning; if he crows at night, it is a sailor's delight.
Cited in Hyatt, p. 25.
3003. If a rooster crows on a rainy night, look for good weather next day; if on a clear night, wet weather.
Cited in Hyatt, p. 25.
3004. If a rooster crows on the ground, it is a sign of rain; if he crows on the fence, it is a sign of fair weather. Var.: When the rooster

crows on the ground, the rain will fall down; when he crows on the fence, the rain will depart hence.
Cited in Dunwoody, p. 39; Lee, p. 62.

3005. If a rooster crows while on the ground, it is a sign of foul weather; if while off the ground, nice weather.
Cited in Hyatt, p. 25.

3006. If a rooster in February stands on a cow-manure pile and crows, the weather will change within twenty-four hours.
Cited in Hyatt, p. 25.

3007. If the rooster crows before midnight, it will rain before daylight.
Cited in Wurtele, p. 300.

3008. If the rooster crows before midnight there will be falling weather.
Cited in Smith, p. 1.

3009. If the rooster crows on the nest at night, it will rain before morninng.
Cited in Smith, p. 5.

3010. In winter the weather becomes colder after a rooster has crowed about nine o'clock at night.
Cited in Hyatt, p. 24.

3011. Prepare for a long dry spell after a rooster crows while it is raining.
Cited in Hyatt, p. 25.

3012. The crowing of a rooster about four o'clock in the morning is a storm token.
Cited in Hyatt, p. 24.

3013. The crowing of a rooster before noon indicates a change of weather.
Cited in Hyatt, p. 24.

3014. When a rooster crows before going to bed, a change in the weather is ahead.
Cited in Smith, p. 1.

3015. When the rooster crows at night, he tells you that a rain's in sight.
Cited in Sloane 2, p. 52.

3016. When the rooster crows at noon, rain will come soon.
Cited in Reeder, no pp.

3017. When the roosters go crowing to bed, they will rise with watery head. Vars.: (a) If a rooster crows when he goes to bed, he will

get up with a wet head. (b) Rooster crows before going to bed, he will get up in the morning with a rainy head.
Cited in Dunwoody, p. 39; Smith, p. 5. Recorded in Bryant (Miss.)

3018. You are warned of rain by a rooster crowing on the roof of your house.
Cited in Hyatt, p. 25.

rope

3019. Ropes difficult to untwist indicate bad weather. Vars.: (a) Ropes are more difficult to untwist before bad weather. (b) Ropes being difficult to untwist indicate bad weather.
Cited in Dunwoody, p. 116; Garriott, p. 21; Inwards, p. 218; Lee, p. 74.

3020. Ropes shorten before a rain. Vars.: (a) Sailors note the tightening of ropes on a ship before rain. (b) When ropes are tight, it's going to rain.
Cited in Freier, p. 38.

3021. Ropes shorten with an increase of humidity.
Cited in Garriott, p. 21; Hand, p. 21.

3022. The rigging rope on vessels and clothes lines grow slack before rain.
Cited in Dunwoody, p. 116.

rubbers (overshoes)

3023. It never rains if you wear your rubbers.
Recorded in Bryant (Wash.).

sage

3024. Plant your sage and rue together, the sage will grow in any weather.
Cited in Northall, p. 482.

salmon

3025. Salmon and trout, plentiful in the river (Columbia), show an abundance of rain in the surrounding country by which the river has risen. Var.: When salmon and trout are plentiful in the Columbia river, it is a sign that there has been abundance of rain in the surrounding country.
Cited in Dunwoody, p. 51; Inwards, p. 199.

salt

3026. A farmer's wife says when her cheese salt is soft, it will rain; when getting dry, fair weather may be expected. Var.: When cheese salt is soft, expect rain.
Cited in Inwards, p. 219; Hand, p. 21; Freier, p. 35.

3027. Either to spill salt at the table or to drop it on the ground betokens rain.
Cited in Hyatt, p. 31.

3028. Salt becomes moist before rain. Vars.: (a) Salt becomes damp before rain. (b) Salt becoming damp and lumpy in the saltcellar is a sign of rain.
Cited in Steinmetz 1, p. 113; Dunwoody, p. 116; Hyatt, p. 31; Lee, p. 74.

3029. Salt increases in weight before rain. Vars.: (a) If salt is sticky, and gains in weight; it will rain before too late. (b) Salt increases in weight before a shower.
Cited in Garriott, p. 21; Inwards, p. 219; Hand, p. 21; Freier, p. 35.

sand

3030. When the sand doth feed the clay, England woe and well a day; but when the clay doth feed the sand, then it is well with England. Comment: Sand feeds clay in a wet summer while clay feeds sand in a dry summer. Because there is more clay than sandy ground in England a dry summer means better growing

conditions.
Cited in Swainson, p. 16; Cheales, p. 25.

sandwort

3031. When the corona of red sandwort contracts, expect rain. Var.:
Purple sandwort expands its beautiful pink flowers only when the
sun shines, but closes them before the coming shower.
Cited in Dunwoody, p. 67; Inwards, p. 215; Sloane 2, p. 112.

sap

3032. Sap from the maple tree flows faster before a rain shower.
Cited in Freier, p. 26.

Saturday

3033. A fine Saturday, a fine Sunday; a fine Sunday, a fine week.
Cited in Marvin, p. 203.

3034. There is never a Saturday without some sunshine. Var.: Never
a Saturday without some sunshine.
Cited in Dunwoody, p. 101; Marvin, p. 203; Wilson, p. 699;
Lee, p. 146.

scorpion

3035. If scorpions crawl with tails up, it will rain soon.
Cited in Reeder, no pp.

3036. When scorpions crawl, expect dry weather.
Cited in Dunwoody, p. 58; Inwards, p. 207.

sea

3037. A heavy inshore swell (sea) during calm weather is believed to
foretell wind.
Cited in Mitchell, p. 214.

3038. Calm seas do not make good sailors. Var.: Smooth seas make
poor sailors.
Cited in Mieder, p. 528.

3039. In a calm sea every man is a pilot. Var.: Every man is a pilot
when the sea is calm.
Cited in Mieder, p. 528.

3040. Just before a storm, the sea heaves and sighs.
Cited in Inwards, p. 153; Alstad, p. 110.

3041. Never go to the sea when a storm is coming.
Cited in Mieder, p. 528.

3042. When the surface of the sea is rough without any wind blowing
at the time, expect a gale before long. Var.: If, during the
absence of wind, the surface of the sea becomes agitated by a
long rolling swell, a gale may be expected.
Cited in Dunwoody, p. 116; Inwards, p. 153.

sea bird

3043. If sea birds fly towards land and land birds toward the sea,
expect wind without rain.
Cited in Dunwoody, p. 40.

3044. When a strong wind and stormy weather are forthcoming, sea
birds, gulls, etc, hang about the land or over it, sometimes flying
inward. Var.: The landward flight and flocking of seagulls
presage wind.
Cited in Mitchell, p. 227; Steinmetz 1, p. 110.

sea gull

3045. If sea gulls fly inland, expect storm. Var.: When sea gulls fly to
land, a storm is at hand.
Cited in Dunwoody, p. 40.

3046. If sea gulls stay on shore, a storm is coming. Var.: Sea gulls
stay on land before a storm.
Cited in Freier, p. 54. Recorded in Bryant (Calif.).

3047. Sea gull in the field indicate a storm from southeast.
Cited in Inwards, p. 197.

3048. Sea gull, sea gull, sit on the sand, it's never good weather while
you're on the land. Var.: Sea gull, sea gull, sit on the sand, it's
never good weather when you're on the land.
Cited in Swainson, p. 239; Wright, p. 78; Whitman, p. 47;
Inwards, p. 197; Alstad, p. 71; Lee, p. 61.

3049. Sea gulls, early in the morning making a gaggling more than
ordinary, foretoken stormy and blustering weather.
Cited in Swainson, p. 240; Inwards, p. 197.

3050. When sea gulls fly out early and far to seaward, moderate winds
and fair weather may be expected.
Cited in Swainson, p. 240; Dunwoody, p. 20.

3051.   When sea mews (gulls) appear in unwonted numbers, expect rain
        and high southwest winds.
        Cited in Inwards, p. 197.
3052.   When the sea gulls cry is frequently repeated, when it is more
        lengthened and dismal than usual, and when it is heard in an
        inland place, rain or snow is prognosticated.
        Cited in Mitchell, p. 227.
3053.   When the sea gulls fly high, there's a storm in the sky.
        Recorded in Bryant (Wash.).

sea urchin

3054.   Sea urchins thrusting themselves into the mud, or striving to
        cover their bodies with sand, foreshow a storm. Var.: Sea
        urchins trying to dig in mud or to cover their bodies with sand,
        foreshadow a storm.
        Cited in Dunwoody, p. 51; Inwards, p. 199; Lee, p. 59.

seafowl

3055.   If seafowl retire to the shore or marshes, a storm is approaching.
        Cited in Garriott, p. 24; Inwards, p. 197; Hand, p. 24.

seaman

3056.   A good seaman is known in bad weather.
        Cited in Mieder, p. 528.

season

3057.   Oregon has two seasons: the rainy and August.
        Recorded in Bryant (Oreg.).

seaweed

3058.   A piece of seaweed hung up will become damp previous to rain.
        Var.: A piece of kelp or seaweed hung up will become damp
        before rain.
        Cited in Swainson, p. 261; Dunwoody, p. 68; Garriott, p. 21;
        Hand, p. 21.
3059.   Seaweed becomes damp and expands before wet weather.
        Cited in Dunwoody, p. 67.
3060.   Seaweed dry, sunny sky; seaweed wet, rain you'll get.
        Cited in Lee, p. 53.

September

3061. A wet September, next summer drought, no crops and famine.
Vars.: (a) A wet September, drought for next summer. (b) A wet
September, drought for next summer, famine, and no crops. (c)
A wet September, drought next year. (d) Heavy September rains
bring drought.
Cited in Dunwoody, p. 99; Garriott, p. 45; Inwards, p. 62;
Hand, p. 45; Freier, p. 81.

3062. As September, so the coming March.
Cited in Dunwoody, p. 98; Garriott, p. 45; Whitman, p. 43;
Inwards, p. 62; Hand, p. 45; Freier, p. 83.

3063. If the storms in September clear off warm, all the storms of the
following winter will be warm.
Cited in Dunwoody, p. 99; Inwards, p. 62.

3064. In September after burning stubble, ponds and streams begin to
bubble.
Cited in Page, p. 33.

3065. Rain in September is good for the farmer, but poison to the vine
growers. Var.: September rain is much liked by the farmer.
Cited in Dunwoody, p. 69; Inwards, p. 62.

3066. September blow soft, till the fruit's in the loft.
Cited in Denham, p. 57; Swainson, p. 126; Cheales, p. 22;
Northall, p. 442; Wright, p. 102; Inwards, p. 62; Wilshere, p.
13; Simpson, p. 200.

3067. September dries up ditches or breaks down bridges.
Cited in Wilshere, p. 13.

3068. September rain is good for crops and vines.
Cited in Dunwoody, p. 99; Inwards, p. 62.

3069. Thunder in September indicates a good crop of grain and fruit
for next year.
Cited in Inwards, p. 62.

3070. When a cold spell occurs in September and passes without a
frost, a frost will not occur until the same time in October.
Cited in Dunwoody, p. 99; Inwards, p. 62.

September 1

3071. Fair on the first of September, fair the entire month. Vars.: (a)
Fair on September the first, fair for the month. (b) If the first of
September be fair, it will be fair for a month.

Cited in Dunwoody, p. 99; Whitman, p. 40; Inwards, p. 62; Wilshere, p. 13.

September 8
3072. As the weather on the eighth of September, so it will be for the next four weeks. Var.: As on September 8th, so for the next four weeks.
Cited in Dunwoody, p. 99; Inwards, p. 62.

September 15
3073. September 15th is said to be a fine day six years out of seven. Var.: September 15th is said to be fine in six years out of seven.
Cited in Garriott, p. 43; Wright, p. 103; Hand, p. 43.

September 19
3074. If on the nineteenth of September there is a storm from the south, a mild winter may be expected.
Cited in Inwards, p. 63.

September 20-21
3075. If the wind blows from the southeast during September 20th and 21st, the weather from the middle of February to the middle of March will be warm.
Cited in Dunwoody, p. 87.

September 20-22
3076. The three days of September (20th, 21st, 22nd) rule the weather for October, November, and December. Var.: September 20th, 21st, and 22nd rule the weather for October, November, and December.
Cited in Dunwoody, p. 100; Inwards, p. 63.

September 21
3077. If the wind blows from the south on the twenty-first of September, it indicates a warm autumn. Var.: South wind on the twenty-first of September indicates that the rest of the autumn will be warm.
Cited in Dunwoody, p. 87; Inwards, p. 63.

3078. On September 21st a south wind indicates a light winter; a north wind, a heavy winter.
Cited in Hyatt, p. 9.

3079. The quarter of the wind on September 21st governs its prevailing direction for the next six months.
Cited in Hyatt, p. 9.

shad

3080. Shad run south when the weather changes to cold.
Cited in Dunwoody, p. 51.

shark

3081. Sharks go out to sea at the approach of a wave of cold weather. Var.: Sharks swim out to sea when a wave of cold weather approaches.
Cited in Dunwoody, p. 51; Inwards, p. 198; Lee, p. 59.

sheep

3082. After sheep turn their backs to the wind, a cold spell may be predicted.
Cited in Hyatt, p. 29.

3083. Before a storm comes sheep become frisky, leap, and butt or "box" each other. Vars.: (a) Before a rain, sheep are frisky and box each other. (b) Before a storm sheep frisk, leap, and butt each other.
Cited in Garriott, p. 23; Inwards, p. 182; Hand, p. 23; Lee, p. 69; Freier, p. 49.

3084. If old sheep turn their backs towards the wind, and remain so for some time, wet and windy weather is coming.
Cited in Inwards, p. 182.

3085. If sheep ascend hills and scatter, expect clear weather. Var.: When sheep go to the hills and scatter, expect nice weather.
Cited in Dunwoody, p. 33; Lee, p. 69.

3086. If sheep feed uphill in the morning, sign of fine weather.
Cited in Inwards, p. 183.

3087. If sheep gambol and fight, or retire to shelter, it presages a change in the weather.
Cited in Swainson, p. 233; Inwards, p. 182.

3088.   If sheep wool is heavy, predict a severe winter; if light, a mild winter. Var.: When sheep have an unusual amount of wool, a cold winter is ahead.
         Cited in Hyatt, p. 29; Smith, p. 14.
3089.   Old sheep and ewes eat greedily before a storm and sparingly before a thaw. Var.: Old sheep eat greedily before a storm, and sparingly before a thaw.
         Cited in Mitchell, p. 228; Swainson, p. 233; Inwards, p. 183.
3090.   Shear your sheep in May, and shear them all away.
         Cited in Northall, p. 440.
3091.   Sheep bleat and seek shelter before snow.
         Cited in Dunwoody, p. 33.
3092.   Sheep crowd together near a fence before a storm. Var.: When sheep or oxen cluster together as if seeking shelter, expect a storm.
         Cited in Hyatt, p. 29; Lee, p. 69.
3093.   Sheep fighting for their food more than usual indicates a storm.
         Cited in Sloane 2, p. 120.
3094.   Sheep leaving the pasture and seeking the fold foretell a storm.
         Cited in Hyatt, p. 29.
3095.   When sheep begin to go up to the mountains, shepherds say it will be fine weather.
         Cited in Wright, p. 78; Inwards, p. 183.
3096.   When sheep, cattle, or horses turn their backs to the wind, it is a sign of rain. Var.: When sheep turn their backs to the wind, it is a sign of rain.
         Cited in Dunwoody, p. 18; Inwards, p. 183; Sloane 2, p. 120.
3097.   When sheep do huddle by tree and bush, bad weather is coming with wind and slush.
         Cited in Wright, p. 80; Inwards, p. 183; Lee, p. 69.
3098.   When sheep feed down the hill in winter, a snowstorm is looked for; when they feed up the burn, wet weather is near.
         Cited in Swainson, p. 233; Inwards, p. 183.
3099.   When sheep leave the high grounds and bleat much in the evening and during the night, severe weather is expected.
         Cited in Swainson, p. 233; Inwards, p. 183.
3100.   When the sheep collect and huddle, tomorrow will become a puddle.
         Cited in Sloane 2, p. 53.

3101. You may shear your sheep when the elder blossoms peep. Var.: When the white pinks (of the elder blossoms) begin to appear, then is the time your sheep to shear.
Cited in Swainson, p. 12; Dunwoody, p. 33; Northall, p. 482; Wright, p. 67; Page, p. 10.

shoe

3102. If your shoes squeak as you walk, a storm is approaching.
Cited in Hyatt, p. 32.

shower

3103. A hasty shower of rain, falling when the wind has raged some hours, soon allays it.
Cited in Inwards, p. 157.

3104. A heavy shower is soon over.
Cited in Mieder, p. 538.

3105. A sunshiny shower will last but half an hour. Vars.: (a) A sunshiny shower lasts half an hour. (b) Sunshine and shower, rain for an hour.
Cited in Cheales, p. 24; Wurtele, p. 302; Freier, p. 45.

3106. A sunshiny shower, won't last half an hour. Vars.: (a) A sunshine shower, never lasts an hour. (b) A sunshiny shower, never lasts half an hour. (c) A sunshiny shower, won't last an hour. (d) A sunshiny shower, won't last out the hour. (e) Sunshine shower won't last half an hour. (f) Sunshining shower won't last half an hour.
Cited in Denham, p. 8; Swainson, p. 213; Dunwoody, p. 79; Northall, p. 464; Wright, p. 79; Whitman, p. 29; Inwards, p. 158; Sloane 2, p. 59; Hyatt, p. 11; Smith, p. 8; Lee, p. 94; Page, p. 6; Mieder, p. 538.

3107. If a shower be approaching from the west, it may be seen shooting forth white feathery rays from its upper edge.
Cited in Inwards, p. 129; Alstad, p. 147.

3108. If we stand the shower we shan't flinch for the drops.
Cited in Whiting 1, p. 393.

3109. Short, slight showers, during dry weather, are called a hardening of the drought.
Cited in Mitchell, p. 229.

3110. Showers are most frequent at the turn of the tide. Var.: Showers occur more frequently at the turn of the tide.
Cited in Humphreys 2, p. 71; Whitman, p. 33; Inwards, p. 153.

3111. Small showers last long, but sudden storms are short.
Cited in Inwards, p. 156; Mieder, p. 538.

3112. When a rain spatters, it is merely a shower.
Cited in Hyatt, p. 11.

Shrove Tuesday

3113. As the weather is on Shrove Tuesday, so will it be to the end of Lent. Comment: Shrove Tuesday is the day before Ash Wednesday that begins the season of Lent.
Cited in Swainson, p. 67.

3114. So much as the sun shineth on Shrove Tuesday, the like will shine every day in Lent. Vars.: (a) So much as the sun shines on Pancake (Shrove) Tuesday, the like will shine every day in Lent.
Cited in Swainson, p. 67; Whitman, p. 38; Inwards, p. 70.

3115. Thunder on Shrove Tuesday foretelleth wind, store of fruit, and plenty.
Cited in Inwards, p. 70.

Shrovetide

3116. When the sun is shining on Shrovetide, it is meant well for rye and peas. Comment: Shrovetide is the three days, Shrove Sunday, Shrove Monday, Shrove Tuesday, preceding Ash Wednesday.
Cited in Dunwoody, p. 103; Inwards, p. 70; Lee, p. 151.

sign

3117. A prediction is seldom founded on a single sign.
Cited in Mitchell, p. 215.

3118. All signs fail in dry weather. Vars.: (a) All signs fail in a dry time. (b) Signs fail in dry weather.
Cited in Taylor 1, p. 113; Taylor 2, p. 333; Hyatt, p. 13; Whiting 1, p. 394. Recorded in Bryant (US).

3119. No sign is considered infallible.
Cited in Mitchell, p. 215.

skate

3120. Skate jump in the direction that the next wind will come from.
Cited in Dunwoody, p. 51.

skunk

3121. A skunk nest deep in the ground forecasts a harsh winter with deep snows.
Cited in Hyatt, p. 29.

3122. A skunk odor in the air is a rain omen.
Cited in Hyatt, p. 29.

3123. When skunks are real fat, we'll have a long winter coming.
Cited in Freier, p. 83.

sky

3124. Neither heat nor cold abides always in the sky.
Cited in Denham, p. 6.

sky (black)

3125. The blacker the sky, the greener the grass.
Recorded in Bryant (Can. [Ont.]).

sky (blazing)

3126. To have a blazing sky reflect against clouds in the south at sunset denotes rain.
Cited in Hyatt, p. 5.

sky (blue)

3127. A blue and white sky, never four and twenty hours dry.
Cited in Northall, p. 459; Inwards, p. 135. Recorded in Bryant (N.Y.).

3128. A dark, gloomy, blue sky indicates wind; a bright, blue sky clear fine weather. Vars.: (a) A dark gloomy blue sky indicates wind. (b) A dark, gloomy blue sky is windy. (c) A dark gloomy sky is windy, but a light, bright blue sky indicates fine weather.
Cited in Steinmetz 1, p. 123; Dunwoody, p. 42; Inwards, p. 149. Recorded in Bryant (Mich.).

3129. A deep-blue sky is always an indication of beautiful weather for the rest of the day.
Cited in Hyatt, p. 3.

3130. A light, bright blue sky indicates fine weather.
Cited in Steinmetz 1, p. 123.

3131. After a storm, if you can find enough blue sky to make a Dutchman's jacket, the rain is gone. Vars.: (a) After a storm, if you can find enough blue sky to make a sailor's breeches, the rain is gone. (b) After a storm, if you can find enough blue sky to make an apron, the rain is gone.
Cited in Wurtele, p. 297.

3132. After it has been raining some time, a blue sky in the southeast indicates that there will be fair weather soon.
Cited in Dunwoody, p. 85.

3133. Blue in the north sky means fair weather.
Recorded in Bryant (N.Dak.).

3134. Blue sky at night is a sailor's delight.
Recorded in Bryant (Wis.).

3135. Enough blue sky in the northwest to make a Scotchman a jacket is a sign of approaching clear weather. Var.: Enough blue sky in the northwest to make a pair of Dutchman's breeches is a sign of approaching fair weather.
Cited in Dunwoody, p. 43; Inwards, p. 148; Hand, p. 12; Freier, p. 43.

3136. If a blue color (of the sky) should predominate, the air is clearing.
Cited in Inwards, p. 161.

3137. If on a gloomy day there is a patch of blue sky large enough to make a pair of britches for a Dutchman, the weather will soon clear. Vars.: (a) If on a gloomy day there is a patch of blue sky large enough to make a shirt for a sailor, the weather will soon clear. (b) If on a gloomy day there is a patch of blue sky the size of a handkerchief, the weather will soon clear.
Cited in Hyatt, p. 3.

3138. If the sky beyond the clouds is blue, be glad, there is a picnic for you.
Cited in Dunwoody, p. 44; Marvin, p. 204; Lee, p. 91. Recorded in Bryant (Mich.).

3139. If the sky is of a deep, clear blue or sea-green color near the horizon, rain will follow in showers.
Cited in Inwards, p. 148.

3140. If there be a light-blue sky with thin, light flying clouds, whilst the wind goes to the south, without much increase in force, or a dirty blue sky when no clouds are to be seen, expect storm.
Cited in Dunwoody, p. 47.

3141. If there's enough blue in the sky to make a Dutchman a pair of britches, it won't rain. Vars.: (a) If there's enough blue in the sky to cut a Dutchman a shirt, it won't rain. (b) If there's enough blue in the sky to make a Dutchman a pair of pants, it won't rain. (c) If you can see enough blue in the sky to patch a Dutchman's breeches, it won't rain. (d) It won't rain if there is blue enough in the sky to make a sailor suit.
Cited in Sloane 2, p. 54. Recorded in Bryant (Ga., Ill., Miss., Nebr., Ohio, Wash.).

3142. Sailors call just enough blue sky to wipe one's face with a precursor of fine weather.
Cited in Steinmetz 2, p. 277.

3143. When as much blue sky is seen as will make a Dutchman a jacket, the weather may be expected to clear up. Vars.: (a) If there is enough blue in the sky to make a Scot's breeches, the weather will clear. (b) If there is enough blue in the sky to make an old woman's apron, the weather will clear. (c) If you can see enough blue in the sky to make a duck's pants, it will fair off. (d) When as much blue is seen in the sky as will make a Dutchman's jacket, the weather will clear. (e) When as much blue sky is seen as will make a sailor's breeches, the weather may be expected to clear up.
Cited in Dunwoody, p. 14; Inwards, p. 148; Lee, p. 91. Recorded in Bryant (Md.).

sky (bright)

3144. Bright sky at night, sailors delight.
Recorded in Bryant (Oreg.).

sky (Carle)

3145. The Carle sky keeps not the head dry. Comment: From Dumfries to Gretna in Scotland, a lurid, yellowish sky in the east or southeast is called a Carlisle or Carle sky, and is regarded as a

sure sign of rain.
Cited in Inwards, p. 149.

sky (clear)

3146. A clear sky, and sun setting in a well-defined form, without dazzling the eye, and a deep salmon-color, foreshows a brilliant and very hot day to succeed.
Cited in Steinmetz 1, p. 120.

3147. A very clear sky without clouds is not to be trusted, unless the barometer be high.
Cited in Inwards, p. 148.

3148. Clear sky Friday night, rain before Monday night.
Recorded in Bryant (Ill.).

3149. It will not rain much as long as the sky is clear before the wind, but when clouds fall in against the wind, rain will soon follow.
Cited in Hand, p. 12.

3150. When there is enough clear sky to patch a Dutchman's breeches, expect fair weather.
Cited in Dunwoody, p. 44.

sky (cloudy)

3151. A mottled sky never lets the well dry. Comment: A mottled sky refers to cumulus clouds.
Recorded in Bryant (Can. [Ont.]).

3152. Every sky has its cloud.
Cited in Mieder, p. 545.

3153. If the sky after fine weather becomes heavy with small clouds, expect rain. Var.: If the sky, from being clear, becomes fretted or spotted all over with bunches of clouds, rain will soon fall.
Cited in Dunwoody, p. 44; Inwards, p. 125.

3154. Trace in the sky the painter's brush, the winds around you soon will rush. Comment: Painter's brush refers to a cirrus cloud.
Cited in Humphreys 2, p. 51; Whitman, p. 23; Inwards, p. 132; Wurtele, p. 293; Lee, p. 95; Wilshere, p. 3.

sky (curdly)

3155. A curdly sky will not be twenty-four hours dry. Vars.: (a) A curdled sky leaves nothing dry. (b) A curdled sky never lets the world go dry. (c) A curdled sky seldom leaves the earth dry. (d)

A curdled sky will never stay dry. (e) A curdly sky will not leave the earth long dry. (f) With curdly sky the land will never be dry. Comment: Curdly or curdled sky refers to cumulus clouds.
Cited in Dunwoody, p. 43; Inwards, p. 134. Recorded in Bryant (N.Y., Ohio, Wis. Can. [Ont.]).

sky (dappled)

3156. The dappled sky is a certain sign of rain.
Cited in Steinmetz 2, p. 83.

sky (dark)

3157. If the sky becomes darker without much rain and divides into two layers of clouds, expect sudden gusts of wind. Var.: If there be a cloudy sky and dark clouds driving fast under higher clouds, expect violent gusts of wind.
Cited in Dunwoody, p. 43; Inwards, p. 149.

sky (dominicker)

3158. Dominicker sky, storm close by. Comment: Dominicker means barred.
Cited in Hyatt, p. 6.

sky (fiery)

3159. If a fiery sky at sunset is reflected on clouds in the north, storms and high winds can be expected.
Cited in Hyatt, p. 5.

sky (fleecy)

3160. If there be a fleecy sky, unless driving northwest, expect rain.
Cited in Dunwoody, p. 44.

sky (gray)

3161. Gray evening sky, not one day dry.
Cited in Lee, p. 79.

3162. If there be a dark, gray sky, with a south wind, expect frost.
Cited in Dunwoody, p. 54; Inwards, p. 149.

sky (green)

3163.   A green sky above the sunset foretells rain next day. Comment: This appears to be one of the most reliable omens in the British Isles. A green sky at any time usually indicates bad weather.
Cited in Inwards, p. 148.

3164.   Rain is in the air when a faint greenish hue overspreads the sky.
Cited in Hyatt, p. 4.

3165.   When the sky during rain is tinged with sea green, the rain will increase; if with deep blue, rain will be showery. Var.: When the sky in rainy weather is tinged with sea-green, the rain will increase; if with deep blue, it will be showery.
Cited in Dunwoody, p. 79; Inwards, p. 148.

3166.   When the sky is of a sickly-looking greenish hue, wind or rain may be expected.
Cited in Inwards, p. 149.

sky (leaden)

3167.   A leaden sky at daybreak in summer will be replaced by intense heat.
Cited in Hyatt, p. 5.

sky (mackerel)

3168.   A mackerel sky denotes fair weather for that day, but predicts rain a day or two after. Vars.: (a) A mackerel sky denotes fair weather for that day, but rain a day or two after. (b) Mackerel sky, fair today, wet tomorrow.
Cited in Mitchell, p. 22; Steinmetz 2, p. 283; Marvin, p. 205; Brunt, p. 73; Inwards, p. 134.

3169.   A mackerel sky foretells rain.
Cited in Smith, p. 7.

3170.   A mackerel sky is as much for wet as 'tis for dry.
Cited in Marvin, p. 205; Inwards, p. 134.

3171.   A mackerel sky is never dry. Vars.: (a) Mackerel clouds in the sky expect more wet than dry. (b) Mackerel sky never leaves the earth dry.
Cited in Hyatt, p. 6. Recorded in Bryant (Ky., Minn., N.J., Utah).

3172.   A mackerel sky, never holds three days dry. Var.: A mackerel sky never three days dry.

Cited in Northall, p. 459; Marvin, p. 204; Hyatt, p. 6.

3173. A mackerel sky, not twenty-four hours dry. Var.: A mackerel sky, never twenty-four hours dry.
Cited in Dunwoody, p. 45; Garriott, p. 12; Marvin, p. 205; Whitman, p. 23; Inwards, p. 134; Hyatt, p. 6; Lee, p. 96.

3174. Mackerel sky and mare's tails, make lofty ships carry low sails. Vars.: (a) Mackerel scales, furl your sails. (b) Mackerel skies and mare's tails make tall ships carry low sails.
Cited in Dunwoody, p. 15; Northall, p. 459; Humphreys 1, p. 441; Marvin, p. 205; Taylor 1, p. 121; Inwards, p. 134; Wilson, p. 497; Lee, p. 96; Page, p. 25; Freier, p. 45. Recorded in Bryant (Md., Mich.).

3175. Mackerel sky, mackerel sky, never long wet, never long dry. Var.: Mackerel sky, mackerel sky, never long wet and never long dry.
Cited in Dunwoody, p. 45; Marvin, p. 204; Taylor 1, p. 111; Whitman, p. 23; Brunt, p. 73; Inwards, p. 134; Page, p. 25; Wilshere, p. 3.

3176. Mackerel sky, sign of dry.
Recorded in Bryant (Ky., Minn., N.J., Utah).

3177. Mackerel sky, wind'll be high.
Cited in Wurtele, p. 293.

sky (pink)

3178. A pinkish sky in the west at night is an omen of rain.
Cited in Hyatt, p. 5.

3179. Pink sky at night, sailor's delight; pink sky in morning, sailor's warning.
Recorded in Bryant (Ill., Ind.).

sky (red)

3180. A red and lowering sky at sunrise indicates a wet day.
Cited in Steinmetz 1, p. 118.

3181. A red evening (sky) indicates fine weather; but if the red extends far upwards, especially in the morning, it indicates wind or rain. Var.: The sky being red at evening, foreshows a fair and clear morning; but if the morning rise red, of wind and rain we shall be sped.
Cited in Dunwoody, p. 78; Inwards, p. 82.

3182.   A red sky at night is the shepherd's delight, a red sky in the morning is the shepherd's warning. Vars.: (a) If the sky is red at night, shepherd's delight, but red in the morning is a shepherd's warning. (b) Red at night, shepherd's delight; red in the morning, shepherd's warning.
Cited in Whitman, p. 32.

3183.   A red sky at night means good weather.
Cited in Smith, p. 3.

3184.   A red sky in the morning means bad weather ahead.
Cited in Smith, p. 1.

3185.   A red sky in the morning signifies blustery winds.
Cited in Hyatt, p. 5.

3186.   A very red sky in the east at sunset indicates stormy winds.
Cited in Dunwoody, p. 78.

3187.   If at morning the sky be red, it bids the traveler stay in bed.
Cited in Hyatt, p. 4.

3188.   Red in the east, neither good for man nor beast.
Recorded in Bryant (Oreg.).

3189.   Red in the morning, soldiers are mourning.
Cited in Hyatt, p. 5.

3190.   Red skies in the evening precede fine morrows.
Cited in Dunwoody, p. 78; Garriott, p. 26; Hand, p. 26.

3191.   Red (sky) in the west, the lamb and I'll go safe to rest.
Recorded in Bryant (Miss.).

3192.   Sky red in the morning is a sailor's sure warning; sky red at night is the sailor's delight. Vars.: (a) A red sky at night is the sailors delight, a red sky in the morning is the sailor's warning. (b) Red at night, sailor's delight; red in the morning, sailor take warning. (c) Red in the morning, sailors take warning; red at night, sailors' delight.
Cited in Swainson, p. 179; Northall, p. 460; Humphreys 1, p. 432; Whitman, p. 32; Sloane 2, p. 97; Hyatt, p. 4; Wilson, p. 741; Wurtele, p. 295; Dundes, p. 94; Whiting 2, p. 573; Mieder, p. 545. Recorded in Bryant (Ill., La., Utah, Wash.).

sky (rosy)

3193.   A rosy sky at sunset is a sign of fair weather.
Cited in Walton, no pp.

3194. After you have seen a rosy sky, make preparations for a hailstorm.
Cited in Hyatt, p. 5.

sky (scaly)
3195. A scaly sky means a change in the weather.
Cited in Smith, p. 1.

sky (starry)
3196. If the sky is starry three nights in a row, there will be rain.
Cited in Smith, p. 6.

sky (streak)
3197. Red or yellow streaks from west to east indicate rain in forty-
eight hours.
Cited in Dunwoody, p. 79.

sky (transparent)
3198. At sunset a transparent sky with a scattering of small red clouds
is a promise of fair weather.
Cited in Hyatt, p. 5.

sky (yellow)
3199. A bright yellow sky at sunset, presages wind; a pale yellow,
wet; and a greenish, sickly-looking color, wind and rain. Var.:
A bright yellow sky in the evening foretells wind, a pale yellow
twilight extending far up the sky indicates threatening weather.
Cited in Steinmetz 1, p. 123.
3200. A light yellow sky at sunset presages wind. Vars.: (a) A pale
yellow sky at sunset presages rain. (b) In the sky at sunset pale
yellow portends high winds.
Cited in Dunwoody, p. 48; Hyatt, p. 5.
3201. Yellow in the sky at sunset is a portent of rain.
Cited in Hyatt, p. 5.
3202. Yellow streaks in sunset sky, wind and day-long rain is nigh.
Cited in Sloane 1, p. 73.

skylark

3203. The skylark is a bird that people usually associate with good weather; if it hovers and glides during its descent, the weather will remain fine, but if it drops straight down to the ground, it will rain.
Cited in Page, p. 19.

sleet

3204. Much sleet in winter will be followed by a good fruit year.
Cited in Dunwoody, p. 76.

sloe

3205. Many slones, many groans. Comment: When there is abundant fruit (plumlike) on the sloe (blackthorn), there will follow a hard winter with much poverty and suffering. Sloe is spelled slone here because of the rhyme.
Cited in Northall, p. 493; Marvin, p. 204.

3206. When the sloe tree is white as a sheet, sow your barley whether it be dry or wet. Comment: The sloe tree is the blackthorn.
Cited in Denham, p. 36; Swainson, p. 12; Dunwoody, p. 90; Northall, p. 479.

smell

3207. Cesspools, dunghills, and water closets smell more before rain.
Cited in Steinmetz 1, p. 112.

3208. Ditches and drains smelling unpleasantly warn of rain.
Cited in Page, p. 7.

3209. Drains, ditches, and dunghills are more offensive (smell) before rain.
Cited in Garriott, p. 22; Inwards, p. 223; Hand, p. 22.

3210. Manure piles smell stronger before a rain.
Cited in Freier, p. 30.

3211. When the perfume (smell) of flowers is unusually perceptible, rain may be expected.
Cited in Dunwoody, p. 65; Hand, p. 21.

smoke

3212. Chimney smoke clinging to the ground in the morning brings a storm before night.

Cited in Hyatt, p. 31.

3213. Fair weather will continue or come, if smoke pours in white clouds from a railroad engine.
Cited in Hyatt, p. 31.

3214. If during calm, smoke does not ascend readily, expect rain.
Cited in Inwards, p. 220.

3215. If the smoke goes straight up it is going to snow.
Cited in Smith, p. 10.

3216. Smoke ascending indicates clear weather. Var.: If smoke rises into the air, we will have clear weather.
Cited in Dunwoody, p. 117; Hyatt, p. 31.

3217. Smoke falling indicates rain. Vars.: (a) If smoke falls to the ground, it is likely to rain. (b) Smoke falling instead of rising is a sign of rain. (c) Smoke falling to the ground indicates rain. (d) Smoke falls to the ground preceding rain.
Cited in Steinmetz 2, p. 293; Dunwoody, p. 117; Garriott, p. 19; Inwards, p. 220; Hand, p. 19; Page, p. 7; Freier, p. 39.

3218. Smoke going down a stream will be followed by rain.
Cited in Hyatt, p. 31.

3219. To have smoke puff from a stovepipe into the room is a forecast of snow.
Cited in Hyatt, p. 31.

3220. When smoke descends, good weather ends.
Cited in Sloane 2, p. 55.

3221. When smoke from the chimney of a cottage descends (flops down) upon the roof and passes along the eaves, expect rain within twenty-four hours. Var.: If the smoke from a chimney doesn't rise straight up, but instead hugs the roof, it will rain.
Cited in Inwards, p. 220; Smith, p. 8.

3222. When smoke goes west, gude weather is past; when the smoke goes east, gude weather comes neist. Comment: Neist means next.
Cited in Denham, p. 17; Northall, p. 471; Humphreys 1, p. 441; Whitman, p. 35; Inwards, p. 118; Freier, p. 60.

3223. When smoke rises but not too high, clouds won't grow and you'll keep dry.
Cited in Freier, p. 45.

3224. When the smoke clings to the chimney, there is bad weather ahead.

Recorded in Bryant (Oreg.).

3225. When the smoke in clear weather rises vertically from the chimney, the weather will remain clear. Vars.: (a) When the smoke rises straight up from the chimney, good weather will follow. (b) If smoke rises high into the air, the weather will be fair.
Cited in Dunwoody, p. 117. Recorded in Bryant (N.Y., Oreg., Utah).

3226. When the sun presses the smoke out of chimneys, foul weather follows.
Cited in Dunwoody, p. 117.

smoke (pipe)

3227. If the smoke of a morning pipe hangs a long while in the air, a good hunting day always follows. Var.: If the smoke from a morning pipe hangs a long while in the air, a good hunting day. Comment: The underbrush will not be dry and crackle; game will be active and moving about.
Cited in Inwards, p. 220; Alstad, p. 51.

smoke stack

3228. The factory smoke stack is more of a nuisance before a rain.
Cited in Freier, p. 40.

snail

3229. If snails and slugs come out abundantly, it is a sign of rain. Vars.: (a) If snails are abundant, it is a sign of rain. (b) Snails seen in large numbers are a token of rain.
Cited in Inwards, p. 201; Sloane 2, p. 122; Hyatt, p. 18.

3230. Snailie, snailie, shoot out your horn, and tell us if it will be a bonnie day in the morn.
Cited in Swainson, p. 252; Inwards, p. 201.

3231. Snails moving on bushes or grass are signs of rain.
Cited in Dunwoody, p. 72.

3232. When black snails cross your path, black clouds much moisture hath.
Cited in Swainson, p. 252; Dunwoody, p. 72; Northall, p. 474; Inwards, p. 201; Sloane 2, p. 122.

3233. When black snails on the road you see, then on the morrow rain will be.
Cited in Whitman, p. 49; Inwards, p. 201; Sloane 2, p. 123; Wilson, p. 747; Lee, p. 60.

3234. When snails and slugs crawl up evergreens and remain there during the whole day, expect rain. Var.: When snails crawl up an evergreen and remain there all day, expect rain.
Cited in Inwards, p. 201; Sloane 2, p. 123; Reeder, no pp.

snake

3235. Along the river, especially in the sloughs, when one sees snakes on tree limbs over-hanging the water or on partly submerged logs, rain may be expected soon.
Cited in Hyatt, p. 20.

3236. As a weather divination, kill a snake: if it remains on its belly, the weather will be fair; if it rolls over on its back, the weather will be wet. Vars.: (a) As a weather divination, kill a snake and throw it up into the air: if the reptile falls, or after falling remains on its belly, the weather will be fair, if on its back, wet. (b) When you kill a snake and it rolls over on its back to die, it will rain; if the snake flips it will drip.
Cited in Hyatt, p. 21; Reeder, no pp.

3237. Bury a snake, good weather to make, hanging it high brings storm clouds nigh. Var.: If you kill a snake and want fair weather, be sure to bury it.
Cited in Sloane 2, p. 55; Wurtele, p. 300.

3238. Hanging a dead snake on a tree will bring rain in a few hours. Var.: Hanging a dead snake on a tree will produce rain in a few hours.
Cited in Dunwoody, p. 72; Inwards, p. 202.

3239. If snakes abandon the water, rain soon follows.
Cited in Hyatt, p. 20.

3240. If snakes along the river abandon the water for high ground, high water is denoted; the higher the ground, the higher the flood stage.
Cited in Hyatt, p. 20.

3241. If snakes appear before February 1st, you are warned of an early spring.
Cited in Hyatt, p. 20.

3242. If you hang a snake on its back on a fence, it will cause rain.
       Cited in Reeder, no pp.
3243. If you kill a black snake in dry weather and hang it up, it will
       rain before morning.
       Cited in Hyatt, p. 21.
3244. If you kill a black snake, skin it and hang the skin on a fence, it
       will rain.
       Cited in Hyatt, p. 21.
3245. If you kill a snake and hang it on a fence, rain will come.
       Cited in Smith, p. 5.
3246. If you kill a snake and hang it up on a fence with his belly up,
       it will storm like hell in five hours.
       Cited in Hyatt, p. 21.
3247. If you kill a snake and lay it on its back, it will rain soon. Vars.:
       (a) If you kill a snake and turn it on its back, it will rain. (b)
       Kill a snake and turn it upside down, it will cause it to rain. (c)
       Kill a snake and turn its belly to the sky for rain.
       Cited in Sloane 2, p. 42; Hyatt, p. 20; Smith, p. 5.
3248. If you kill a snake and let it lie on the ground, it will rain before
       morning.
       Cited in Hyatt, p. 20.
3249. If you kill any kind of snake, skin it and nail the skin to the
       barn; it will rain.
       Cited in Hyatt, p. 21.
3250. In Oregon the approach of snakes indicates that a spell of fine
       weather will follow. Var.: In Oregon the approach of snakes
       indicates that a spell of fine weather will continue.
       Cited in Dunwoody, p. 73; Inwards, p. 202.
3251. Kill a black snake and hang it up in a tree to make it rain. Var.:
       To bring rain kill a black snake and hang him in a tree with his
       belly up.
       Cited in Smith, p. 5; Wurtele, p. 300.
3252. Rain is foretold by the appearance and activity of snakes. Vars.:
       (a) More snakes than usual during the day is an indication of
       rain. (b) Snakes expose themselves on the approach of rain.
       Cited in Swainson, p. 252; Dunwoody, p. 73; Garriott, p. 22;
       Inwards, p. 202; Hand, p. 22; Hyatt, p. 20.
3253. Snakes and snake trails may be seen near houses and roads
       before rain. Vars.: (a) Snake tracks in the dust of a road mean

rain within twenty-four hours. (b) Snake trails may be seen near houses before rain. (c) Snakes and snake trails are often seen near houses and roads before rain.
Cited in Dunwoody, p. 73; Inwards, p. 202; Hyatt, p. 20; Lee, p. 58.

3254. Snakes in great numbers during the spring foretell a dry summer.
Cited in Hyatt, p. 20.

3255. The tracks of a snake that has zigzagged back and forth across a road signify rain within several days.
Cited in Hyatt, p. 20.

3256. To have a snake cross your path betokens rain.
Cited in Hyatt, p. 20.

3257. To see at anytime or anywhere a stretched-out live snake is a storm warning.
Cited in Hyatt, p. 20.

3258. When snakes hunt food, rain may be expected; after a rain, they cannot be found.
Cited in Dunwoody, p. 73; Inwards, p. 202; Lee, p. 58.

snipe

3259. The drumming of the snipe in the air, and the call of the partridge, indicate dry weather and frost at night.
Cited in Mitchell, p. 227; Inwards, p. 190.

3260. Winter is broken by the first cry of the snipe.
Cited in Hyatt, p. 23.

snow

3261. A fall of snow in May is worth a ton of hay.
Recorded in Bryant (S.C.).

3262. A heavy November snow will last until April.
Cited in Dunwoody, p. 75.

3263. A year of snow, a year of plenty.
Cited in Wright, p. 11.

3264. April snow is as good as lamb's manure.
Cited in Freier, p. 84.

3265. As many days as the snow remains on the trees, just so many days will it remain on the ground.
Cited in Dunwoody, p. 76.

3266. As many days as there are between the first snow and Christmas, so many will be the snows that winter.
Cited in Hyatt, p. 11.

3267. As many days old as the moon is at the first snow, there will be as many snows before crop planting time.
Cited in Dunwoody, p. 63; Sloane 2, p. 114.

3268. Heavy snows in winter favor the crops of the following summer.
Var.: More snow in winter, more wheat in spring.
Cited in Dunwoody, p. 75; Sloane 2, p. 113.

3269. If a snow stays on the ground longer than seven days, it will be a long winter.
Cited in Smith, p. 10.

3270. If all snow melts on reaching the ground, except occasional patches in fence corners and other sheltered places, it will soon be snowing again.
Cited in Hyatt, p. 11.

3271. If all snow melts on reaching the ground, the storm will be a flurry only.
Cited in Hyatt, p. 11.

3272. If it is snowing and the sun comes out, expect snow next day.
Cited in Hyatt, p. 11.

3273. If it snows in May, look for an early summer and a late winter.
Cited in Hyatt, p. 11.

3274. If it snows on February 2nd only as much as to be seen on a black ox, then summer will come soon.
Cited in Dunwoody, p. 90.

3275. If snow falls in flakes which increase in size, expect a thaw.
Cited in Dunwoody, p. 117.

3276. If the first snow falls on moist, soft earth, it indicates a small harvest; but if upon hard, frozen soil, a good harvest.
Cited in Dunwoody, p. 99; Inwards, p. 64.

3277. If the first snow sticks to the trees, it foretells a bountiful harvest.
Cited in Dunwoody, p. 75; Sloane 2, p. 113.

3278. If the snow comes down crisscrossed it will be a bad winter.
Recorded in Bryant (Mich.).

3279. If there be neither snow nor rain, then will be dear (expensive) all sorts of grain.
Cited in Marvin, p. 211.

3280. If there is no snow before January, there will be more snow in March and April.
Cited in Dunwoody, p. 75; Garriott, p. 44; Hand, p. 44; Freier, p. 81.

3281. In the winter, a heavy snow is predicted if the barometer falls and the temperature rises.
Cited in Freier, p. 25.

3282. It will snow again after a snowfall if the snow remains on the ground for three days.
Cited in Smith, p. 10.

3283. Much February snow a fine summer doth show.
Cited in Wilshere, p. 7.

3284. Much snow, much hay.
Recorded in Bryant (Ohio).

3285. Never eat the first snow, it is full of germs, thereafter snow is pure enough to eat.
Cited in Smith, p. 10.

3286. Snow clears the air of all germs.
Cited in Smith, p. 10.

3287. Snow coming two or three days after new moon will remain on the ground some time, but that falling just after full moon will soon go off.
Cited in Dunwoody, p. 63.

3288. Snow feather, fair weather.
Recorded in Bryant (Wis.).

3289. Snow for se'nnight (seven nights) is a mother to the earth, for ever after a stepmother.
Cited in Wilson, p. 748.

3290. Snow hanging in ditches and along hedges is waiting for more.
Var.: Snow hanging in ditches and along hedges, there'll be more to fetch it away.
Cited in Wilshere, p. 16.

3291. Snow in April is manure; snow in March devours.
Cited in Dunwoody, p. 95; Inwards, p. 52.

3292. Snow in March is bad for fruit and grapevine.
Cited in Garriott, p. 42.

3293. Snow is generally preceded by a general animation of man and beast which continues until the snowfall ends.
Cited in Dunwoody, p. 74.

3294. Snow is the poor man's fertilizer, and good crops will follow a winter of heavy snowfall. Vars.: (a) Snow takes the place of manure. (b) Spring snow is poor man's fertilizer.
Cited in Dunwoody, p. 75; Whiting 1, p. 404; Whiting 2, p. 580.

3295. Snow never comes until the streams are full.
Cited in Smith, p. 10.

3296. The date of the first snowfall indicates the number of snowfalls there will be that winter.
Cited in Smith, p. 10.

3297. The first snow comes six weeks after the last thunderstorm in September.
Cited in Freier, p. 84.

3298. The more snow, the more healthy the season.
Cited in Dunwoody, p. 75; Sloane 2, p. 113.

3299. The number of days the last snow remains on the ground indicates the number of snowstorms which will occur during the following winter.
Cited in Dunwoody, p. 75; Sloane 2, p. 114.

3300. The snows are heaviest on the leeward side of a lake.
Cited in Alstad, p. 35.

3301. There will be as many snows as there are fogs during the month of August.
Recorded in Bryant (N.Y.).

3302. Under water, dearth; under snow, bread. Var.: Under water, famine; under snow. bread. Comment: Dearth means scarcity or famine.
Cited in Denham, p. 26; Cheales, p. 20; Northall, p. 434; Wright, p. 11; Inwards, p. 35.

3303. Whatever day it is that the first snow falls on will indicate just how many snow falls there will be within that year. Var.: The date of the month on which we have the first snowfall will be the number of winter snows.
Cited in Hyatt, p. 11; Smith, p. 10.

3304. When in the ditch the snow doth lie, 'tis waiting for more by and by.
Cited in Dunwoody, p. 74. Recorded in Bryant (Wis.).

3305. When it snows the Old Lady is plucking her white goose.
Cited in Smith, p. 9.

3306. When snow falls in the mud it remains all winter.
Cited in Dunwoody, p. 75.

3307. When the first snow remains on the ground some time, in places
not exposed to the sun, expect a hard winter.
Cited in Dunwoody, p. 75.

3308. When the snow falls dry it means to lie, but flakes light and soft
bring rain oft.
Cited in Dunwoody, p. 74; Sloane 2, p. 113; Lee, p. 30.

3309. Whenever there is much snow, a fruitful season generally
follows. Var.: When there's lots of snow, a fruitful crop will
often grow.
Cited in Whitman, p. 44; Freier, p. 82.

3310. You'll have as many snows in the winter as there are fogs in the
autumn.
Cited in Smith, p. 9.

snowball

3311. Cut a snowball in halves, if it is wet inside, the snow will pass
off with rain; if it is dry inside, the snow will be melted by the
sun. Var.: Cut a snowball in half, wet center means rain; dry
center can only be melted by the sun.
Cited in Dunwoody, p. 75; Lee, p. 30.

snowbird

3312. Snowbirds flying along a fence or near a grove of trees are a
sign of colder weather within twenty-four hours.
Cited in Hyatt, p. 23.

3313. When snowbirds gather in flocks and light on fences and hedges,
expect rain.
Cited in Dunwoody, p. 40.

snowdrop

3314. The snowdrop, in purest white array, first rears her head on
Candlemas Day. Comment: Candlemas Day is February 2nd.
Cited in Swainson, p. 262.

snowflake

3315. If snowflakes increase in size, a thaw will follow.
Cited in Dunwoody, p. 75.

3316. Large snowflakes, short snowstorm; small snowflakes, long snowstorm. Vars.: (a) Large snowflakes foretell a small snow fall. (b) Small snowflakes mean a big snowfall.
Cited in Hyatt, p. 11; Smith, p. 10.

snowstorm

3317. A snowstorm, in the middle of spring, not infrequently proves to be the forerunner of the first hot weather, which is developed in ten days, or at most two weeks after it.
Cited in Steinmetz 2, p. 227.

3318. If a snowstorm begins when the moon is young, the rising of the moon will clear away the snow. Var.: If a snowstorm begins when the moon is young, it will cease at moonrise.
Cited in Dunwoody, p. 75; Inwards, p. 92.

3319. If after a snowstorm you find snow sticking to the sides of trees, it will snow again within a few hours.
Cited in Hyatt, p. 11.

3320. There will be as many snowstorms during the season as there are days remaining in the month after the time of the first snow.
Cited in Dunwoody, p. 74.

3321. There will be as many snowstorms during the winter as the moon is days old at the first snowstorm.
Cited in Dunwoody, p. 63.

soap

3322. Make soap on the growing of the moon to make it thick.
Recorded in Bryant (Miss.).

3323. Soap covered with moisture indicates bad weather.
Cited in Dunwoody, p. 117; Garriott, p. 22; Inwards, p. 220; Hand, p. 22.

3324. Soap gets slippery before a rain.
Cited in Freier, p. 35.

soot

3325. If on lifting up a stove lid the soot on the bottom burns off, look for rain in summer and snow in winter.

Cited in Hyatt, p. 31.

3326. If soot drops to the ground or back down the chimney, rain is coming in summer and snow in winter.
Cited in Hyatt, p. 31.

3327. Soot burning in the chimney means it will rain.
Recorded in Bryant (Wash.).

3328. Soot burning on the back of the chimney indicates storms.
Cited in Inwards, p. 222.

3329. Soot falls down before a rain.
Cited in Freier, p. 26.

3330. Soot frequently falling down chimneys is an indication of coming rain. Vars.: (a) If soot falls down the chimney, rain will ensue. (b) Soot falling down the chimney, rain.
Cited in Steinmetz 2, p. 292; Inwards, p. 222; Page, p. 7.

3331. Soot hanging from the bars of the grate, a sign of wind.
Cited in Inwards, p. 222.

3332. When in cold weather the soot falls from the chimney, the weather will change.
Cited in Dunwoody, p. 117.

3333. When the soot burns on the chimney, it is going to snow.
Cited in Smith, p. 10.

3334. When the soot sparkles on pots over the fire, rain follows.
Cited in Inwards, p. 222.

sound

3335. A good hearing day is a sign of wet.
Cited in Dunwoody, p. 19; Inwards, p. 151; Alstad, p. 107; Sloane 2, p. 118.

3336. Bells are heard at greater distances before rain.
Cited in Dunwoody, p. 106.

3337. If a train whistle sounds dull, look for rain; if sharp, nice weather.
Cited in Hyatt, p. 12.

3338. If the noise (sound) of a steamer or a railway train is heard at a great distance, bad weather is predicted.
Cited in Mitchell, p. 225.

3339. If the noise (sound) of a train is heard at a great distance, rain is predicted. Vars.: (a) A distant train whistle heard clearly is a

sign of rain. (b) If you hear a train whistling at a greater distance than usual, it portends rain.
Cited in Dunwoody, p. 18; Hyatt, p. 12; Smith, p. 8.

3340. In winter, when the sound of the breakers on the shore is unusually distinct, frost is indicated.
Cited in Mitchell, p. 225; Dunwoody, p. 21.

3341. Much sound in the air is a sign of rain. Vars.: (a) Distant sounds heard with distinctness during the day indicate rain. (b) When on calm days the sound is carried far, rain follows.
Cited in Dunwoody, p. 19.

3342. Sound travelling far and wide, a stormy day will betide. Var.: Sounds are heard with unusual clearness before a storm.
Cited in Dunwoody, p. 117; Humphreys 1, p. 442; Humphreys 2, p. 68; Whitman, p. 35; Inwards, p. 152; Alstad, p. 105; Sloane 2, p. 118; Lee, p. 37; Freier, p. 65. Recorded in Bryant (N.J., Oreg.).

3343. Stringed instruments giving forth clear, ringing sounds indicate fair weather. Var.: When stringed instruments give forth clear, ringing sounds, there will be fair weather.
Cited in Dunwoody, p. 117; Inwards, p. 218; Lee, p. 74.

3344. The sound of breakers from a distant shore foretells rain.
Recorded in Bryant (Mich.).

sow

3345. Sow dry and plant wet.
Cited in Denham, p. 11; Cheales, p. 25; Northall, p. 482.

sow thistle

3346. The non-closing of the flower heads of the sow thistle warn us that it will rain next day, whilst the closing of them denotes fine weather.
Cited in Inwards, p. 215.

spaniel

3347. When the spaniel sleeps, it indicates rain.
Cited in Dunwoody, p. 33.

spark

3348. If sparks fly when a pot of water is put on the fire, it will rain.
Cited in Lee, p. 75.

sparrow

3349. If sparrows chirp a great deal, wet weather will ensue. Vars.: (a)
If sparrows chirp a great deal, expect a storm. (b) If sparrows
make a lot of noise, rain will follow. (c) If the sparrow makes
a lot of noise, rain will follow. (d) The chirping of the sparrow
in the morning signifies rain.
Cited in Swainson, p. 245; Inwards, p. 195; Sloane 2, p. 130;
Lee, p. 64; Freier, p. 22.

3350. If sparrows mate in March, there will be six more weeks of cold
weather.
Cited in Hyatt, p. 23.

3351. If the hedge sparrow is heard before the grapevine is putting
forth its buds, it is said that a good crop is in store.
Cited in Dunwoody, p. 37; Inwards, p. 195.

3352. Sparrows collecting into large flocks mean an early autumn.
Cited in Hyatt, p. 23.

3353. Sparrows group together and keep up a discordant chirping
before rain.
Cited in Dunwoody, p. 39.

3354. Sparrows will collect into large flocks just before weather turns
colder.
Cited in Hyatt, p. 23.

3355. When sparrows perch and fly together in clusters, it will rain.
Cited in Lee, p. 64.

speedwell

3356. When the corona of the speedwell and stichwort contract, expect
rain. Var.: The germander speedwell closes its blue petals before
rain, and opens them again when it has ceased.
Cited in Dunwoody, p. 67; Inwards, p. 216.

spider

3357. Before rain, spiders "put their house in order" by drawing in
their tackle, "shutting up" completely. Var.: Before rain or wind

spiders fix their frame lines unusually short.
Cited in Steinmetz 2, p. 290; Inwards, p. 205.

3358. Flies come down and fly low in a moisture-laden atmosphere, so expect rain when spiders are busy making their webs at high speed near the ground.
Cited in Inwards, p. 206.

3359. If garden spiders forsake their webs, rain is at hand. Vars.: (a) If garden spiders forsake their webs, it indicates rain. (b) If spiders leave their webs, expect rain. (c) Spiders desert their webs before a rain.
Cited in Swainson, p. 256; Garriott, p. 24; Inwards, p. 206; Hand, p. 24; Sloane 2, p. 123; Hyatt, p. 18; Freier, p. 31.

3360. If in a rainy spell you find spiders making webs between sundown and dark, it will not rain that night.
Recorded in Bryant (Miss.).

3361. If spiders are indolent, rain generally soon follows. Vars.: (a) If spiders are indolent, rain will soon come. (b) If spiders are totally indolent, rain generally soon follows.
Cited in Garriott, p. 24; Inwards, p. 206; Hand, p. 24; Sloane 2, p. 123.

3362. If spiders are many, and spinning their webs, the weather is fair.
Cited in Freier, p. 31.

3363. If spiders break off and remove their webs, the weather will be wet.
Cited in Dunwoody, p. 58; Inwards, p. 206.

3364. If spiders in spinning their webs make the terminating filaments long, we may in proportion to their lengths expect rain.
Cited in Dunwoody, p. 58; Inwards, p. 206.

3365. If spiders make new webs and ants build new hills, the weather will be clear. Var.: When spiders build new webs, the weather will be clear.
Cited in Dunwoody, p. 58; Inwards, p. 206; Sloane 2, p. 123.

3366. If spiders mend their webs between 6:00 and 7:00 p.m., it is a sign of a serene night.
Cited in Inwards, p. 206.

3367. If spiders undo their webs, tempests follow.
Cited in Inwards, p. 206.

3368. If the garden spiders break and destroy their webs and creep away, expect continued rain.

Cited in Dunwoody, p. 56; Inwards, p. 207.

3369. If the spider works during rain, it is an indication that the weather will be clear. Vars.: (a) If spiders are many and spinning their webs, the spell (rain) will soon be very dry. (b) If the spider works during rain, it is an indication that the weather will soon be clear.
Cited in Dunwoody, p. 58; Inwards, p. 206; Freier, p. 31.

3370. If you kill a spider, it will rain within twenty-four hours.
Cited in Hyatt, p. 18.

3371. If you kill a spider on Friday, it will rain on Sunday.
Cited in Hyatt, p. 18.

3372. If you kill a spider on Sunday, it will rain on Monday.
Cited in Hyatt, p. 18.

3373. If you notice outdoor spiders mending their webs, there will not be any rain that day.
Cited in Hyatt, p. 18.

3374. If you step on a spider it will rain.
Cited in Smith, p. 4.

3375. Large spiders trying to get into the house all summer signify an extremely cold winter.
Cited in Hyatt, p. 19.

3376. Many spiders in the house; much rain soon. Var.: Spiders are more evident inside house before rain.
Cited in Hyatt, p. 18; Wilshere, p. 23.

3377. Spiders are restless and uneasy, and frequently drop from the wall, before rain, the moist air getting into their webs and making them heavy. Vars.: (a) If spiders fall from their webs or from the walls, it signifies rain. (b) Spiders fall from their webs before a rain.
Cited in Steinmetz 1, p. 111; Inwards, p. 206; Alstad, p. 100.

3378. Spiders creep out of their holes and narrow receptacles against wind or rain.
Cited in Swainson, p. 256.

3379. Spiders generally change their webs once every twenty-four hours. If they make the change between 6:00 and 7:00 p.m. expect a fair night. If they change in the morning, a fine day may be expected. If they work during rain, expect fine weather soon, and the more active and busy the spider the finer will be

the weather.
Cited in Dunwoody, p. 58; Inwards, p. 206.
3380.   Spiders strengthening their webs indicate rain.
Cited in Dunwoody, p. 57; Garriott, p. 24; Hand, p. 24.
3381.   Spiders, when they are seen crawling on the walls more than usual indicate that rain will probably ensue. Var.: Spiders in motion indicate rain.
Cited in Dunwoody, p. 58; Inwards, p. 206.
3382.   The ample working of the spinner (spider) in the air shows a tempest.
Cited in Inwards, p. 205.
3383.   When spiders weave their webs by noon, fine weather is coming soon.
Cited in Lee, p. 51.
3384.   When spiders work at their webs in the morning expect a fair day.
Cited in Dunwoody, p. 58.
3385.   When the spider cleans its web fair weather is indicated.
Cited in Dunwoody, p. 58; Inwards, p. 206.

spider web
3386.   If spider webs fly in the autumn with a south wind, expect east winds and fine weather.
Cited in Dunwoody, p. 58.
3387.   If spiders' frame lines are very long, the weather will usually be fine for fourteen days. Vars.: (a) If spider frames are long, they predict fair weather. (b) Spiders spin long webs for hot, dry weather.
Cited in Inwards, p. 205; Sloane 2, p. 123; Wilshere, p. 23.
3388.   In September more spider webs than usual presages an early winter with cold weather.
Cited in Hyatt, p. 19.
3389.   Long single, separate spider webs on grass is a sign of frost next night. Var.: Long, single, separate spiders' webs on grass indicate frost next night.
Cited in Dunwoody, p. 58; Inwards, p. 206.
3390.   Many spider webs on the grass is a prediction of dry weather.
Cited in Hyatt, p. 19.

3391. Spider webs floating at autumn sunset, bring a night frost, this you may bet.
Cited in Dunwoody, p. 55; Inwards, p. 206.

3392. Spider webs scattered thickly over a field covered with dew glistening in the morning sun indicate rain. Var.: Many spider webs on the grass is a prediction of rain.
Cited in Dunwoody, p. 57; Inwards, p. 206; Hyatt, p. 19.

3393. When, after a long drought, you observe in hedges some very dense woven spiders' webs, funnel-shaped, there will be a change of weather within three days.
Cited in Inwards, p. 206.

3394. When spider webs are seen floating about in the air, farmers regard it as a sign of coming rain. Var.: Spider webs floating in the air mean rain.
Cited in Mitchell, p. 231; Dunwoody, p. 14; Hyatt, p. 19.

3395. When spiders' webs in air do fly, the spell will soon be very dry.
Cited in Dunwoody, p. 57; Whitman, p. 50; Inwards, p. 206; Freier, p. 31.

3396. When you see the ground covered with spider webs which are wet with dew, and there is no dew on the ground, it is a sign of rain before night, for the spiders are putting up umbrellas.
Cited in Dunwoody, p. 58; Garriott, p. 24; Hand, p. 24.

spring

3397. A cold spring kills the roses.
Cited in Dunwoody, p. 89.

3398. A dry spring, rainy summer.
Cited in Garriott, p. 45; Whitman, p. 44; Hand, p. 45; Freier, p. 77. Recorded in Bryant (Utah).

3399. A January spring is worth nothing. Var.: A January spring is nothing worth.
Cited in Denham, p. 25; Swainson, p. 20; Cheales, p. 20; Dunwoody, p. 94; Northall, p. 431; Brunt, p. 67; Inwards, p. 36; Page, p. 43; Wilshere, p. 7; Freier, p. 81.

3400. A late spring is a great blessing. Var.: A late spring, a great blessing.
Cited in Denham, p. 39; Swainson, p. 11; Dunwoody, p. 89; Northall, p. 476; Garriott, p. 45; Whitman, p. 44; Hand, p. 45.

3401. A late spring is bad for cattle, and an early spring for corn.
Vars.: (a) A late spring is good for corn, but bad for cattle. (b)
A late spring is good for corn but not cattle.
Cited in Dunwoody, p. 89; Garriott, p. 45; Whitman, p. 44;
Hand, p. 45; Freier, p. 79.

3402. A late spring never deceives.
Cited in Dunwoody, p. 89; Garriott, p. 45; Humphreys 1, p.
430; Humphreys 2, p. 9; Whitman, p. 44; Hand, p. 45;
Wurtele, p. 298; Wilshere, p. 17.

3403. A wet spring is a sign of dry weather for harvest. Var.: A wet
spring, a dry harvest.
Cited in Denham, p. 32; Swainson, p. 11; Dunwoody, p. 91;
Inwards, p. 30.

3404. A wet spring, a well stocked-cellar.
Cited in Hyatt, p. 34.

3405. Better a late spring and bear, than early blossom and blast.
Cited in Garriott, p. 45; Whitman, p. 44; Hand, p. 45; Freier,
p. 78.

3406. If spring begins between midnight and morning, the spring will
be mild.
Cited in Boughton, p. 124.

3407. If the spring is cold and wet, the autumn will be hot and dry.
Cited in Dunwoody, p. 90; Garriott, p. 45; Whitman, p. 44;
Inwards, p. 30; Hand, p. 45; Freier, p. 78.

3408. If there's spring in winter, and winter in spring, the year won't
be good for anything. Var.: When there is a spring in the winter,
or a winter in the spring, the year is never good.
Cited in Swainson, p. 15; Inwards, p. 30. Recorded in Bryant
(Wis.).

3409. In spring a tub of rain makes a spoonful of mud; in autumn a
spoonful of rain makes a tub of mud. Var.: In spring a tub of
rain makes a spoonful of mud.
Cited in Dunwoody, p. 90; Wright, p. 36; Inwards, p. 30.

3410. Many changes take place on a spring day.
Recorded in Bryant (Miss.).

3411. The first spring day is in the devil's pay.
Recorded in Bryant (N.Y.).

3412. The spring is not always green.
Cited in Denham, p. 31; Swainson, p. 15; Inwards, p. 30.

3413. Through all the sad and weary hours which cold and dark and storms will bring, we scarce believe in what we know, that time drags on at last to spring.
Cited in Wright, p. 10.

3414. Weather the first day of spring governs weather the next three months.
Cited in Smith, p. 11.

spring (water)

3415. Springs running flusher are an indication of rain. Vars.: (a) Springs increase their flow just before a rain. (b) Springs rise against rain. (c) Springs start to flow just before a rain. (d) When spring water rises, it's a sign of rain.
Cited in Inwards, p. 222; Smith, p. 8; Freier, p. 26; Reeder, no pp.

3416. When the spring that's low begins to flow; then sure, we know, 'twill rain or snow.
Cited in Humphreys 2, p. 75; Whitman, p. 50.

squall

3417. Generally squalls are preceded, or accompanied, or followed by clouds; but the dangerous white squall of the West Indies is indicated only by a rushing sound and by white wave crests to windward.
Cited in Inwards, p. 128; Hand, p. 13.

3418. Squalls making up on the flood tide will culminate about high water; those making on ebb tide will culminate about low water.
Cited in Dunwoody, p. 87.

3419. When rain squalls break to the westward, it is a sign of foul weather; when they break to leeward, it is a sign of fair weather.
Cited in Dunwoody, p. 70.

squirrel

3420. Baby squirrels found in open nests during the latter part of February betoken an early spring.
Cited in Hyatt, p. 30.

3421. If squirrels are active or chase each other up and down trees, it warns you of unsettled weather; if they are inactive, settled weather.
Cited in Hyatt, p. 30.

3422. If squirrels bury their nuts deep in the ground, a cold winter is presaged; if under the fallen leaves or near the surface of the ground, a warm winter.
Cited in Hyatt, p. 30.

3423. If the squirrel has hoarded a small quantity of nuts, there will be an open winter; if a large quantity, a bitter winter.
Cited in Hyatt, p. 30.

3424. It will be a very cold January and February when squirrels build their nests real large and very deep in the fall. When they build small ones and not deep, it is a sign of a warm January and February.
Cited in Hyatt, p. 30.

3425. The gathering of nuts by squirrels is followed by bad weather.
Cited in Hyatt, p. 30.

3426. When squirrels and small animals lay away a larger supply of food than ususal, it indicates that a long and severe winter will follow. Var.: As a rule squirrels hoard nuts in the autumn, but you will see them also hoarding grain before a hard winter.
Cited in Dunwoody, p. 33; Garriott, p. 40; Hand, p. 40; Hyatt, p. 30.

3427. When squirrels are scarce in the autumn, it indicates a cold winter.
Cited in Dunwoody, p. 33; Garriott, p. 40; Hand, p. 40; Freier, p. 79.

3428. When squirrels eat nuts on the tree, weather as warm as warm can be. Var.: When a squirrel eats nuts in a tree, weather as warm as warm can be.
Cited in Inwards, p. 185; Lee, p. 67.

3429. When squirrels gather a large number of hickory nuts a severe winter season is coming. Var.: When unusual quantities of walnuts are gathered by squirrels, a severe season is ahead.
Cited in Smith, p. 13.

3430. When squirrels lay in a winter supply of nuts, expect a cold winter. Vars.: (a) When squirrels lay in a large supply of nuts,

expect a cold winter. (b) When squirrels lay in a large supply of nuts, there will be a severe winter.
Cited in Dunwoody, p. 33; Inwards, p. 185; Freier, p. 79.

3431. When squirrels store nuts high in trees, we will have a winter of deep snow.
Cited in Freier, p. 82.

3432. When the flying squirrels sing in midwinter it indicates an early spring.
Cited in Garriott, p. 40; Hand, p. 40.

3433. When the ground squirrel is seen in winter, it is a sign that snow is about over.
Cited in Dunwoody, p. 31; Garriott, p. 40; Hand, p. 40.

St. Andrew's Day

3434. If on St. Andrew's Day in the evening, much dew or wet remains on the grass, it betokens a wet season to follow; if dry, the contrary. Comment: St. Andrew's Day is November 30th.
Cited in Swainson, p. 149.

3435. On St. Andrew's Day the night is twice as long as the day.
Cited in Swainson, p. 149; Northall, p. 455.

St. Anne's Day

3436. If it rains on St. Anne's Day, it will rain for one month and one week. Var.: Rain on St. Anne's will continue for a month and a week. Comment: St. Anne's Day is July 26th.
Cited in Sloane 2, p. 94; Lee, p. 150.

St. Barnabas' Day

3437. Barnaby bright, the longest day and the shortest night. Vars.: (a) Barnaby bright; all day, no night. (b) Barnaby bright, Barnaby bright, the longest day and the shortest night. (c) On St. Barnabas' Day, the sun comes to stay. Comment: St. Barnabas' Day is June 11th.
Cited in Swainson, p. 104; Cheales, p. 22; Northall, p. 452; Wright, p. 73; Sloane 2, p. 93; Wilson, p. 31; Simpson, p. 10.

3438. Rain on St. Barnabas' Day is a good harvest in grapes. Var.: Rain on Barnabas is good for grapes.
Cited in Dunwoody, p. 97; Garriott, p. 42; Inwards, p. 57; Hand, p. 42; Lee, p. 149.

3439. When St. Barnaby bright smiles night and day, poor ragged robin blooms in the hay.
Cited in Swainson, p. 263.

St. Bartholomew's Day
3440. As Bartholomew's Day, so the whole autumn. Comment: St. Bartholomews' Day is August 24th.
Cited in Dunwoody, p. 98; Whitman, p. 39; Inwards, p. 62.
3441. At St. Bartholomew there comes cold dew. Vars.: (a) At St. Barthol'mew, then comes cold dew. (b) St. Bartholomew brings the cold dew. Comment: Because the nights begin to be cold on August 24th.
Cited in Denham, p. 55; Swainson, p. 124; Northall, p. 453; Inwards, p. 61; Lee, p. 61; Wilshere, p. 13.
3442. If Bartlemy's Day be fair and clear, we may hope for a prosperous autumn that year. Var.: If St. Bartholomews' Day be clear, a prosperous autumn comes that year.
Cited in Northall, p. 453; Inwards, p. 62; Page, p. 39; Wilshere, p. 12.
3443. If it rains on St. Bartholomew's Day, it rains for forty days. Var.: If it rains on St. Bartholomew's Day, it will rain for forty days.
Cited in Swainson, p. 125; Dunwoody, p. 103; Whitman, p. 39.
3444. If St. Bartholomew's Day be misty, the morning beginning with a hoar frost, the cold weather will soon come, and a hard winter.
Cited in Inwards, p. 61; Sloane 2, p. 95.
3445. St. Barthelemy's mantle wipes dry all the tears that St. Swithin can cry. Comment: St. Swithin's Day is July 15th.
Cited in Garriott, p. 43; Whitman, p. 40; Inwards, p. 62; Hand, p. 42.
3446. Thunderstorms after Bartholomew's Day are mostly violent. Var.: Thunderstorms after Bartholomew's Day are more violent.
Cited in Dunwoody, p. 98; Inwards, p. 62.

St. Benedict's Day
3447. St. Benedict, sow thy pease or keep them in thy rick. Comment: St. Benedict's Day is March 21st.
Cited in Northall, p. 451.

St. Benoit's Day

3448. If it rains on St. Benoit's Day, it will rain for forty days.
Comment: St. Benoit's Day is March 21st.
Cited in Lee, p. 149; Freier, p. 91.

St. Bridget's Day

3449. As long as the sunbeam comes in on Bridget's feast day, the
snow comes before May Day. Comment: St. Bridget's Day is
February 1st.
Cited in Inwards, p. 41.

3450. Bridget's feast day white, every ditch full. Var.: Snow on St.
Bridget's Day indicates full ditches come spring.
Cited in Inwards, p. 41; Lee, p. 148.

St. Bullion's Day

3451. If Bullion's Day be dry, there will be a good harvest. Comment:
St. Bullion's Day is July 4th.
Cited in Marvin, p. 207; Inwards, p. 50; Sloane 2, p. 95.

3452. If St. Martin Bullion's Day brings rain, it will rain for forty days
and nights. Comment: In Scotland this day is called St. Martin
of Bullion's Day, for what reason it is uncertain.
Cited in Lee, p. 150.

3453. If the deer rise dry and lie down dry on Bullion's Day, there will
be a good harvest.
Cited in Swainson, p. 113; Inwards, p. 58.

3454. St. Bullion's Day, if ye be fair, for forty days 'twill rain nae
mair.
Cited in Swainson, p. 113; Marvin, p. 207; Whitman, p. 39;
Inwards, p. 58.

St. Catherine's Day

3455. At Catherine foul or fair, so will be the next February. Var.: At
St. Catherine's fair or foul, so it will be next February. Com-
ment: St. Catherine's Day is November 25th.
Cited in Dunwoody, p. 99; Inwards, p. 67; Wilshere, p. 15.

3456. Sts. Catherine and Clement's Day inaugurate winter. Comment:
St. Clement's Day is November 23rd.
Cited in Sloane 2, p. 96.

St. Chad's Day

3457. On or before St. Chad, every goose lays, both good and bad.
Vars.: (a) On St. Chad's Day every goose lays both good and
bad. (b) Before St. Chad every goose lays, both good and bad.
Comment: St. Chad's Day is March 2nd.
Cited in Denham, p. 40; Northall, p. 450; Wilson, p. 40.

St. Clement's Day

3458. St. Clement gives the winter. Comment: St. Clement's Day is
November 23rd.
Cited in Wilshere, p. 15.

St. David's Day

3459. First comes David, then comes Chad, and then comes Winneral
as though he was mad. Comment: St. David's Day is March 1st,
St. Chad's Day is March 2nd and St. Winneral's Day is March
3rd.
Cited in Inwards, p. 49; Freier, p. 90.

3460. Sow peas and beans on David and Chad, be the weather good or
bad. Var.: David and Chad sow peas, good or bad.
Cited in Denham, p. 40; Swainson, p. 61.

3461. Upon St. David's Day put oats and barley in the clay. Var.: On
St. David's Day put oats and barley in the clay.
Cited in Swainson, p. 61; Northall, p. 450; Wright, p. 24;
Wilson, p. 693; Lee, p. 148.

St. Dorothea's Day

3462. St. Dorothea's Day gives the most snow. Var.: St. Dorothea
brings the most snow. Comment: St. Dorothea's Day is February
6th.
Cited in Dunwoody, p. 94; Inwards, p. 44; Lee, p. 148; Freier,
p. 90.

St. Dunstan's Day

3463. On St. Dunstan's Day the devil may bring a frost to blast the
apple crop. Comment: St. Dunstan's Day is May 19th.
Cited in Sloane 2, p. 92.

St. Edmund's Day

3464. Set garlike and pease St. Edmund to please. Comment: St. Edmund's Day is November 20th.
Cited in Wright, p. 119.

St. Elmo's star

3465. Last night I saw Saint Elmo's stars, with their glittering lanterns all at play, on the tops of the masts and the tips of the spars, and I knew we should have foul weather that day. Comment: Refers to the phenomena known as St. Elmo's fire.
Cited in Swainson, p. 193; Humphreys 2, p. 70; Whitman, p. 36; Sloane 2, p. 127; Lee, p. 86.

St. Eulalie's Day

3466. If the sun smiles on St. Eulalie's Day, it is good for apples and cider. Comment: St. Eulalie's Day is February 12th.
Cited in Lee, p. 148; Freier, p. 90.

St. Gallus' Day

3467. No rain on Gallus, a dry spring will follow. Comment: St. Gallus' Day is October 16th.
Cited in Lee, p. 151.

3468. The weather on St. Gallo's will prevail for forty days. Var.: The weather on St. Gallo's Day will last for forty days.
Cited in Sloane 2, p. 94; Lee, p. 150.

St. George's Day

3469. If it rains on St. George's, he eats all the cherries. Comment: St. George's Day is April 23rd.
Cited in Sloane 2, p. 92.

3470. If on St. George's Day rye has grown so high as to hide a crow, a good harvest may be expected.
Cited in Dunwoody, p. 96; Lee, p. 149.

St. Godelieve's Day

3471. If it rains on St. Godelieve's Day, it will rain for forty days after. Var.: Rain on St. Godelieve's Day will continue for forty days. Comment: St. Godelieve's Day is July 27th.
Cited in Sloane 2, p. 94; Lee, p. 150.

St. Gorgen's Day

3472. At St. Gorgen's Day the meadow turns to hay. Comment: St. Gorgen's Day is April 24th.
Cited in Dunwoody, p. 96.

St. Hilary's Day

3473. St. Hilary's Day is the coldest day of the year. Var.: January 14th, St. Hilary, the coldest day of the year. Comment: St. Hilary's Day is January 14th.
Cited in Swainson, p. 30; Inwards, p. 38; Lee, p. 147; Freier, p. 89.

St. Jacob's Day

3474. Clear on St. Jacob's Day promises plenty of fruit. Comment: St. Jacob's Day is July 20th.
Cited in Dunwoody, p. 98.

3475. If three days previous to St. Jacob's are clear, then the rye will be good.
Cited in Dunwoody, p. 98.

3476. Rain on Jacob's Day, expect a fertile year.
Cited in Lee, p. 149.

St. James' Day

3477. St. James' Day, Oyster Day. Who eats oysters on St. James' Day will never want. Comment: St. James' Day is July 25th.
Cited in Wright, p. 95.

3478. Till St. James' Day be come and gone, you may have hops, or you may have none. Var.: Till St. James is past and gone, there may be hops and there may be none.
Cited in Denham, p. 54; Cheales, p. 22.

St. John's Day

3479. Before St. John's Day we pray for rain; after that we get it anyhow. Comment: St. John the Baptist's Day is June 24th.
Cited in Dunwoody, p. 103; Garriott, p. 42; Inwards, p. 57; Hand, p. 42.

3480. Cut your thistles before St. John, you will have two instead of one.
Cited in Swainson, p. 107.

3481. If it rains on St. John's Eve, the filberts will be spoiled.
Cited in Lee, p. 150.
3482. No crop before St. John's Day is worthy of praising.
Cited in Lee, p. 150.
3483. Previous to St. John's Day we dare not praise barley.
Cited in Dunwoody, p. 97.
3484. Rain on St. John's Day, and we may expect a wet harvest. Var.:
Rain on St. John's Day, expect a wet harvest.
Cited in Dunwoody, p. 97; Inwards, p. 57; Sloane 2, p. 93.
3485. Rain on St. John's will damage the nuts. Var.: Rain on St.
John's Day, damage to nuts.
Cited in Dunwoody, p. 97; Garriott, p. 42; Inwards, p. 57;
Hand, p. 42.

St. Joseph's Day
3486. If on St. Joseph's Day clear, so follows a fertile year. Comment:
St. Joseph's Day is March 19th.
Cited in Dunwoody, p. 94; Garriott, p. 42; Whitman, p. 38;
Inwards, p. 49; Hand, p. 42; Lee, p. 148; Freier, p. 91.

St. Laurence's Day
3487. If the weather is fair on St. Laurence's Day, fine autumn and
good wine. Var.: Fine weather on St. Lawrence's Day indicates
a good autumn. Comment: St.Laurence's Day is August 10th.
Cited in Sloane 2, p. 95; Lee, p. 150.

St. Lucy's Day
3488. Lucy light, the shortest day and the longest night. Comment: St.
Lucy's Day was the shortest day of the year in the old calendar,
December 13th.
Cited in Swainson, p. 152; Northall, p. 455.

St. Luke's Day
3489. There is often, about October 18th, a spell of fine, dry weather,
and this has received the name of St. Luke's little summer.
Cited in Hand, p. 43; Sloane 2, p. 95; Lee, p. 151; Page, p. 39.

St. Mamertus' Day

3490. St. Mamertus, St. Pancras, St. Servatius do not pass without a frost. Comment: Sts. Mamertus', Pancras' and Servatius' Days are May 11th, 12th, and 13th respectively.
Cited in Sloane 2, p. 92.

St. Margaret's Day

3491. "Margaret's Flood"; expect rain. Comment: St. Margaret's Day is July 20th.
Cited in Lee, p. 150.

3492. So much rain often falls about St. Margaret's Day that people often speak of "Margaret's flood." Var.: So much rain falls on St. Margaret's Day that they call it "Margaret's flood."
Cited in Inwards, p. 60; Sloane 2, p. 94.

St. Mark's Day

3493. As long frogs are heard before St. Marc's (Mark's) Day, that long will they keep quiet afterward. Var.: As long before St. Mark's Day as the frogs are heard croaking, so long will they keep quiet afterwards. Comment: St. Mark's Day is April 25th.
Cited in Dunwoody, p. 72; Inwards, p. 53.

3494. Rain on St. Mark's Day speaks ill for fruit crops.
Cited in Lee, p. 149.

St. Martin's Day

3495. At St. Martin's Day winter is on his way. Comment: St. Martin's Day is November 11th.
Cited in Swainson, p. 145.

3496. If it is cold, fair, and dry at Martinmas, the cold in winter will not last long. Var.: If St. Martin's Day is dry and cold, the winter will not be long lasting.
Cited in Dunwoody, p. 92; Garriott, p. 43; Wright, p. 118; Inwards, p. 66; Hand, p. 43; Lee, p. 151.

3497. If the geese at Martin's Day stand on ice, they will walk in mud on Christmas.
Cited in Dunwoody, p. 92; Marvin, p. 211; Inwards, p. 66.

3498. If the leaves of the trees and grapevines do not fall before St. Martin's Day, a cold winter may be expected. Var.: If the leaves do not fall before St. Martin's, expect a cold winter.

Cited in Dunwoody, p. 92; Garriott, p. 43; Inwards, p. 66; Hand, p. 43; Sloane 2, p. 96.

St. Mary's Day

3499. A clear St. Mary's Day, a fruitful year ahead. Comment: St. Mary's Annunciation Day is March 25th.
Cited in Lee, p. 149.

3500. If it rains on St. Mary's Day, it will rain for four weeks.
Cited in Whitman, p. 38; Lee, p. 150; Page, p. 38.

3501. If on St. Mary's Day it is bright and clear, fertile is said to be the year. Var.: St. Mary's bright and clear, fertile is said to be the year.
Cited in Dunwoody, p. 94; Garriott, p. 42; Inwards, p. 49; Hand, p. 42; Freier, p. 91.

3502. On St. Mary's Day sunshine brings much and good wine. Var.: On St. Mary's, sunshine, much good wine.
Cited in Dunwoody, p. 98; Inwards, p. 61; Sloane 2, p. 95.

St. Matthew's Day

3503. After Matthew's no fox will run over the ice. Comment: St. Matthew's Day is September 21st.
Cited in Dunwoody, p. 92.

3504. If it freezes on St. Matthew's Day, it will freeze for a month together.
Cited in Dunwoody, p. 103; Whitman, p. 37; Freier, p. 90.

3505. If the ground is frozen on St. Matthias' Day, it will soon thaw; if the ground has thawed by that day, it will freeze again. Var.: If there is ice on St. Matthias' Day it will break it; if no ice, it will make it. Comment: St. Matthias' Day is September 21st.
Cited in Hyatt, p. 12; Lee, p. 148.

3506. Matthew's Day bright and clear, brings good wine in next year.
Cited in Inwards, p. 63; Lee, p. 150.

3507. St. Matthew breaks the ice; if he finds none he will make it. Var.: St. Matthew breaks the ice, if he finds none he'll have some.
Cited in Denham, p. 27; Dunwoody, p. 103.

3508. St. Matthew brings on the cold dew. Var.: St. Matthew brings the rain and cold dew.
Cited in Inwards, p. 63; Wilshere, p. 13.

3509. St. Matthew's Day makes the days and nights equal.
Cited in Garriott, p. 43; Hand, p. 43.

3510. St. Matthew's rain fattens pigs and goats.
Cited in Sloane 2, p. 95.

St. Medard's Day

3511. If it rains on St. Medard's Day, it will rain again forty days later.
Cited in Whitman, p. 38; Sloane 2, p. 93.

St. Michael's Day

3512. If it does not rain on St. Michael's and Gallus, the farmer will promise a dry spring. Vars.: (a) If it does not rain on St. Michael's and Gallus, a dry spring is indicated for the next year. (b) If it does not rain on St. Michael and St. Gallus the following spring will be dry and propitious. Comment: St. Michael's Day is September 29th and St.Gallus' Day is October 16th. Propitious means presenting favorable circumstances.
Cited in Dunwoody, p. 91; Inwards, p. 64; Page, p. 39; Wilshere, p. 5.

3513. If Michaelmas Day be fair, the sun will shine much in the winter, though the wind at northeast will frequently reign long and be sharp and nipping.
Cited in Swainson, p. 130; Inwards, p. 63.

3514. If on St. Michael's Day the winds blow from the north and east, a cold winter may be expected.
Cited in Dunwoody, p. 92.

3515. If St. Michael brings many acorns, Christmas will cover the fields with snow.
Cited in Dunwoody, p. 99; Garriott, p. 43; Inwards, p. 63; Hand, p. 43.

3516. Light rain on St. Michael's Day is followed by mild winter.
Cited in Dunwoody, p. 92.

3517. So many days old the moon is on Michaelmas Day, so many floods after.
Cited in Denham, p. 56; Swainson, p. 133; Sloane 2, p. 95; Wilson, p. 529; Lee, p. 151.

3518. The Michaelmas moon rises nine nights alike soon. Comment: The nearest moon to the autumnal equinox is called "the Harvest

Moon," rising nearer to the same time each succeeding night, at this time of year, than it does at any other.
Cited in Swainson, p. 133.

St. Pancras' Day
3519. St. Pancras does not pass without frost. Comment: St. Pancras' Day is May 12th.
Cited in Lee, p. 149.

St. Patrick's Day
3520. On St. Patrick's Day the warm side of a stone turns up, and the broadback goose begins to lay. Comment: St. Patrick's Day is March 17th.
Cited in Dunwoody, p. 104; Garriott, p. 42; Inwards, p. 49; Hand, p. 42; Lee, p. 148; Freier, p. 91.

St. Paul's Day
3521. Fair on St. Paul's Day is favorable to all fruits. Comment: St. Paul's Day is January 25th.
Cited in Dunwoody, p. 93; Inwards, p. 38.
3522. If on Saint Paul's it rains and snows, the grain will be costly.
Cited in Dunwoody, p. 93.
3523. If St. Paul's Day be fair and clear, it doth betide a happy year; but if by chance it then should rain, it will make dear all kinds of grain, and if the clouds make dark the sky, then cattle and fowls this year shall die; if blustering winds do blow aloft, then wars shall trouble the realm full oft. Vars.: (a) If St. Paul's Day is fair and clear, it does betide a happy year. (b) If St. Paul's be fair and clear, it doth betide a happy year; if it chances to snow or rain, there will be dear all sorts of grain, or if the wind doth blow aloft, great stirs will vex the world full oft. (c) If St. Paul be fair and clear, then betides a happy year, but if it chance to snow or rain, dear will be all sorts o' grain.
Cited in Denham, p. 24; Dunwoody, p. 103; Northall, p. 444; Brunt, p. 68; Lee, p. 148; Page, p. 36; Wilshere, p. 7; Simpson, p. 196; Freier, p. 89.
3524. If St. Paul's Day be fine, the year will be the same.
Cited in Whitman, p. 37.

3525. If St. Paul's Day is bright and clear, one does hope for a good year.
Cited in Dunwoody, p. 93; Garriott, p. 41; Whitman, p. 37; Inwards, p. 38; Hand, p. 41.

3526. If the sun shine on St. Paul's Day, it betokens a good year; if rain or snow, indifferent; if misty, it predicts great dearth; if thunder, great winds and death of people that year.
Cited in Inwards, p. 39.

3527. Nor Paul nor Swithin rules the clouds and winds. Comment: St. Swithin's Day is July 15th.
Cited in Mitchell, p. 218.

3528. Paul's Day stormy and windy, famine on earth and much death on people; Paul's Day beautiful and fair, abundance on the earth of corn and meal.
Cited in Inwards, p. 39.

3529. St. Paul fair with sunshine, brings fertility to rye and wine.
Cited in Dunwoody, p. 93; Inwards, p. 38.

3530. Upon St. Paul's Day put oats and barley in the clay.
Cited in Dunwoody, p. 104.

St. Peter's Day

3531. If cold at St. Peter's Day, it will last longer. Comment: St. Peter's Day is February 22nd.
Cited in Dunwoody, p. 91; Marvin, p. 208; Inwards, p. 44.

3532. If it rains on St. Peter's Day, the bakers will have to carry double flour and single water; if dry, they will carry single flour and double water. Comment: May refer to Sts. Peter and Paul's Day, June 29th.
Cited in Dunwoody, p. 97; Inwards, p. 57; Sloane 2, p. 93.

3533. The night of St. Peter's shows what weather we will have for the next forty days.
Cited in Dunwoody, p. 91; Inwards, p. 44; Lee, p. 148; Freier, p. 90.

St. Philip's Day

3534. When it rains on St. Philips, the poor will need no help from the rich. Var.: If it rains on St. Philip's Day the poor will not need help from the rich. Comment: St. Philip's Day is May 26th.
Cited in Sloane 2, p. 93; Lee, p. 149.

Sts. Philip & James' Day

3535. If it rains on Sts. Philip and James', a fertile year may be expected. Comment: Sts. Philip's and James' Day is May 1st.
Cited in Dunwoody, p. 97; Inwards, p. 55; Sloane 2, p. 92.

St. Processus' Day

3536. If it rains on the feast of Processus and Martinian, it suffocates the corn. Comment: St. Processus' Day is July 14th, St. Martin of Bullion's Day is July 4th.
Cited in Sloane 2, p. 94.

St. Protasius' Day

3537. If it rains on St. Protasius' Day, it will continue raining throughout forty days. Comment: St. Protasius' (St. Protase) Day is June 19th.
Cited in Whitman, p. 38; Sloane 2, p. 93.

St. Romanus' Day

3538. Romanus bright and clear indicates a goodly year. Comment: St. Romanus Day is February 28th.
Cited in Inwards, p. 44; Lee, p. 148; Freier, p. 90.

St. Servatius' Day

3539. He who shears his sheep before St. Servatius loves more his wool than his sheep. Comment: St. Servatius' Day is May 13th.
Cited in Inwards, p. 55; Sloane 2, p. 92; Lee, p. 149.

St. Simon's Day

3540. St. Simon and St. Jude have a reputation for bad weather. It is a time when the weather breaks up, gales begin, and sailors make for harbor at the first sign of wind. Var.: On St. Simon's and Jude's Day gales begin, St. Luke's Little Summer is ended. Comment: Sts. Simon's and Jude's Day is October 28th, St. Luke's Little Summer is about October 18th.
Cited in Page, p. 40; Wilshere, p. 14.

3541. St. Simon's is a rainy day. Comment: St. Simon's Day is October 28th.
Cited in Sloane 2, p. 95.

3542.   St. Simon's is never dry.
        Cited in Lee, p. 151.

St. Stephen's Day
3543.   If the wind blows much on Stephen's Day, the grape will be bad
        in next year. Vars.: (a) St. Stephen's Day windy, bad for next
        year's grapes. (b) Wind on St. Stephen's Day foretells of bad
        grapes the following year. Comment: St. Stephen's Day is
        December 26th.
        Cited in Dunwoody, p. 100; Inwards, p. 69; Sloane 2, p. 96;
        Lee, p. 151.

St. Sulpicius' Day
3544.   Frost on St. Sulpicius' Day augurs well for the spring. Com-
        ment: St. Sulpicius' Day is January 17th.
        Cited in Freier, p. 89.

St. Swithin's Day
3545.   Against St. Swithin's hasty showers, the lily white reigns queen
        of flowers. Comment: St. Swithin's Day is July 15th.
        Cited in Swainson, p. 263.
3546.   All the tears St. Swithin can cry, St. Bartholomew's dusty
        mantle wipes dry. Comment: St. Bartholomew's Day is August
        24th.
        Cited in Swainson, p. 125; Inwards, p. 60; Wilshere, p. 11.
3547.   If it rains on St. Swithin's Day it will rain for forty days. Var.:
        If on St. Swithin's Day it rains, the forty days thereafter will be
        wet.
        Cited in Whitman, p. 39; Sloane 2, p. 53; Hyatt, p. 10; Smith,
        p. 8.
3548.   If on St. Swithin's Day it proves fair, a temperate winter will
        follow, but if rainy, stormy, or windy, then the contrary.
        Cited in Swainson, p. 115.
3549.   If St. Swithin weeps, the weather will be foul for forty days.
        Vars.: (a) If St. Swithin weeps, it will rain for forty days. (b) If
        St. Swithin weep, that year, the weather will be foul for forty
        days.
        Cited in Swainson, p. 115; Garriott, p. 43; Whitman, p. 39;
        Inwards, p. 59; Sloane 2, p. 94; Hand, p. 43.

3550. St. Swithin is christening the apples, when it rain on this day.
Var.: St. Swithin christens the apples. Comment: St. Swithin's
Day is July 15th.
Cited in Swainson, p. 115; Wright, p. 89; Inwards, p. 59.

3551. St. Swithin's Day, if thou dost rain, for forty days it will
remain: St. Swithin's Day, if thou be fair, for forty days 'twill
rain na mair. Vars.: (a) St. Swithin's Day, and ye do rain, for
forty days it will remain; St. Swithin's Day, and ye be fair, for
forty days, 'twill rain nae mair. (b) St. Swithun's Day, if it do
rain, for forty days it will remain; St. Swithun's Day an' it be
fair, for forty days 'twill rain nae mair. (c) St. Swithin's Day,
if thou be fair, for forty days it will remain; St. Swithin's Day,
if thou bring rain, for forty days it will remain. (d) St. Swithin's
Day, if thou be fair, 'twill rain for forty days nae mair; St.
Swithin's Day, if thou dost rain, for forty days it will remain.
Cited in Denham, p. 52; Steinmetz 2, p. 304; Swainson, p. 115;
Cheales, p. 25; Northall, p. 452; Wright, p. 89; Whitman, p.
39; Inwards, p. 59; Wilson, p. 696; Lee, p. 150; Page, p. 38;
Wilshere, p. 5; Simpson, p. 196.

St. Thomas' Day

3552. Look at the weathercock on St. Thomas' Day at 12 o'clock
(noon), and see which way the wind is; there it will stick for the
next three months. Vars.: (a) Look at the weathercock on St.
Thomas' Day at noon, to see which direction the wind will
remain for the next three months. (b) Look at the weathercock
on St. Thomas at 12 o'clock (noon) and see which way the wind
is; there it will stick for the next lunar quarter. Comment: St.
Thomas' Day is December 21st.
Cited in Swainson, p. 154; Wright, p. 125; Inwards, p. 67;
Sloane 2, p. 96; Wilshere, p. 15.

3553. St. Thomas gray, St. Thomas gray, the longest night and the
shortest day.
Cited in Swainson, p. 154; Northall, p. 455; Wright, p. 124;
Wilson, p. 696; Page, p. 40.

St. Urban's Day

3554. St. Urban's Day inaugurates summer. Comment: St. Urban's
Day is May 25th.

Cited in Sloane 2, p. 92; Lee, p. 149.

St. Valentine's Day

3555. Birds of a feather on St. Valentine's Day will meet together.
Comment: St. Valentine's Day is February 14th.
Cited in Wright, p. 19.

3556. In Valentine, March lays her line.
Cited in Northall, p. 448.

3557. On St. Valentine's Day, beans should be in the clay.
Cited in Northall, p. 448.

3558. Spring is a near neighbor on Valentine's Day.
Cited in Lee, p. 148.

3559. The crocus was dedicated to St. Valentine, and ought to blossom about this time.
Cited in Inwards, p. 44.

3560. Winter breaks her back on St. Valentine's Day.
Cited in Freier, p. 90.

St. Vincent's Day

3561. If St. Vincent's Day is fair, there will be more water than wine.
Comment: St. Vincent's Day is January 22nd.
Cited in Lee, p. 149; Freier, p. 91.

3562. If St. Vincent's has sunshine, one hopes much rye and wine.
Var.: If St. Vincent's has sunshine, we get much rye and wine.
Cited in Dunwoody, p. 93; Garriott, p. 41; Whitman, p. 37; Inwards, p. 38; Hand, p. 41; Freier, p. 89.

3563. On St. Vincent's Day rains cease and winds come.
Cited in Sloane 2, p. 94.

3564. On St. Vincent's Day, winter waxes or wanes.
Cited in Sloane 2, p. 96.

3565. Sun on St. Vincent's Day means good wine crops next season.
Cited in Lee, p. 147.

3566. Sunny on St. Vincent's, we shall have more wine than water.
Cited in Freier, p. 89.

St. Vitus' Day

3567. If St. Vitus' Day be rainy weather, it will rain for thirty days together. Var.: If St. Vitus' Day be rainy, it will rain for thirty more days. Comment: St. Vitus' Day is June 15th.

Cited in Denham, p. 49; Dunwoody, p. 97; Northall, p. 452; Whitman, p. 38; Inwards, p. 57; Sloane 2, p. 93; Wilson, p. 696; Lee, p. 149; Page, p. 38; Wilshere, p. 11.

St. Winwaloe's Day

3568. Winwaloe comes as if he were mad. Comment: St. Winwaloe's Day is March 3rd.
Cited in Lee, p. 148.

star

3569. A multitude of stars means pretty weather and a scarcity of stars means falling weather.
Cited in Hyatt, p. 4.

3570. A star dogging the moon foretells bad weather.
Cited in Inwards, p. 98.

3571. Count the number of stars inside the ring around the moon, the number tells how many days until the next snow. Var.: Count the stars in the circle around the moon to see how many days before the next rain.
Cited in Smith, p. 10. Recorded in Bryant (N.Y.).

3572. Excessive twinkling of stars indicates very heavy dews, rain, and snow. Vars.: (a) Excessive twinkling of stars indicates heavy dews, rain, or snow, or stormy weather in the near future. (b) Excessive twinkling of the stars indicates foul weather. (c) When the stars twinkle very brightly, expect stormy weather in the near future.
Cited in Dunwoody, p. 74; Garriott, p. 28; Hand, p. 28; Freier, p. 55.

3573. If a big star is dogging the moon, wild weather may be expected.
Cited in Inwards, p. 98.

3574. If shooting stars fall in the south in winter, there will be a thaw.
Cited in Dunwoody, p. 74.

3575. If stars are shining at night, it won't rain the next day.
Cited in Smith, p. 7.

3576. If stars do not come out by night, then sunshine won't come out in daylight.
Cited in Smith, p. 14.

3577. If stars twinkle brightly, radiant weather is at hand.
Cited in Hyatt, p. 4.

3578. If there are no falling stars on a bright summer night, expect fine weather.
Cited in Dunwoody, p. 73.

3579. If there are three stars close to the moon, it will be foggy the next day.
Cited in Smith, p. 14.

3580. In summer, when many stars twinkle, clear weather is indicated.
Cited in Dunwoody, p. 73.

3581. Many shooting stars on summer nights indicate hot weather.
Cited in Dunwoody, p. 74; Inwards, p. 100.

3582. Many stars in winter indicate frost.
Cited in Dunwoody, p. 73.

3583. More than usual twinkling of the stars significant of increasing wind. Var.: Stars twinkle, we cry wind.
Cited in Steinmetz 1, p. 116; Alstad, p. 168.

3584. Numerous falling stars presage wind next day. Vars.: (a) In whatever direction a star shoots, the wind will blow next day. (b) Shooting stars denote wind and its direction. (c) When stars shoot precipitant through the sky, approaching wind.
Cited in Mitchell, p. 223; Steinmetz 1, p. 116; Inwards, p. 100; Hyatt, p. 3.

3585. Stars appear clearest in a frosty sky.
Cited in Wilshere, p. 21.

3586. Stars are not seen by sunshine.
Cited in Denham, p. 5; Mieder, p. 561.

3587. Stars that sparkle and seem larger than usual in summer are forecasting rain; in winter, a sharper temperature or frost.
Cited in Hyatt, p. 4.

3588. The dimness of the stars and other heavenly bodies is one of the surest signs of rainy weather. Var.: When dimmer stars disappear, expect rain or snow. Comment: The increase in humidity and haze make stars appear to fade before rain.
Cited in Steinmetz 1, p. 117; Humphreys 2, p. 45; Whitman, p. 32; Sloane 2, p. 58; Freier, p. 55.

3589. When a star tows the moon and another chases her astern, tempestuous weather will follow. Var.: One star ahead of the moon, towing her, and another astern, chasing her, is a sure sign of a storm. Comment: The phenomenon is styled a big star

chasing the moon.
Cited in Dunwoody, p. 74; Inwards, p. 98.

3590. When it gets dark the stars come out.
Cited in Mieder, p. 561.

3591. When the North Star flickers strangely, or appears closer than usual, expect rain.
Cited in Dunwoody, p. 73; Garriott, p. 28; Hand, p. 28.

3592. When the sky is very full of stars, expect rain. Var.: When the sky seems very full of stars, expect frost.
Cited in Dunwoody, p. 73; Garriott, p. 28; Inwards, p. 98; Hand, p. 28; Freier, p. 55.

3593. When the stars begin to hide, soon the rain will betide.
Cited in Alstad, p. 168.

3594. When the stars begin to huddle, the earth will soon become a puddle. Comment: When a mist forms over the sky, as it does before a storm, the smaller stars cease to be visible, while the brighter ones shine dimly with a blur of light about them looking like a cluster of stars.
Cited in Dunwoody, p. 73; Garriott, p. 28; Humphreys 1, p. 440; Humphreys 2, p. 44; Whitman, p. 32; Inwards, p. 98; Hand, p. 28; Hyatt, p. 3; Page, p. 28; Wilshere, p. 21; Alstad, p. 167.

3595. When the stars flicker in a dark background, rain or snow follows soon.
Cited in Dunwoody, p. 73; Garriott, p. 28; Humphreys 1, p. 440; Humphreys 2, p. 46; Whitman, p. 33; Inwards, p. 98; Hand, p. 28; Freier, p. 54.

3596. When the stars look bigger than usual, and pale and dull and without rays, this undoubtedly indicates that the clouds are condensing into rain, which will very soon fall. Var.: When stars appear to be numerous, very large, and dull, and do not twinkle, expect rain.
Cited in Steinmetz 1, p. 117; Dunwoody, p. 73; Garriott, p. 28; Hand, p. 28.

starling

3597. If starlings and crows congregate together in large numbers, expect rain. Vars.: (a) If starlings or crows congregate, expect

rain. (b) When starlings and crows group together in large numbers, expect rain.
Cited in Swainson, p. 246; Inwards, p. 194; Sloane 2, p. 130; Lee, p. 65.

stoat

3598. When stoats put on their white coats, severe weather will last long and snow is certain. Comment: Refers to the European ermine.
Cited in Inwards, p. 185.

stomach

3599. An uneasy stomach always tells you of an advancing storm.
Cited in Hyatt, p. 30.

stone

3600. A sweating stone indicates rain. Vars.: (a) When stones sweat in the afternoon, it indicates rain. (b) When stones sweat, rain you'll get.
Cited in Dunwoody, p. 118; Lee, p. 75; Freier, p. 34.

3601. Quarries of stone and slate indicate rain by a moist exudation from the stones.
Cited in Dunwoody, p. 117; Garriott, p. 21; Inwards, p. 219; Hand, p. 21; Freier, p. 34.

3602. When walls built of stones which have been quarried below high water mark become damp, wet weather is at hand.
Cited in Inwards, p. 219.

stork

3603. If storks and cranes fly high and steady, expect fair weather.
Cited in Dunwoody, p. 40; Lee, p. 63.

storm

3604. A coming storm makes his first announcement down the chimney. Var.: A storm makes its first announcement down the chimney.
Cited in Alstad, p. 106; Sloane 2, p. 23.

3605. A Friday storm will reappear before Monday.
Cited in Hyatt, p. 7.

3606. A southwest storm, twelve hours long.
Recorded in Bryant (Mich.).

3607. After a storm comes a calm. Vars.: (a) After a bustle in the air,
the weather's very calm and fair. (b) After a storm, there must
be a calm. (c) Always a calm after a storm. (d) There is a lull
after a storm. (e) There's always quietness after a storm.
Cited in Denham, p. 2; Swainson, p. 214; Inwards, p. 122;
Wilson, p. 6; Whiting 1, p. 418; Simpson, p. 2; Mieder, p. 566.
Recorded in Bryant (Ill., Ind., Kans., Ky., N.J., N.Dak., Ohio,
Tex., Wis., Can. [Ont.]).

3608. Any port in a storm. Vars.: (a) Any old port in a storm. (b)
Sailors have a port in every storm.
Cited in Wilson, p. 15; Whiting 1, p. 343; Simpson, p. 4;
Mieder, p. 474.

3609. High regions are never without storms.
Cited in Mieder, p. 502.

3610. If a storm subsides before sunset, next day will be fair; if during
the night, next day will be cloudy.
Cited in Hyatt, p. 7.

3611. If the first spring storm is from the north or southwest, all
subsequent storms will come from the same direction.
Cited in Hyatt, p. 7.

3612. If the wind is from the northwest or southwest, the storm will be
short; if from the northeast, it will be a hard one; if from the
northwest, a cold one, and from the southwest, a warm one.
Cited in Dunwoody, p. 68.

3613. It is always calm before a storm. Vars.: (a) The calm before the
storm. (b) The lull before the storm. (c) There is always a calm
before a storm.
Cited in Dunwoody, p. 87; Whiting 2, p. 390; Mieder, p. 566.

3614. Ride the whirlwind and direct the storm.
Cited in Mieder, p. 566.

3615. Severe storms in winter are from east to northeast.
Cited in Hyatt, p. 7.

3616. Storms make oaks take deeper roots.
Cited in Mieder, p. 566.

3617. The rising or the setting of the sun or moon, especially the
moon, will be followed by a decrease of a storm which is then

prevailing.
Cited in Dunwoody, p. 63.

3618.   The sharper the storm, the sooner it's over. Var.: The harder the
storm, the sooner it's over.
Cited in Wilson, p. 721; Simpson, p. 201; Mieder, p. 566.

3619.   The storm alights on the mountain, and walks into the valley
before the rain arrives.
Cited in Sloane 2, p. 62.

3620.   The sudden storm lasts not three hours.
Cited in Swainson, p. 214; Sloane 2, p. 115.

3621.   The worst storms follow an east wind.
Cited in Hyatt, p. 7.

3622.   When it storms on the first Sunday of the month, it will storm
every Sunday. Var.: When it storms on the first Sunday in the
month, it will storm every Sunday during that month.
Cited in Dunwoody, p. 101; Garriott, p. 43; Whitman, p. 40;
Inwards, p. 74; Hand, p. 43.

storm (hurricane)

3623.   June, too soon: July, stand by; August, look out you must;
September, remember; October, all over. Var.: June, too soon:
July, stand by; August, look out! September you will remember;
October, all over. Comment: This refers to tropical hurricanes
of the Atlantic and notes the season of their probable occurence.
Cited in Garriott, p. 10; Hand, p. 10; Sloane 2, p. 40.

storm (norther)

3624.   If the first norther of the year is wet, the winter will be a wet
one.
Cited in Reeder, no pp.

stormy

3625.   Birds or chickens flying low and lighting every few feet is a sign
of stormy and rainy weather.
Recorded in Bryant (N.Y.).

3626.   Stormy March, pleasant May.
Recorded in Bryant (N.Y.).

3627.   Stormy weather on Friday; clear weather on Saturday.
Cited in Hyatt, p. 7.

stormy petrel

3628. The stormy petrels, or Mother Carey's chickens, so ominous to sailors, either forecast a storm, or show that one has been raging in the latitutde where they appear so mysteriously, as if they were "walking on water." Var.: The stormy petrel is found to be a sure token of stormy weather. When these birds gather in numbers in the wake of a ship, the sailors feel sure of an impending tempest.
Cited in Steinmetz 1, p. 111; Garriott, p. 24; Hand, p. 24.

stovelid

3629. A change of weather is indicated, if the stovelid turns red immediately after a fire has been started.
Cited in Hyatt, p. 31.

straw

3630. Small straws will show which way winds blow. Var.: Straws show which way the wind blows.
Cited in Inwards, p. 209; Wilson, p. 779.

stream

3631. Winter never really sets in until the streams are full.
Cited in Smith, p. 13.

string (violin)

3632. Strings of violins and guitars shrink and snap against rain.
Cited in Steinmetz 1, p. 113.

summer

3633. A cool summer, and a light weight in the bushel.
Cited in Cheales, p. 20; Inwards, p. 32.
3634. A dry summer never begs its bread.
Cited in Inwards, p. 31.
3635. A dry summer never leaves famine in its wake.
Cited in Inwards, p. 31.
3636. A dry summer ne'er made a dear peck.
Cited in Cheales, p. 22; Inwards, p. 31.
3637. A hot, dry summer means a cold, wet winter.
Cited in Smith, p. 13.

3638. A hot summer, a cold winter; a cool summer, a mild winter. Vars.: (a) Hot summer, cold winter. (b) If the summer is extremely hot, the winter will be unusually cold.
Cited in Hyatt, p. 14; Smith, p. 14; Mieder, p. 572.

3639. A very hot and dry summer is sometimes followed by a severe winter.
Cited in Inwards, p. 31.

3640. An English summer, two hot days and a thunderstorm.
Cited in Swainson, p. 17.

3641. If summer lasts till the New Year there will be winter till May.
Cited in Inwards, p. 32.

3642. If the summer be rainy the following winter will be severe. Var.: A wet summer almost always precedes a cold, stormy winter.
Cited in Dunwoody, p. 88; Inwards, p. 31.

3643. If the summer is moist, the winter will be dry.
Cited in Smith, p. 14.

3644. In the summer, if the sky is red at sunset, the next day will be hot.
Cited in Smith, p. 11.

3645. Summer in winter and summer's flood never boded an Englishman good.
Cited in Dunwoody, p. 91; Northall, p. 477; Inwards, p. 33.

3646. The summer shall be good and dry, corn and beasts shall multiply.
Cited in Northall, p. 456.

3647. There's no summer, but has a winter.
Cited in Denham, p. 48.

3648. When the summer is dry, the winter will be moist with much snow, sleet, and rain.
Cited in Smith, p. 14.

sun

3649. Although its appearance may be brief, you will always see the sun on Easter.
Cited in Hyatt, p. 2.

3650. Although the sun shine, leave not your cloak at home.
Cited in Wilson, p. 12; Mieder, p. 573.

3651. Behind the clouds, the sun is shining, Vars.: (a) Somewhere behind the clouds the sun is shining. (b) The sun is always

shining behind the clouds.
Cited in Mieder, p. 573. Recorded in Bryant (Ind., Vt.).

3652. Better be bitten by a snake, than to feel the sun in March.
Cited in Marvin, p. 207.

3653. February sun is dearly won.
Cited in Mieder, p. 573.

3654. If the sun burn more than usual, or there be a halo around the sun in fine weather, "wet."
Cited in Dunwoody, p. 77.

3655. If the sun comes up like a ball of fire and immediately disappears behind clouds, it is a sign of rain before ten o'clock that morning; if such a sun disappears behind clouds later in the morning, it is a sign of rain anytime that day.
Cited in Hyatt, p. 1.

3656. If the sun goes behind a cloud on Friday, it will rain before Sunday evening.
Cited in Smith, p. 6.

3657. If the sun goes behind clouds, rain within three days.
Cited in Smith, p. 7.

3658. If the sun shines during rain and if the cock crows, the weather will improve.
Cited in Wurtele, p. 302.

3659. If the sun shines during rain spit under a rock. Look under the rock next day and you will see a hair of the man you are going to marry.
Cited in Wurtele, p. 302.

3660. If the sun shines during rain turn your back to the sun and you will see the rainbow.
Cited in Wurtele, p. 302.

3661. If the sun shines while it is raining summer is coming soon.
Cited in Wurtele, p. 301.

3662. In the summer when the sun burns more than usual, expect thunderstorms.
Cited in Garriott, p. 20; Hand, p. 20.

3663. No matter how much rain or how overcast the sky during a rainy season, the sun will appear every fourth day, if only for a minute. Comment: That is, the sun never hides for more than three days.
Cited in Hyatt, p. 2.

3664. No sun, no showers, no summer flowers.
Cited in Mieder, p. 573.

3665. Somewhere the sun is shining.
Cited in Mieder, p. 573.

3666. Sun follows the rain. Var.: Sunshine follows rain.
Cited in Mieder, p. 573.

3667. The Lord sends sun at one time, rain at another.
Recorded in Bryant (Wis.).

3668. The morning sun never lasts the day. Vars.: (a) For age and want save while you may; no morning's sun lasts all the day. (b) No morning sun lasts a whole day.
Cited in Cheales, p. 19; Wilson, p. 544; Whiting 1, p. 424; Mieder, p. 573.

3669. The sun breaking through a stormy day is making holes for the wind to blow through.
Cited in Alstad, p. 152.

3670. The sun doesn't shine on both sides of the hedge at once.
Cited in Mieder, p. 573.

3671. The sun getting up its back-stays indicates foul weather. Comment: This phenomenon is referred to as the sun drawing water.
Cited in Sloane 2, p. 116.

3672. The sun is none the worse for shining on a dunghill. Var.: The sun is not less bright for sitting on a dunghill.
Cited in Denham, p. 5; Mieder, p. 573.

3673. The sun is the mother of the earth.
Cited in Alstad, p. 24.

3674. The sun will shine through the darkest clouds.
Cited in Mieder, p. 573.

3675. There is never a Saturday in the year, but what the sun it doth appear. Vars.: (a) Each Saturday during the year the sun shines long enough for a virgin (the Virgin?) to dry her shirt. (b) Each Saturday during the year the sun shines long enough for a workingman to dry his shirt. (c) The sun always comes out on Saturday.
Cited in Northall, p. 467; Hyatt, p. 1; Whiting 2, p. 605.

3676. When the sun draws water, rain follows soon. Vars.: (a) Expect rain when the sun draws water. (b) If a morning sun draws water, rain will fall that night. (c) If an afternoon sun draws water, rain will fall the next day. (d) If the sun draws water in the morning, it will rain before night. (e) Sun drawing water, indicates rain. (f) The sun drawing water, indicates the coming of rain. (g) The sun draws water, rain will follow. (h) When the sun is observed "drawing water," rain is coming.

Cited in Dunwoody, p. 77; Garriott, p. 26; Whitman, p. 25; Hand, p. 26; Sloane 2, p. 58; Hyatt, p. 1; Smith, p. 6. Recorded in Bryant (Okla).

3677. When the sun enters Leo, the greatest heat will then arise. Comment: Leo is a constellation in the Northern Hemisphere near Cancer and Virgo, containing the bright stars Regulus and Denebola.

Cited in Dunwoody, p. 91; Inwards, p. 57.

3678. When the sun in the morning is breaking through the clouds and scorching, a thunderstorm follows in the afternoon.

Cited in Dunwoody, p. 78; Garriott, p. 26; Hand, p. 26.

3679. When the sun is in the west, lazy folk work best. Var.: The lazy man works best when the sun is in the west.

Cited in Mieder, p. 574. Recorded in Bryant (Can. [Ont.]).

3680. When the sun is scorching, rain follows soon.

Cited in Dunwoody, p. 78.

3681. While the sun shines it is day.

Recorded in Bryant (N.Y.).

3682. Yearly there are three Saturdays on which the sun will not shine.

Cited in Hyatt, p. 1.

sun (eclipse)

3683. An eclipse of the sun is followed by five successive days of rain.

Cited in Hyatt, p. 2.

sun (halo)

3684. A bright circle (halo) around the sun denotes a storm, and cooler weather.

Cited in Dunwoody, p. 77.

3685. A halo around the sun indicates the approach of a storm, within three days, from the side which is more brilliant.

Cited in Dunwoody, p. 77.

3686. A solar halo indicates bad weather.
Cited in Dunwoody, p. 77; Garriott, p. 26; Hand, p. 26; Freier, p. 144. Recorded in Bryant (Mich.).

3687. If there be a ring or halo around the sun in bad weather, expect fine weather soon.
Cited in Dunwoody, p. 77.

3688. Ring around sun, rain none. Var.: Circle or ring around the sun, rain none.
Cited in Hyatt, p. 2. Recorded in Bryant (Ind.).

3689. The circle of the sun wets a shepherd. Vars.: (a) A ring around the sun or moon, means that rain will come real soon. (b) Sun or moon halos indicate a coming rain or snow, the larger the halo, the nearer the precipitation.
Cited in Whitman, p. 27; Sloane 2, p. 40; Lee, p. 84; Freier, p. 46.

3690. When the sun is in his house it will rain soon. Comment: Sun in his house means a circle or halo around the sun.
Cited in Whitman, p. 25.

sun (hazy)

3691. A blur or haziness about the sun indicates a storm.
Cited in Dunwoody, p. 77; Garriott, p. 26; Hand, p. 192; Freier, p. 26.

3692. A hazy sun early in the morning indicates rain; a clear sun, fair weather.
Cited in Hyatt, p. 1.

sun (mock)

3693. Mock suns in winter are usually followed by intense cold.
Cited in Dunwoody, p. 77.

sun (orange)

3694. Orange sun usually means foul weather.
Cited in Dunwoody, p. 77.

sun (pale)

3695. When the sun appears of a light pale color, or goes down into a bank of clouds, it indicates the approach or continuance of bad

weather.
Cited in Inwards, p. 82.

sun (red)

3696. A red sun has water in his eye. Var.: A reddish sun has water in his eye; before long you won't be dry.
Cited in Dunwoody, p. 78; Garriott, p. 21; Humphreys 1, p. 431; Marvin, p. 209; Humphreys 2, p. 13; Whitman, p. 25; Hand, p. 21; Hyatt, p. 1; Freier, p. 32. Recorded in Bryant (Mich.).

3697. If the sun in red should set, the next day surely will be wet; if the sun should set in gray, the next will be a rainy day. Vars.: (a) If the sun goes red to bed, 'twill rain tomorrow it is said. (b) Sun down gray and sun up red will send rain down upon his head. (c) When the sun goes red to bed, 'twill rain tomorrow, it is said.
Cited in Denham, p. 10; Swainson, p. 221; Northall, p. 466; Humphreys 1, p. 436; Whitman, p. 25; Hyatt, p. 1; Wilson, p. 787; Wilshere, p. 2; Freier, p. 33; Mieder, p. 573. Recorded in Bryant (Ill., Kans., N.Y., S.C.).

3698. Red sun indicates fair weather.
Cited in Dunwoody, p. 77.

3699. When the sun appears red or green, expect rain.
Cited in Dunwoody, p. 78.

3700. When the sun burns more than usual, rain may be expected.
Cited in Dunwoody, p. 16.

3701. When the sun sets red, it will be a clear day tomorrow. Var.: If the sun goes down red, it will be dry.
Cited in Whitman, p. 25; Reeder, no pp.

sun (rise)

3702. If the sun at rising appears enlarged there will shortly be sudden and sharp showers, if in summer; but in winter settled and moderate weather.
Cited in Steinmetz 1, p. 119.

3703. If the sun rise cloudy, and it soon decrease, certain fair weather. Var.: If at sunrise the clouds are driven away and retire as it were to the west, this denotes fair weather.
Cited in Mitchell, p. 232; Steinmetz 1, p. 119.

3704. If the sun rises clear, then shadowed by a cloud, and comes out again clear, it will rain before night.
Cited in Dunwoody, p. 79.

3705. In every country the sun rises in the morning. Vars.: (a) The sun shines on all the world. (b) The sunrise never failed us yet.
Cited in Wilson, p. 787; Mieder, p. 573. Recorded in Bryant (N.Y.).

3706. When the sun rises with dim, murky clouds, expect rain.
Cited in Dunwoody, p. 78.

sun (rise pale)

3707. A pale sun in the morning announces rain. Vars.: (a) If the sun rises pale, a pale red, or even dark blue, there will be rain during the day. (b) When the sun appears a pale or dull color, expect rain.
Cited in Steinmetz 2, p. 83; Dunwoody, p. 78.

sun (rise red)

3708. At dawn in summer a red sun means a sultry day.
Cited in Hyatt, p. 1.

3709. If red the sun begins his race, expect that rain will flow apace. Vars.: (a) If red the sun begins his race, be sure that rain will fall apace. (b) If the sun comes up red, it will rain. (c) In fiery red the sun doth rise, then wades through clouds to mount the skies.
Cited in Denham, p. 11; Steinmetz 1, p. 118; Northall, p. 466; Humphreys 1, p. 431; Marvin, p. 209; Humphreys 2, p. 14; Whitman, p. 24; Wilson, p. 669; Lee, p. 79; Wilshere, p. 3; Reeder, no pp.

3710. If the sun rise red and fiery, wind and rain.
Cited in Mitchell, p. 232; Whitman, p. 24.

3711. If the sun rises red, tomorrow will be a hot day.
Cited in Smith, p. 11.

3712. Sun red at morning, sailor's warning; sun red at night, sailor's delight.
Recorded in Bryant (Ill.).

3713. When the sun comes up with a glare, it is a warning.
Cited in Smith, p. 4.

3714. When the sun rises in a red cloud there will be rain.
Cited in Smith, p. 7.

sun (set)

3715. As the sun descends, the temperature ascends.
Cited in Humphreys 2, p. 113.

3716. If the sun goes down clear on Friday night, it will rain before Sunday night. Vars.: (a) If the sun sets clear on Friday night, rain before Monday night. (b) Sun sets clear on Friday, rain before Monday night.
Recorded in Bryant (N.Y. Can. [Ont.]).

3717. If the sun set behind a straight skirting of cloud, be sure of wind from the point where the sun is setting.
Cited in Steinmetz 1, p. 120.

3718. If the sun sets behind a bank, it will rain before Wednesday night.
Recorded in Bryant (Ill.).

3719. If the sun sets behind a cloud on Sunday evening, the weather will be rough the next week.
Cited in Smith, p. 1.

3720. If the sun sets behind a rugged, rocky or mixed bank of clouds, very stormy, wet, or showery will be the morrow.
Cited in Steinmetz 1, p. 120.

3721. If the sun sets clear, a fair day is near.
Recorded in Bryant (Ind.).

3722. If the sun sets clear Friday evening, it will rain before Monday night.
Cited in Dunwoody, p. 77.

3723. If the sun sets clear on Friday, it will storm on Sunday.
Cited in Sloane 2, p. 38.

3724. If the sun sets clear on Sunday evening, it will be a good week ahead.
Cited in Smith, p. 3.

3725. If the sun sets in a general sheet of haziness of a dusky or leaden hue, bad weather is near.
Cited in Steinmetz 1, p. 120.

3726. If the sun sets in dark, heavy clouds, expect rain the next day.
Var.: If the sun sets behind a cloud, it forebodes rain the next day.

Cited in Garriott, p. 26; Sloane 2, p. 126. Recorded in Bryant (Mich.).

3727. Sun set in a clear, easterly wind's near; sun set in a bank, westerly will not lack.
Cited in Wright, p. 31.

3728. Sun setting double indicates much rain.
Cited in Dunwoody, p. 77.

3729. The sun never sets on the British Isles.
Cited in Mieder, p. 573.

3730. The sun sets weeping in the lowly west, witnessing storms to come, woe and unrest.
Cited in Whitman, p. 25.

3731. The sun setting after a fine day behind a heavy bank of clouds, with a falling barometer, is generally indicative of rain or snow, according to the season, either in the night or next morning.
Cited in Garriott, p. 26; Hand, p. 26.

3732. The sun setting behind clouds on Wednesday is a token of rain before Sunday.
Cited in Hyatt, p. 1.

3733. The sun setting in clouds, after fine weather, is a sign of coming rain.
Cited in Steinmetz 2, p. 83.

3734. When the sun sets bright and clear, an easterly wind you need not fear. Var.: When the sun sets in a clear, an easterly wind you need not fear.
Cited in Denham, p. 20; Swainson, p. 178; Dunwoody, p. 83; Northall, p. 467; Garriott, p. 26; Whitman, p. 25; Hand, p. 26; Wilson, p. 787; Freier, p. 32. Recorded in Bryant (Mich.).

3735. When the sun sets in a bank, a westerly wind we shall not want. Vars.: (a) A sunset and a cloud so black, a westerly wind you shall not lack. (b) When the sun sets in a bank, a westerly wind we shall not lack. Comment: Bank indicates a heavy dark cloud.
Cited in Denham, p. 12; Northall, p. 467; Inwards, p. 82; Sloane 2, p. 126; Wilson, p. 787; Freier, p. 59.

3736. When (the sun) setting in part clear, but among curly locks of thin cloud, like tufts of hair or the strippings of goose quills, expect fog or rain next morning.
Cited in Steinmetz 1, p. 120.

sun (set pale)

3737. If the sun goes pale to bed, 'twill rain tomorrow, it is said. Vars.: (a) If the sun sets pale, it will rain tomorrow. (b) Last night the sun went pale to bed, the moon in halos hid her head. Comment: The advent of a moon halo after a pale sunset is a fairly certain rain sign.
Cited in Swainson, p. 178; Dunwoody, p. 78; Northall, p. 467; Humphreys 1, p. 436; Whitman, p. 24; Inwards, p. 82; Sloane 2, p. 40; Hyatt, p. 1; Wilson, p. 786; Lee, p. 78; Page, p. 7; Wilshere, p. 2. Recorded in Bryant (Ill., N.Y., S.C.).

sun (set red)

3738. If on Friday the sun sets in a blaze, it will bring rain before Monday morning say some; before Monday night say others. The contrary is also believed: if on Friday the sun sets in clouds, it will bring rain before Monday morning say some; before Monday night say others.
Cited in Hyatt, p. 1.

3739. If the sun sets with very red sky in the east, expect wind; in the southeast, expect rain.
Cited in Dunwoody, p. 78.

3740. Red sun at night is a sailor's delight.
Cited in Mieder, p. 545. Recorded in Bryant (Wis.).

3741. The flame of the setting sun tells you there are better days to come.
Recorded in Bryant (Wis.).

sun dog

3742. A sun dog at night is a sailor's delight; a sun dog in the morning is the sailor's warning. Var.: A dog in the morning, sailor, take warning; a dog at night is the sailor's delight. Comment: Sun dog (dog) refers to a mock sun or a fragment of a rainbow.
Cited in Dunwoody, p. 79; Whitman, p. 31; Inwards, p. 160; Sloane 2, p. 119.

3743. A sun dog on each side of the sun in the morning is a portent of milder weather; in the afternoon, harsher weather.
Cited in Hyatt, p. 2.

3744. If a sun dog is seen on each side of the sun, a severe storm will arrive during the night.

Cited in Hyatt, p. 2.

3745. In summer a sun dog warns you of cooler weather; in winter, chillier weather or a blizzard.
Cited in Hyatt, p. 2.

3746. Morning sun dog, colder weather; afternoon sun dog, warmer weather.
Cited in Hyatt, p. 2.

3747. Sun dogs in the morning will bark before night; but sun dogs in the evening are sailors' delight.
Recorded in Bryant (Mich.).

3748. Sun dogs indicate cold weather in winter or storm in summer.
Cited in Dunwoody, p. 79.

3749. Sun dogs mean cold weather.
Recorded in Bryant (Wis.).

3750. The meaning of a sun dog north of the sun is rain from the northwest; south of the sun, rain from the southwest.
Cited in Hyatt, p. 2.

3751. Two sun dogs in the east denote cold weather.
Cited in Hyatt, p. 2.

sun spot

3752. Wet seasons occur in years when sun spots are frequent.
Cited in Dunwoody, p. 78.

sunbeam

3753. A single sunbeam drives away many shadows.
Cited in Mieder, p. 574.

Sunday

3754. A wet Sunday, a fine Monday, wet the rest of the week. Var.: A wet Sunday, a fine Monday.
Cited in Wright, p. 118; Inwards, p. 74; Lee, p. 146.

3755. Sunday clearing, clear till Wednesday.
Cited in Dunwoody, p. 101; Whitman, p. 40; Inwards, p. 74.

3756. The last Sunday of the month indicates the weather of the next month.
Cited in Dunwoody, p. 101; Inwards, p. 74.

sunflower

3757. Sunflower raising its head indicates rain. Var.: When sunflowers raise their heads, it is a sign of rain.
Cited in Dunwoody, p. 67; Hand, p. 22; Lee, p. 73.

3758. Tall sunflowers, deep snow.
Recorded in Bryant (Utah).

sunset

3759. A bright yellow sunset indicates wind, a pale yellow, wet.
Cited in Dunwoody, p. 79.

3760. A clear orange-colored sunset foretells a very fine day to follow, and more surely if with a rising barometer and a calm dewy evening.
Cited in Steinmetz 1, p. 120.

3761. A cloudy sunset on Friday turns the weather colder.
Cited in Hyatt, p. 1.

3762. A cloudy sunset on Monday, rain before Friday.
Cited in Hyatt, p. 1.

3763. A concealed sunset on Thursday denotes rainless weather until Sunday.
Cited in Hyatt, p. 1.

3764. A dull sunset is attended by bad weather.
Cited in Hyatt, p. 1.

3765. A flaming sunset on Friday; a rain before Tuesday night.
Cited in Hyatt, p. 1.

3766. A gray sunset indicates wet.
Cited in Steinmetz 1, p. 119.

3767. A pale sunset, a golden sunset, or a green sunset, indicates rain.
Cited in Dunwoody, p. 78.

3768. A pink sunset means bad weather is on the way.
Cited in Smith, p. 1.

3769. A ruddy sunset, especially if small horizontal lines of clouds lie as shoals of fish about the horizon, betokens windy weather.
Cited in Steinmetz 1, p. 120.

3770. A sunset and a cloud so black, a westerly wind you shall not lack. Var.: At sunset with a cloud so black, a westerly wind you shall not lack.
Cited in Cheales, p. 26; Inwards, p. 82.

3771. A very clear sunset, of a pale gold color, is a sign of fine weather, if there be a calm and dewy evening with it.
Cited in Steinmetz 1, p. 120.

3772. A yellow sunset indicates wet.
Cited in Steinmetz 1, p. 119.

3773. After a cloudy sunset there will be three rainy days; after a cloudless sunset, three sunny days.
Cited in Hyatt, p. 1.

3774. Clear sunset, clear tomorrow.
Recorded in Bryant (N.Y., S.C.).

3775. If at sunset a ruddy sky reflects from clouds in the east, a change of weather is near.
Cited in Hyatt, p. 5.

3776. If red covers the whole sky at sunset, it will be windy the next day.
Cited in Smith, p. 12.

3777. If Sunday sunset is obscured, expect rain before Wednesday.
Vars.: (a) An unclear sunset on Sunday is a forecast of rain before Wednesday. (b) If sunset on Sunday is cloudy, it will rain before Wednesday. (c) If the sun sets behind the cloud on Sunday, it will rain by Wednesday. (d) When the sun sets behind a cloud on Sunday night, it will rain before Wednesday.
Cited in Dunwoody, p. 46; Inwards, p. 74; Hyatt, p. 1; Reeder, no pp. Recorded in Bryant (Miss).

3778. On Sunday a murky sunset means rain before morning.
Cited in Hyatt, p. 1.

3779. Red sunset indicates a fine day tomorrow. Vars.: (a) A red sunset, a good day tomorrow; a red dawn, a bad day. (b) Red sunset, no rain you get.
Cited in Steinmetz 1, p. 119; Smith, p. 8. Recorded in Bryant (Ill.).

3780. Red sunset, sailor's delight; red sunrise, cry before night.
Recorded in Bryant (Ill., La., Utah, Wash.).

3781. Red sunset, wind tomorrow.
Recorded in Bryant (Can. [Ont.]).

3782. Rose tints at sunset and gray dawn, a fine day to follow.
Cited in Inwards, p. 82.

3783. The significance of a glowing sunset is a storm.
Cited in Hyatt, p. 1.

3784. When the sunset is a red glow in the sky, snow is coming.
Cited in Smith, p. 10.

sunshine

3785. After sunshine come showers; after pleasure comes sorrow.
Cited in Mieder, p. 574.

3786. All sunshine makes a desert.
Cited in Mieder, p. 574.

3787. Sunshine and shower, rain again tomorrow. Vars.: (a) If the sun
and rain appear in the same day, it will rain again the next day.
(b) If the sun shines during a rain, it signifies rain next day. (c)
If the sun shines while it is raining, it will rain again tomorrow
at the same hour.
Cited in Humphreys 2, p. 55; Whitman, p. 25; Hyatt, p. 11;
Smith, p. 7; Wurtele, p. 301.

3788. Sunshine and showers make up life's hours.
Cited in Mieder, p. 574.

3789. Sunshine on Monday; sunshine all week.
Cited in Hyatt, p. 1.

3790. Take the sunshine with the rain.
Cited in Mieder, p. 574.

3791. There is never any sunshine on Good Friday.
Cited in Hyatt, p. 1.

3792. We never appreciate the sunshine and the rainbow until the storm
clouds hang low.
Cited in Mieder, p. 574.

supper

3793. It will be a clear day if everything has been eaten from supper.
Vars.: (a) If everything has been eaten at supper, it will not rain
the following day. (b) It will not rain if everything has been
eaten from supper. (c) It is going to be a clear day tomorrow if
there is little or no food remaining on the table after a meal.
Recorded in Bryant (Ohio, W.Va., Wis.).

suspenders

3794. Wear your suspenders twisted during a storm and lightning will
not strike you.
Cited in Hyatt, p. 33.

swallow

3795. Circling swallows indicate rain. Var.: If flying swallows undulate in circles near the ground, rain is near.
Cited in Dunwoody, p. 40; Hyatt, p. 23.

3796. If swallows fly low, it means wet weather; if high, fair weather. Vars.: (a) Swallows fly high, clear blue sky; swallows fly low, rain we shall know. (b) Swallows fly high, sun in the sky; swallows fly low, rain you must know.
Cited in Hyatt, p. 23; Lee, p. 63; Wilshere, p. 22. Recorded in Bryant (Mich.).

3797. If swallows touch the water as they fly, rain approaches.
Cited in Swainson, p. 246; Inwards, p. 195.

3798. One swallow does not make a spring, nor a woodcock a winter.
Cited in Denham, p. 31.

3799. One swallow does not make a summer. Vars.: (a) It takes more than two swallows to make a summer. (b) Three swallows never make a summer.
Cited in Denham, p. 40; Inwards, p. 31; Page, p. 21; Simpson, p. 218; Dundes, p. 95; Whiting 2, p. 607; Mieder, p. 575.

3800. Rain is indicated when: Low o'er the grass the swallows wing, and crickets, too, how sharp they sing.
Cited in Dunwoody, p. 70.

3801. Swallows and swifts fly close to the ground or water before rain. Vars.: (a) Swallows fly close to the ground before a rain. (b) Swallows flying low indicate rain. (c) Swallows skimming along the ground indicate rain.
Cited in Mitchell, p. 227; Dunwoody, p. 40; Sloane 2, p. 32.

3802. Swallows fly low, in pursuit of the insects they feed on, seeking the warmer air in the change of weather before rain.
Cited in Steinmetz 2, p. 290.

3803. The first swallow is a messenger of spring.
Cited in Hyatt, p. 23.

3804. When swallows fleet, soar high, and sport in air, the welkin will be clear. Comment: Welkin means sky.
Cited in Garriott, p. 19; Inwards, p. 195; Hand, p. 19; Lee, p. 46; Freier, p. 21.

3805. When swallows fly high, following the insects on which they feed, it is safe to say that the day will be fine.
Cited in Page, p. 21.

3806. When swallows in the evenings fly high and chirp, fair weather follows; when low, rain follows.
Cited in Dunwoody, p. 40; Garriott, p. 19; Hand, p. 19.

3807. When the swallow's nest is high, the summer is very dry; when the swallow buildeth low, you can safely reap and sow.
Cited in Dunwoody, p. 40.

3808. When the swallows fly low or when the geese fly, expect storm or cold.
Cited in Dunwoody, p. 40.

swan

3809. If the swan flies against the wind, it is a certain indication of a hurricane within twenty-four hours, generally within twelve.
Cited in Swainson, p. 246; Inwards, p. 189.

3810. Swans are hatched in thunderstorms.
Cited in Inwards, p. 189.

3811. The swan builds its nest high before high waters, but low when there will not be the unusual rains.
Cited in Dunwoody, p. 40; Garriott, p. 40; Inwards, p. 189; Hand, p. 40.

3812. When swans fly against the wind, rain is coming.
Cited in Lee, p. 61.

3813. When swans fly, it is a sign of rough weather.
Cited in Inwards, p. 189.

3814. When swans fly, it is a sign of wet weather.
Cited in Sloane 2, p. 130.

sweat

3815. If you are sweating before seven, it will rain before the day ends.
Cited in Smith, p. 7.

swift

3816. When there are many more swifts than swallows in the spring, expect a hot and dry summer.
Cited in Wright, p. 43; Inwards, p. 195.

swine

3817. If swine be restless and grunt loudly, if they squeal and jerk up their ears, there will be much wind.
Cited in Dunwoody, p. 33.

3818. Swine carry straw in their mouths and toss about their bedding before a storm.
Cited in Mitchell, p. 228.

3819. Swine make lairs on the south side of a shelter before cold weather. Comment: Lair means bed or resting place of an animal.
Cited in Dunwoody, p. 33.

sycamore

3820. Sycamore tree peeling off white in the fall, indicates a cold winter.
Cited in Dunwoody, p. 67.

3821. Sycamore trees with smooth white bark in the autumn indicate an open winter.
Cited in Hyatt, p. 16.

3822. The looseness or tightness of sycamore bark in the autumn shows what kind of weather we shall have: peeled easily, a loose winter; peeled with difficulty, a tight winter.
Cited in Hyatt, p. 16.

table

3823. The cracking of tables and chairs indicates rain or frost.
Cited in Dunwoody, p. 118.

tarantula

3824. When tarantulas crawl by day, rain will surely come. Var.: If tarantulas are moving, it will rain soon.
Cited in Dunwoody, p. 58; Inwards, p. 207; Reeder, no pp.

teakettle

3825. A singing teakettle will warn you of rain.
Cited in Hyatt, p. 32.

3826. By the sweating of a teakettle you are warned of rain.
Cited in Hyatt, p. 31.

3827. Sparks of fire on the bottom of the teakettle denote rain.
Cited in Hyatt, p. 32.

3828. Teakettle water evaporates quicker than usual just before a rain.
Cited in Hyatt, p. 32.

telephone

3829. The replacing of a telephone receiver upside down on its hook signifies rain.
Cited in Hyatt, p. 32.

3830. When telephone wires hum, a change in the weather is coming.
Cited in Smith, p. 1.

3831. Whistling telephone lines tell of a storm to come.
Cited in Sloane 2, p. 117.

temperature

3832. A sudden and extreme change of temperature of the atomosphere, either from heat to cold, or cold to heat, is generally followed by rain within twenty-four hours.
Cited in Inwards, p. 154.

3833. If during the night the temperature fall and the barometer rise, we shall have fine weather and clear skies.
Cited in Denham, p. 11.

3834. If the temperature increase between 9:00 p.m. and midnight, when the sky is cloudless, expect rain; and if, during a long and severe period of low temperature, the temperature increases

between midnight and morning, expect a thaw.
Cited in Hand, p. 20.

termite

3835. If wood lice (termites) run about in great numbers, expect rain.
Cited in Dunwoody, p. 58; Inwards, p. 207; Lee, p. 57.

3836. When the wood lice (termites) fly following a shower, the rainy spell is over.
Cited in Reeder, no pp.

terrapin

3837. If a terrapin moves across the road, it is going to rain.
Cited in Smith, p. 5.

3838. Whenever you see terrapins crawling, it is a sign of rain.
Cited in Smith, p. 5.

thaw

3839. A January thaw betokens a wet July.
Cited in Hyatt, p. 12.

3840. Always expect a thaw in January.
Cited in Garriott, p. 44; Wurtele, p. 296.

3841. As many days from the first snow to the next new moon, so many times will it thaw during winter.
Cited in Dunwoody, p. 92.

3842. If it thaws enough for water to run down the ruts in a road during the first three days of January, an open winter may be predicted.
Cited in Hyatt, p. 12.

3843. Quick thaw, long frost.
Cited in Swainson, p. 208.

3844. The ground thawing in December indicates a thaw every month during winter.
Cited in Hyatt, p. 12.

3845. When a thaw comes on, the frost goes down.
Cited in Humphreys 2, p. 114.

thistle

3846. Cut thistles in May, they grow in a day; cut thistles in June, that is too soon; cut thistles in July, then they will die.

Cited in Northall, p. 482.

3847. If the flowers of the Siberian sow thistle keep open all night the weather will be wet the next day.
Cited in Swainson, p. 261.

3848. If the hog thistle closes for the night, expect fair weather; if it remains open, expect rain.
Cited in Dunwoody, p. 66.

3849. Teasel or Fuller's thistle hung up will open for fine weather, and close for wet.
Cited in Inwards, p. 215.

thrush

3850. After the thrush has arrived in the spring, frosts are finished.
Cited in Hyatt, p. 23.

3851. Missel thrush have been observed to sing particularly loud just before a storm. Var.: The missel thrush sings particularly loud and long before rain.
Cited in Dunwoody, p. 38; Inwards, p. 191.

3852. When a missel thrush, from the very top of a tree, sings into the wind, it is a bad sign, and this habit has led to the bird acquiring the name "storm cock" or "storm thrush." Even its song signifies rain, for the bird sings "More wet, more wet."
Cited in Page, p. 19.

3853. When the thrush perches itself upon the topmost bough of a tree, and remains there for some time, singing loudly, expect rain.
Cited in Inwards, p. 191.

3854. When the thrush sings at sunset, a fair day will follow.
Cited in Dunwoody, p. 40; Inwards, p. 191.

thunder

3855. A January thunder, a June flood.
Cited in Hyatt, p. 7.

3856. A January thunder, a June frost.
Cited in Hyatt, p. 7.

3857. A spring thunder proclaims a cold spell.
Cited in Hyatt, p. 7.

3858. A winter's thunder, a summer's hunger. Vars.: (a) A winter's thunder makes summer's hunger. (b) Winter's thunder: rich man's food and poor man's hunger.

Cited in Mieder, p. 595.

3859. After heavy thunder, little rain.
Cited in Smith, p. 8.

3860. After much thunder much rain. Var.: After thunder comes rain.
Cited in Swainson, p. 216; Wilson, p. 820.

3861. After thunder and lightning on New Year's Day comes a cold snap.
Cited in Hyatt, p. 7.

3862. An excellent crop year, corn in particular, will follow a November thunder.
Cited in Hyatt, p. 34.

3863. As many thunders as are in February, so many will be the frost of May. Vars.: (a) February thunder denotes a May frost. (b) Thunder in February brings a frost in May.
Cited in Hyatt, p. 7.

3864. August thunder indications do not come alone: one thunderstorm will follow another.
Cited in Marvin, p. 217.

3865. Autumn thunder, warmer weather; winter thunder, colder weather.
Cited in Hyatt, p. 7.

3866. December thunder indicates good weather.
Cited in Marvin, p. 217.

3867. December thunder makes the weather colder.
Cited in Hyatt, p. 7.

3868. Do not expect any peaches during the year in which thunder is heard on February 12th.
Cited in Hyatt, p. 34.

3869. Early March thunder brings cooler weather.
Cited in Hyatt, p. 7.

3870. Early thunder, early spring.
Cited in Dunwoody, p. 80; Garriott, p. 45; Marvin, p. 216; Whitman, p. 44; Inwards, p. 30; Hand, p. 45; Wurtele, p. 298; Freier, p. 78. Recorded in Bryant (Utah).

3871. Early thunder, late hunger.
Cited in Northall, p. 467.

3872. First thunder in the spring, if in the south it indicates a wet season, if in the north it indicates a dry season.
Cited in Dunwoody, p. 82; Inwards, p. 30.

3873. First thunder in winter or spring indicates rain and very cold weather.
Cited in Dunwoody, p. 80.

3874. He who carries a bay leaf shall never take harm from thunder.
Cited in Denham, p. 4.

3875. If it thunders before seven, it will rain before eleven. Vars.: (a) Thunder before seven, rain before eleven. (b) Thunder before seven, rain before eleven; rain before seven, stop before eleven. Cited in Hyatt, p. 6; Mieder, p. 599. Recorded in Bryant (Ark., Fla., Miss.).

3876. If it thunders both in February and March, crops will be abundant, especially fruit crops.
Cited in Hyatt, p. 34.

3877. If it thunders in November, there will not be any cold weather until after Christmas.
Cited in Hyatt, p. 7.

3878. If it thunders much at the beginning of September, much grain will be raised the following year. Var.: Thunder in September indicates a good crop of grain and fruit for next year.
Cited in Dunwoody, p. 81; Marvin, p. 217.

3879. If it thunders on All Fools' Day, it brings good crops of corn and hay. Vars.: (a) If it thunders on All Fools' Day it brings good crops of grain and hay. (b) If it thunders on All Fools' Day, it brings good crops of grass and hay. (c) If it thunders on All Fools' Day, this assures your crop of hay. Comment: All Fools' Day is April 1st.
Cited in Swainson, p. 84; Dunwoody, p. 104; Northall, p. 451; Garriott, p. 42; Wright, p. 37; Whitman, p. 38; Inwards, p. 52; Hand, p. 42; Sloane 2, p. 30; Lee, p. 152.

3880. If the first thunder is in the north, the bear has stretched his left leg in his winter bed.
Cited in Dunwoody, p. 81.

3881. If there be thunder in the evening, there will be much rain and showery weather.
Cited in Dunwoody, p. 80; Inwards, p. 168.

3882. If there is much thunder in May, the months of August and September will be without it.
Cited in Dunwoody, p. 81; Marvin, p. 216.

3883. If there is thunder and lightning before leaves appear on the trees, old and young will be hungry.
Cited in Hyatt, p. 34.

3884. If there is thunder in October, January will be wet.
Cited in Wilshere, p. 14.

3885. If while trees are leafless there is thunder or lightning, or both, six more weeks of cold weather may be expected.
Cited in Hyatt, p. 7.

3886. If you hear thunder before seven in the morning, seven thunder-claps will be heard before night.
Cited in Hyatt, p. 6.

3887. In February if you hear thunder, you will see a summer wonder.
Cited in Swainson, p. 41; Marvin, p. 212.

3888. It never thunders but it rains.
Cited in Marvin, p. 212; Mieder, p. 595. Recorded in Bryant (Wis.).

3889. It's the thunder that frights, but the lightning that smites.
Recorded in Bryant (Can. [Ont.]).

3890. January thunder indicates an April frost.
Cited in Hyatt, p. 7.

3891. January thunder indicates wind, corn, and cattle. Var.: Thunder in January signifies the same year great winds, plentiful of corn and cattle.
Cited in Marvin, p. 216.

3892. Late November or early December thunder does not change the weather.
Cited in Hyatt, p. 7.

3893. March thunder ends wintry weather.
Cited in Hyatt, p. 7.

3894. Morning thunder is followed by a rain the same day.
Cited in Dunwoody, p. 81.

3895. Much thunder in July injures wheat and barley. Var.: July thunder indicates that the wheat and barley will suffer harm.
Cited in Dunwoody, p. 80.

3896. Northwest thunder means rain within forty-eight hours.
Cited in Hyatt, p. 7.

3897. October thunder will be followed by milder weather.
Cited in Hyatt, p. 7.

3898. On whatever day it thunders in February, on a similar date it will thunder in May.
Cited in Hyatt, p. 7.

3899. The date of thunder in January will be the number of spring days during May.
Cited in Hyatt, p. 7.

3900. The distant thunder speaks of coming rain. Var.: Thunder first, then rain.
Cited in Dunwoody, p. 80.

3901. The first thunder of the year awakes the frogs and snakes from their winter sleep. Var.: The first thunder of the year awakes all the frogs and snakes.
Cited in Dunwoody, p. 80; Wright, p. 37; Whitman, p. 36; Freier, p. 77.

3902. Thunder and lightning early in winter or late in fall indicates warm weather.
Cited in Dunwoody, p. 80; Inwards, p. 167.

3903. Thunder and lightning in February or March, poor maple sugar year. Vars.: (a) February thunder indicates poor maple sugar year. (b) Thunder in February or March, poor sugar maple year.
Cited in Dunwoody, p. 80; Garriott, p. 44; Marvin, p. 216; Hand, p. 44.

3904. Thunder and lightning on the northern lakes in November is an indication that the lakes will remain open until the middle of December or until Christmas. Var.: Thunder in November on the northern lakes of the United States is taken as an indication that the lakes will remain open till at least the middle of December.
Cited in Dunwoody, p. 81; Inwards, p. 65.

3905. Thunder before noon, showers in the afternoon.
Cited in Hyatt, p. 7.

3906. Thunder curdles cream.
Cited in Sloane 2, p. 43; Smith, p. 11.

3907. Thunder during Christmas week indicates that there will be much snow during the winter.
Cited in Dunwoody, p. 79; Inwards, p. 69.

3908. Thunder first, then rain.
Cited in Whiting 1, p. 437.

3909. Thunder from the south is followed by warmth, and from the north by cold. When the storm disappears in the east, it is a sign of fine weather.
Cited in Inwards, p. 169.

3910. Thunder from the south or southeast indicates foul weather; from the north or northwest fair weather.
Cited in Dunwoody, p. 82; Inwards, p. 169.

3911. Thunder in April, floods in May.
Cited in Page, p. 46.

3912. Thunder in December presages fair weather.
Cited in Swainson, p. 150; Inwards, p. 67.

3913. Thunder in February frightens the maple syrup back into the ground.
Cited in Sloane 2, p. 37.

3914. Thunder in February means frost in April. Vars.: (a) If it thunders in February, it will frost on or near that same date in April, or rain that day in June. (b) Thunder in February, rain in July.
Cited in Mieder, p. 595. Recorded in Bryant (Miss.).

3915. Thunder in February; snow in May.
Cited in Hyatt, p. 7.

3916. Thunder in March betokens a fruitful year. Var.: Thunder in March, a fruitful harvest.
Cited in Humphreys 2, p. 10; Whitman, p. 36; Freier, p. 77.

3917. Thunder in March, corn to parch.
Cited in Mieder, p. 117.

3918. Thunder in March, floods in May.
Cited in Page, p. 46; Wilshere, p. 9.

3919. Thunder in November signifies that same year to be fruitful and merry, and cheapness of corn. Vars.: (a) November thunder indicates that the coming year will be fertile. (b) Thunder in November indicates a fertile year to come.
Cited in Swainson, p. 141; Inwards, p. 65.

3920. Thunder in spring, cold will bring.
Cited in Swainson, p. 12; Dunwoody, p. 91; Northall, p. 468; Wright, p. 31; Marvin, p. 216; Whitman, p. 36; Boughton, p. 124; Inwards, p. 30; Wilshere, p. 17.

3921. Thunder in the evening indicates much rain.
Cited in Dunwoody, p. 80.

3922. Thunder in the fall indicates a mild, open winter. Var.: Thunder in the fall, no winter at all.
Cited in Dunwoody, p. 80; Marvin, p. 216; Inwards, p. 33; Mieder, p. 595.

3923. Thunder in the morning, all the day storming; thunder at night is the traveller's delight.
Cited in Smith, p. 11.

3924. Thunder in the morning is a sailor's warning. Var.: Thunder in the morning, sailor take warning; thunder at night, sailor's delight.
Cited in Hyatt, p. 7; Smith, p. 11.

3925. Thunder in the morning signifies wind, about noon rain, in the evening great tempest.
Cited in Inwards, p. 168.

3926. Thunder in the north indicates cold weather and rain from the west.
Cited in Dunwoody, p. 81.

3927. Thunder is the water wagon crossing a bridge.
Cited in Smith, p. 9.

3928. Thunder without rain is like words without deeds.
Recorded in Bryant (Calif., Nebr., N.C.).

3929. Thundery weather in March is a sign of a cool summer.
Cited in Hyatt, p. 7.

3930. To have thunder in December is a forecast of frost in May.
Cited in Hyatt, p. 7.

3931. When it is thundering, the angels are moving God's furniture.
Cited in Smith, p. 9.

3932. When it thunders in March, it brings sorrow. Var.: March thunder indicates coming sorrow.
Cited in Swainson, p. 56; Marvin, p. 216; Wilson, p. 820.

3933. When it thunders in the morning, it will rain before night. Var.: Thunder in the morning brings rain in the afternoon.
Cited in Dunwoody, p. 81; Whitman, p. 35; Smith, p. 8.

3934. When it thunders on the day of the moon's disappearance, the crops will prosper and the market will be steady.
Cited in Whitman, p. 36.

3935. When it thunders the angels are bowling.
Cited in Smith, p. 9.

3936. When it thunders, the thief becomes honest.
Cited in Inwards, p. 170.
3937. When milk becomes suddenly and inexplicably sour, a thunder-storm is at hand. Var.: Thunder causes milk to sour sooner than ordinary.
Cited in Mitchell, p. 231. Recorded in Bryant (Calif., Nebr., N.C.).
3938. When the thunder is very loud, there's very little rain.
Cited in Mieder, p. 595.
3939. While trees are bare, thunder or lightning, or both, signifies chillier weather; after trees have leafed, milder weather.
Cited in Hyatt, p. 7.
3940. Winter is ended by the first thunder of spring.
Cited in Hyatt, p. 7.
3941. Winter thunder and summer's flood never boded an Englishman any good. Var.: Winter thunder and summer's flood never boded any good.
Cited in Swainson, p. 10; Dunwoody, p. 92; Northall, p. 468; Inwards, p. 34; Wilson, p. 897.
3942. Winter thunder bodes summer hunger. Var.: Winter thunder, summer hunger.
Cited in Denham, p. 3; Mitchell, p. 230; Swainson, p. 10; Northall, p. 468; Marvin, p. 216; Inwards, p. 34; Wilson, p. 897; Page, p. 27; Wilshere, p. 17; Freier, p. 77.
3943. Winter thunder is to old folks death, and to young folks plunder.
Cited in Dunwoody, p. 80.
3944. Winter thunder, poor man's death, rich man's hunger. Var.: Winter thunder is poor man's death and rich man's hunger.
Cited in Swainson, p. 9; Dunwoody, p. 92; Northall, p. 468; Inwards, p. 34.
3945. Winter thunder, rich man's good and poor man's hunger. Var.: Winter thunder, rich man's food, poor man's hunger.
Cited in Inwards, p. 35; Page, p. 27. Recorded in Bryant (Tex.).
3946. Winter's thunder is summer's wonder. Vars.: (a) Winter's thunder makes summer's wonder. (b) Winter's thunder, summer's wonder.
Cited in Denham, p. 2; Northall, p. 468; Inwards, p. 34; Hyatt, p. 34; Wilson, p. 897.

3947. Winter's thunder is the world's wonder.
Cited in Northall, p. 468.

thundershower

3948. Thundershower in the morning, rain all day.
Recorded in Bryant (Miss., N.Y.).

thunderstorm

3949. A summer thunderstorm which does not much depress the barometer will be very local and of slight consequence.
Cited in Garriott, p. 16; Inwards, p. 174; Freier, p. 25.

3950. A thunderstorm from northwest is followed by fine, bracing weather; but thunder and lightning from northeast indicates sultry, unsettled weather.
Cited in Dunwoody, p. 81.

3951. A thunderstorm from the south is followed by warmth, and from the north, by cold. When the storm disappears in the east, it is a sign of fine weather.
Cited in Mitchell, p. 223.

3952. A thunderstorm in April is the end of hoar frost.
Cited in Dunwoody, p. 95; Inwards, p. 52.

3953. A thunderstorm is a sign of a change in the weather.
Cited in Smith, p. 1.

3954. A thunderstorm often comes up against the surface wind.
Cited in Inwards, p. 168.

3955. Don't talk during a thunderstorm.
Cited in Smith, p. 11.

3956. Don't walk or move around during a thunderstorm.
Cited in Smith, p. 11.

3957. Early thunderstorms bring wonderful crops.
Cited in Hyatt, p. 34.

3958. If there be showery weather, with sunshine and increase of heat in the spring, a thunderstorm may be expected every day, or at least every other day.
Cited in Dunwoody, p. 81.

3959. The thunderstorms of the season will come from the direction of the first thunderstorm.
Cited in Dunwoody, p. 80.

3960. When the wind freshens and blows towards a distant thunder-storm, the storm will come nearer; when the wind blows outwards from it, the storm will drift away.
Cited in Inwards, p. 168.

Thursday

3961. If it storms on the first Thursday, or any subsequent, of a month, count the remaining days of the month, add to this the number of days remaining of the moon, and it will give the number of storms for that season.
Cited in Dunwoody, p. 100.

3962. On Thursday at three, look out, and you'll see, what Friday will be.
Cited in Northall, p. 478; Wright, p. 96; Whitman, p. 41; Inwards, p. 73.

3963. The first Thursday in March, the first Thursday in June, the first Thursday in September, and the first Thursday in December are the governing days for each season. Whatever point of the compass the wind is on these days, that will be the prevailing direction of the wind for that season.
Cited in Dunwoody, p. 100.

tick

3964. Take a tick off a dog, lay it on its back under a rock, and it will cause rain.
Cited in Wurtele, p. 300.

tide

3965. Every tide has its ebb.
Cited in Mieder, p. 595.

3966. Every tide must turn.
Cited in Mieder, p. 595.

3967. If, after the first ebb of the tide, it flows again for a little while, a storm approaches.
Cited in Inwards, p. 153.

3968. If it rains at tide's flow, you may safely go and mow; but if it rains at the ebb, then, if you like, go off to bed.
Cited in Inwards, p. 153.

3969. In threatening weather it is more apt to rain at the turn of the tide, especially at high-water.
Cited in Inwards, p. 153.

3970. It is a long tide that never turns.
Cited in Whiting 1, p. 438.

3971. Rain is likely to commence on the turn of the tide. Var.: Rain is most likely to set in at the turn of the tide.
Cited in Dunwoody, p. 70; Inwards, p. 153.

3972. Some storms cause tides in fresh-water lakes.
Cited in Sloane 2, p. 50.

3973. Storms burst as the tides turn.
Cited in Humphreys 2, p. 71; Whitman, p. 33; Inwards, p. 153.

3974. The highest tides produce lowest ebbs.
Cited in Whiting 1, p. 438.

3975. Time and tide wait for no man. Vars.: (a) The tide stays for no man. (b) Time and tide tarry for no man. (c) Time and tide wait no one's pleasure.
Cited in Whiting 1, p. 440; Whiting 2, p. 628; Mieder, p. 595.

titmouse

3976. The saw-like note of the great titmouse foretells rain.
Cited in Inwards, p. 193.

3977. The titmouse foretells cold, if crying, "Pincher."
Cited in Inwards, p. 193.

toad

3978. If toads come out of their holes in great numbers, rain will fall soon. Vars.: (a) An exceptional number of toads at one time is followed by rain. (b) If toads appear in large number, expect rain.
Cited in Swainson, p. 252; Dunwoody, p. 14; Whitman, p. 48; Inwards, p. 203; Hyatt, p. 20; Freier, p. 41.

toadstool

3979. If toadstools spring up in the night in dry weather, they indicate rain. Var.: Toadstools spring up overnight when dry weather is ending.
Cited in Dunwoody, p. 70; Alstad, p. 93.

3980. If you see toadstools in the morning, expect rain by evening.
Cited in Freier, p. 41.

toes

3981. If in winter your toes are burning, a snowstorm is approaching.
Cited in Hyatt, p. 30.

tortoise

3982. Tortoises creep deep into the ground so as to completely conceal
themselves from view when a severe winter is to follow. When
a mild winter is to follow they go down just far enough to protect
the opening of their shells.
Cited in Dunwoody, p. 118; Inwards, p. 203.

tree

3983. Examine trees after it has snowed: snow only in the forks means
more snow soon; snow only over the top branches, no more
snow.
Cited in Hyatt, p. 11.

3984. If on the trees the leaves still hold, the winter coming will be
cold. Var.: If trees hang onto their leaves, the coming winter will
be cold.
Cited in Northall, p. 474; Freier, p. 79.

3985. North side of trees covered with moss indicate cold weather.
Cited in Dunwoody, p. 68.

3986. Tree putting on new shoots is a sign of rain.
Cited in Reeder, no pp.

3987. Trees grow dark before a storm. Vars.: (a) Trees are light green
when the weather is fair; they turn quite dark when a storm's in
the air. (b) Trees become dark before a storm.
Cited in Dunwoody, p. 68; Garriott, p. 25; Hand, p. 25; Freier,
p. 57.

3988. Trees snapping and cracking in the fall indicates cold weather.
Var.: Trees snapping and cracking in the autumn indicate dry
weather.
Cited in Dunwoody, p. 118; Inwards, p. 209.

3989. Trees warm the house in winter, and cool it in the summer.
Cited in Alstad, p. 20.

3990. When caught by the tempest, wherever it be, if it lightens and thunders beware of a tree!
Cited in Denham, p. 19; Northall, p. 523.

3991. When the leaves of trees curl with the wind from the south, it indicates rain.
Cited in Dunwoody, p. 68.

3992. When the trees begin to bloom, the weather becomes cold.
Cited in Boughton, p. 125.

3993. When tree limbs break off during calm, expect rain.
Cited in Dunwoody, p. 68.

tree frog

3994. The green tree frog becomes very unquiet before rain. Var.: When you hear a tree frog, it's going to rain.
Cited in Inwards, p. 203; Smith, p. 4.

3995. Tree frogs crawl up the branches of trees before a change of weather.
Cited in Dunwoody, p. 72; Inwards, p. 203.

3996. Tree frogs croak at night a day or two before a rain.
Cited in Reeder, no pp.

3997. Tree frogs piping during rain indicates continued rain.
Cited in Dunwoody, p. 72; Inwards, p. 203.

tree toad

3998. A tree toad that trills about ten o'clock at night will bring rain.
Cited in Hyatt, p. 20.

3999. A tree toad trilling in a tree is a forecast of rain.
Cited in Hyatt, p. 20.

trout

4000. Trout bite voraciously before rain.
Cited in Dunwoody, p. 51.

4001. Trout jump and herring school more rapidly before rain.
Cited in Dunwoody, p. 51; Garriott, p. 24; Hand, p. 24.

4002. Trout jump high when a rain is nigh.
Cited in Freier, p. 23.

4003. When trout refuse bait or fly, there ever is a storm nigh.
Cited in Dunwoody, p. 51; Inwards, p. 199; Lee, p. 59.

tulip

4004. Tulips and dandelions close just before rain.
Cited in Dunwoody, p. 68; Inwards, p. 216.

4005. Tulips open their blossoms when the temperature rises, they close
again when the temperature falls.
Cited in Freier, p. 29.

tumblebug

4006. If you see a tumblebug pushing his dung ball, an exceptionally
harsh winter can be expected.
Cited in Hyatt, p. 19.

4007. That year in which tumblebugs are numerous will be a year
having a severe winter.
Cited in Hyatt, p. 19.

turkey

4008. If the turkey's feathers are ruffled, it will rain.
Cited in Freier, p. 52.

4009. If turkeys in winter climb to the highest perch, it will be cold; if
they stop in the middle of the roost, not very cold; and if they
remain on the ground, not cold at all.
Cited in Hyatt, p. 26.

4010. If turkeys roost in the top of a tree, good weather is betokened;
if on the lower limbs, a change of weather.
Cited in Hyatt, p. 26.

4011. Turkeys perched in trees and refusing to descend indicate snow.
Cited in Dunwoody, p. 40; Inwards, p. 188; Sloane 2, p. 130.

4012. Water turkeys flying against the wind indicate falling weather.
Cited in Dunwoody, p. 40; Inwards, p. 188.

4013. When a turkey stands with his back to the wind so that his
feathers become ruffled, a storm is coming.
Cited in Lee, p. 66.

4014. When water turkeys fly against the wind, rain is coming.
Cited in Lee, p. 65.

4015. You will always see turkeys hopping up and down before a rain.
Cited in Hyatt, p. 26.

turkey buzzard

4016. A solitary turkey buzzard at a great altitude indicates rain.

Cited in Dunwoody, p. 34.

turtle

4017. If a turtle bites you, he will not turn loose till it thunders.
Recorded in Bryant (Miss.).
4018. If you kill a turtle, there will soon be lightning and thunder.
Cited in Hyatt, p. 20.
4019. Turtles crawling uphill means it is going to rain; if they crawl downhill it is going to be dry.
Cited in Reeder, no pp.

twilight

4020. Pale, yellow twilight, extending high up, indicates threatening weather.
Cited in Dunwoody, p. 78.
4021. Twilight looming indicates rain.
Cited in Dunwoody, p. 78; Inwards, p. 74.

umbrella

4022. Carry an umbrella on a cloudy day and you will scare rain away.
Var.: If you carry an umbrella, you can be sure it won't rain.
Cited in Hyatt, p. 32; Smith, p. 7.
4023. If you leave your umbrella at home, it is sure to rain.
Cited in Mieder, p. 623. Recorded in Bryant (N.Y.).
4024. The umbrella goes up when the rain comes down.
Cited in Mieder, p. 623.
4025. Two under an umbrealla makes the third a wet fellow.
Cited in Mieder, p. 623.
4026. Watch for rain after somebody opens an umbrella in the house.
Cited in Hyatt, p. 32.
4027. When fine, take your umbrella; when raining, please yourself.
Cited in Inwards, p. 23.

vegetable

4028. Long-tailed vegetables: beets, carrots, parsnips, radishes and turnips, have longer tails before a hard winter.
Cited in Hyatt, p. 17.

4029. When vegetables in the spring begin to wilt, a long dry spell is near.
Cited in Hyatt, p. 17.

violet

4030. Violets flowering in October betoken a mild winter.
Cited in Hyatt, p. 17.

virgin's bower

4031. When Mary left us here below, the virgin's bower begins to blow. Comment: The flower, the virgin's bower, is said to "blow" on the celebration day of the Ascension of the Virgin Mary, August 15th.
Cited in Swainson, p. 263.

vision

4032. Great clearness of vision in the air is a sign of rain. Vars.: (a) Distant objects appear to be closer before rain. (b) If some distant object seems unusually clear, a rain is close; the clearer the object, the closer the rain. (c) Unusual clearness in the atmosphere, objects being seen very distinctly, indicates rain.
Cited in Steinmetz 2, p. 84; Dunwoody, p. 68; Hyatt, p. 12; Freier, p. 65.

4033. If you can see the mountains, it's sure to rain; if you can't see them, it's already raining.
Recorded in Bryant (Wash.).

4034. The further the sight (vision), the nearer the rain. Vars.: (a) The farther the sight, the closer the rain. (b) When distant view is clear, rain will soon come.
Cited in Dunwoody, p. 16; Humphreys 1, p. 442; Whitman, p. 29; Brunt, p. 73; Inwards, p. 150; Alstad, p. 102; Sloane 2, p. 118; Wilson, p. 245; Wilshere, p. 4; Freier, p. 64. Recorded in Bryant (Mich., N.Y.).

vulture

4035. When vultures scent carrion at a great distance, they indicate that state of the atmosphere which is favorable to the perception of smells, which forbodes rain.
Cited in Dunwoody, p. 40.

wall

4036. Brick walls become damp before a rain.
Cited in Dunwoody, p. 106.

4037. When in cold weather the walls begin to show dampness, the weather changes.
Cited in Dunwoody, p. 119; Inwards, p. 219; Lee, p. 74.

4038. When walls are unusually damp, rain is expected. Var.: When walls are wet expect some rain.
Cited in Garriott, p. 21; Inwards, p. 219; Hand, p. 21; Freier, p. 33.

walnut

4039. Great store of walnuts and almonds presage a plentiful year of corn.
Cited in Swainson, p. 261.

4040. If walnut or hickory nut hulls are loose, winter will be open; if they are tight, winter will be closed.
Cited in Hyatt, p. 16.

4041. If walnuts drop early, watch for an early autumn.
Cited in Hyatt, p. 16.

4042. If walnuts fall faster than squirrels can store them away, look for a big wheat crop next year.
Cited in Hyatt, p. 34.

4043. If walnuts or hickory nuts have thin hulls, a thin winter is presaged; thick hulls, a thick winter.
Cited in Hyatt, p. 16.

walnut tree

4044. If the walnut tree has plenty of blossoms, a sign of a fruitful year.
Cited in Freier, p. 82.

4045. When walnut trees bear bountifully, we will have a warm winter.
Cited in Freier, p. 84.

wash

4046. Hang out wash on a line in the sunshine and it will rain before they are dry.
Cited in Smith, p. 7.

wash rag

4047.   If a wash rag or a sponge does not dry out rapidly after it has been used, rain is in the air.
Cited in Hyatt, p. 31.

wasp

4048.   A summer of many wasps will bring a winter of much snow, but by others this is said to apply only when wasps are numerous in autumn.
Cited in Hyatt, p. 19.

4049.   If mud wasps build their nests in sheltered areas, expect a harsh winter.
Cited in Freier, p. 83.

4050.   Wasps and hornets biting more eagerly than usual is a sign of rainy weather.
Cited in Swainson, p. 256; Lee, p. 57.

4051.   Wasps attempt to enter the house as cold weather approaches.
Cited in Hyatt, p. 19.

4052.   Wasps building nests in exposed places indicate a dry season.
Cited in Dunwoody, p. 58; Inwards, p. 205.

4053.   Wasps in great numbers and busy indicate fair and warm weather. Var.: Wasps in great numbers and busy indicate warm weather.
Cited in Dunwoody, p. 58; Inwards, p. 1455.

4054.   When wasps build their nests high on the banks of a stream you may expect a wet summer; but if near the level of the water, a dry summer is indicated.
Cited in Inwards, p. 205.

water

4055.   If standing water be at any time warmer than it was commonly wont to be, and no sunshine help, it foretelleth rain.
Cited in Inwards, p. 222.

4056.   Water rising in wells and springs indicates rain.
Cited in Inwards, p. 222; Freier, p. 26.

4057.   Water seeks its level.
Cited in Whiting 1, p. 473.

4058.   When boiling water rapidly evaporates, expect rain. Var.: When boiling water more rapidly vanishes, expect rain.

Cited in Inwards, p. 219; Freier, p. 25.

4059. When you gently drop water from a feather upon water, and you find that the drops, instead of at once mixing freely with the water, float away on its surface, expect rain.
Cited in Inwards, p. 219.

4060. Witches produce bad weather by stirring water with branches of elder.
Cited in Inwards, p. 212.

water fowl

4061. If the feathers of water fowl be thicker and stronger than usual, expect a cold winter. Var.: If the feathers of water fowl be thicker and stronger, expect a severe winter.
Cited in Inwards, p. 197; Sloane 2, p. 132.

4062. If water fowl scream more than usual and plunge into water, expect rain.
Cited in Dunwoody, p. 40.

watercress

4063. If watercress beds steam on a summer evening, the next day will be hot.
Cited in Inwards, p. 216; Lee, p. 73.

waterfall

4064. If waterfalls roar loudly, bad weather is coming.
Cited in Freier, p. 65.

wave

4065. The waves do not rise but when the winds blow.
Cited in Whiting 1, p. 473.

4066. White horses (cresting waves) warn of rain.
Cited in Page, p. 7.

weasel

4067. The cry of the weasel will bring rain.
Cited in Hyatt, p. 30.

4068. Weasels, stoats, etc., seen running about in the forenoon, foretell rain in the after part of the day. Vars.: (a) If weasels and stoats are seen running about much in the forenoon, it foretells rain in

the after part of the day. (b) When weasels run about in the morning, expect rain later in the day.
Cited in Mitchell, p. 228; Swainson, p. 234; Inwards, p. 185; Lee, p. 66.

weather

4069.   Be it dry, or be it wet, the weather'll always pay its debt.
Cited in Cheales, p. 18; Northall, p. 477; Inwards, p. 25; Wilshere, p. 4.

4070.   Change of weather is the discourse of fools.
Cited in Denham, p. 3; Wilson, p. 114.

4071.   Dry weather follows a shower that threatens and does not keep its threat.
Cited in Hyatt, p. 9.

4072.   Everybody talks about the weather, but nobody does anything about it.
Cited in Mieder, p. 646.

4073.   Expect good weather when the moon is coming, and bad when it is going back.
Recorded in Bryant (Mich.).

4074.   Fair weather after foul.
Cited in Wilson, p. 240.

4075.   Fair weather comes out of the north.
Cited in Sloane 2, p. 117. Recorded in Bryant (Wis.).

4076.   Fine, warm days are called "weather breeders."
Cited in Dunwoody, p. 106; Inwards, p. 154.

4077.   God's weather is good weather.
Cited in Mieder, p. 647.

4078.   He, who is weather-wise, is not otherwise.
Cited in Cheales, p. 18.

4079.   If a change of weather occur when the sun or moon is crossing the meridian, it is for twelve hours at least.
Cited in Inwards, p. 74.

4080.   If it's long fair (weather), it'll be long foul.
Recorded in Bryant (Can. [Ont.]).

4081.   If the weather clears off before noon, it will be pleasant longer.
Recorded in Bryant (Vt.).

4082.   If the weather is fine, put on your cloak; if it is wet do as you please. Var.: One should turn his coat according to the weather.

Cited in Whitman, p. 51. Recorded in Bryant (Calif.).

4083. Ill weather and sorrow come unsent for.
Cited in Denham, p. 3.

4084. Ill weather is seen soon enough when it comes.
Cited in Denham, p. 11.

4085. In fair weather prepare for foul.
Cited in Wilson, p. 240; Mieder, p. 647.

4086. In Texas, if there be dry weather with a light south wind for five or six days, it having previously blown strongly from the same direction, expect fine weather.
Cited in Inwards, p. 119.

4087. Man is never satisfied with the weather.
Cited in Mieder, p. 647.

4088. Never waste time complaining about Arkansas weather; it will soon change. Vars.: (a) If you don't like Oklahoma weather, wait a minute. (b) If you don't like the weather in Chicago, wait a few minutes. (c) If you don't like the weather, wait five minutes. (d) Never waste time complaining about the weather, it will soon change.
Cited in Mieder, p. 647. Recorded in Bryant (Ark., Ill., Okla.).

4089. No weather is ill, if the wind be still.
Cited in Denham, p. 2; Cheales, p. 17; Dunwoody, p. 1883; Wright, p. 84; Marvin, p. 213; Sloane 2, p. 116; Wilson, p. 875; Lee, p. 114; Wilshere, p. 18; Mieder, p. 647. Recorded in Bryant (Wis.).

4090. Only Yankees and fools predict the weather.
Cited in Mieder, p. 685.

4091. Praise a maid in the morning and the weather in the evening.
Recorded in Bryant (Minn.).

4092. So it falls that all men are with fine weather happier far.
Cited in Whitman, p. 51.

4093. Take the weather as it comes.
Cited in Wilson, p. 875.

4094. The general character of the weather during the last twenty days of March, June, September, and December will rule the following season.
Cited in Dunwoody, p. 100.

4095. The kind of weather during the first twelve days of the year indicates what the weather will be like for the year.
Recorded in Bryant (Kans.).

4096. The longer the dry weather has lasted, the less is rain likely to follow the cloudiness of cirrus.
Cited in Garriott, p. 12; Inwards, p. 129; Hand, p. 12.

4097. The weather on the first three days of any season decides the weather for that season.
Cited in Hyatt, p. 14.

4098. The weather usually clears at noon when a southerly wind is blowing.
Cited in Inwards, p. 119.

4099. There is usually fair weather before a settled course of rain.
Cited in Inwards, p. 158.

4100. To talk of the weather, is nothing but folly; for when it rains on the hill, the sun shines in the valley. Var.: The talk of the weather is nothing but folly; when it rains on the hill, the sun's in the valley.
Cited in Denham, p. 17; Wright, p. 80; Inwards, p. 24; Sloane 1, p. 139.

4101. Tomorrow will be fair (weather), if the sun sets clear.
Recorded in Bryant (N.Y.).

4102. When everything is eaten at the table, it indicates continued clear weather.
Cited in Dunwoody, p. 106; Inwards, p. 218.

4103. When fine weather is lost, it will come from the north.
Cited in Marvin, p. 217.

4104. When the landscape looks clear, having your back towards the sun, expect fine weather; but when it looks clear with your face towards the sun, expect showery, unsettled weather.
Cited in Inwards, p. 150.

4105. Whether it's cold or whether it's hot, we must have weather whether or not.
Cited in Whitman, p. 51.

weathercock

4106. A weathercock that swings to the west, proclaims the weather to be the best; a weathercock that swings to the east, proclaims no

good to man or beast.
Cited in Sloane 2, p. 25.

Wednesday

4107. Wednesday clearing, clear till Sunday. Var.: Clearing on
Wednesday, clear 'til Sunday.
Cited in Dunwoody, p. 100; Garriott, p. 43; Whitman, p. 41;
Inwards, p. 73; Hand, p. 43; Lee, p. 146.

4108. When the sun sets clear on Wednesday, expect clear weather the
rest of the week.
Cited in Dunwoody, p. 100; Inwards, p. 73.

weed

4109. As high as the weeds grow, so will be the bank of snow.
Cited in Freier, p. 78.

4110. Tall weeds in autumn; deep snows in winter.
Cited in Hyatt, p. 16.

well

4111. Very well we know, when a well doth well; a rain and a blow it
doth foretell.
Cited in Humphreys 2, p. 75; Whitman, p. 50.

4112. Water rising in wells and springs indicates approaching rain.
Vars.: (a) A rising well and a gushing spring, are two good signs
of the very same thing. (b) If in a seep-hole, spring or well, the
water level rises, or if water is found in a dry seep-hole, rain will
soon appear. (c) When water rises in wells and springs, rain is
approaching.
Cited in Dunwoody, p. 119; Whitman, p. 50; Hyatt, p. 13; Lee,
p. 75.

4113. Wells give murky water before a storm. Vars.: (a) Wells gurgle
and yield muddy water before a storm. (b) Wells gurgle before
a storm.
Cited in Freier, p. 24; Alstad, p. 110.

wet

4114. No one so surely pays his debt as wet to dry and dry to wet.
Var.: No one surely pays his debt as wet to cold and cold to wet.
Cited in Inwards, p. 155; Sloane 2, p. 119.

4115. Open and shet, is a sign of wet. Comment: Shet means shut and refers to blossoms closing.
Cited in Hyatt, p. 5.

4116. Wet causes more damage than frost before Christmas than after Christmas.
Cited in Dunwoody, p. 100.

4117. Wet continues if the ground dries up too soon.
Cited in Inwards, p. 158.

whale

4118. Whales "sounding" foretell a storm.
Cited in Lee, p. 60.

wheat

4119. An abundant wheat crop does not follow a mild winter. Var.: Abundant wheat crops never follow a mild winter.
Cited in Garriott, p. 45; Whitman, p. 45; Inwards, p. 33; Hand, p. 45; Sloane 2, p. 131.

4120. He that goes to see his wheat in May comes weeping away.
Cited in Northall, p. 494.

4121. If the spring is dry, sow wheat; if it is wet, plant corn.
Cited in Dunwoody, p. 119.

4122. If wheat remains the same color as when it was threshed, a light winter is coming; if the grain darkens, a rigorous winter.
Cited in Hyatt, p. 17.

4123. If (wheat) sown in the slop, 'twill be thick on top. Var.: Sow in the slop, 'twill be heavy at top. Comment: Wheat sown when the ground is wet is most productive.
Cited in Inwards, p. 213; Wilson, p. 756.

4124. It is said that there is never too little summer rain for a satisfactory wheat crop in England.
Cited in Inwards, p. 31.

4125. Sow wheat in dirt and rye in dust.
Cited in Denham, p. 31; Inwards, p. 213.

4126. Sow wheat in mud, 'twill stand a flood; barley in dust, be dry that must.
Cited in Cheales, p. 25; Northall, p. 483.

4127. Wheat in the dust, and oats in the dab, that is, sow wheat dry and oats anyhow.

Cited in Dunwoody, p. 119; Inwards, p. 213.

wheel

4128. When the wheel is far, the storm is near; when the wheel is near, the storm is far. Comment: Wheel is the halo around the moon. Cited in Humphreys 1, p. 438; Marvin, p. 210; Humphreys 2, p. 38; Whitman, p. 27; Inwards, p. 89; Sloane 2, p. 126.

whippoorwill

4129. If a whippoorwill sings in a tree at night, it will be clear the next day.
Cited in Hyatt, p. 23.

4130. When the whippoorwill calls in the fall, it is a sign that the first big frost is near.
Cited in Smith, p. 2.

4131. When the whippoorwill hollers, sign of dry weather; unless it chucks just before he hollers whippoorwill, that means rain.
Cited in Hyatt, p. 23.

whirlwind

4132. A whirlwind as a rule will indicate dry weather, but at times it is thought by some to be an indication of rain.
Cited in Hyatt, p. 9.

4133. If you see a whirlwind traveling downstream, rain is imminent.
Cited in Hyatt, p. 9.

4134. When numerous whirlwinds are observed, the rotation being opposite to that of the sun, look for wind and rain.
Cited in Dunwoody, p. 88.

4135. When you see a whirlwind moving away from a river, it will be dry.
Cited in Reeder, no pp.

4136. When you see a whirlwind moving toward a river, it will rain soon.
Cited in Reeder, no pp.

4137. When you see a whirlwind unwinding, it will rain soon.
Cited in Reeder, no pp.

4138. Whirlwinds in the spring mean a droughty summer.
Cited in Hyatt, p. 9.

whitebrush

4139. When the whitebrush blooms, it is a sign of rain.
Cited in Reeder, no pp.

whitethorn

4140. If many whitethorn blossoms or dog-roses are seen, expect a severe winter.
Cited in Inwards, p. 212.

whitlow grass

4141. Look for wet weather if the leaves of the whitlow grass droop, and if lady's bedstraw becomes inflated and gives out a strong odour.
Cited in Inwards, p. 213.

Whitsunday

4142. If Whitsunday bring rain, we expect many a plague. Comment: Whitsunday is the seventh Sunday after Easter.
Cited in Inwards, p. 72.
4143. Whitsunday bright and clear will bring a fertile year.
Cited in Dunwoody, p. 104; Inwards, p. 72.
4144. Whitsunday wet, Christmas fat.
Cited in Dunwoody, p. 104; Inwards, p. 72.

Whitsuntide

4145. Rain on Whitsuntide is said to make the wheat mildewed. Comment: Whitsuntide is the week beginning with Whitsunday or Pentecost.
Cited in Dunwoody, p. 104; Inwards, p. 72.
4146. Strawberries at Whitsuntide indicate good wine.
Cited in Dunwoody, p. 104; Inwards, p. 72.
4147. Whitsuntide rain, blessing for wine.
Cited in Dunwoody, p. 104.

wind

4148. A brisk wind generally precedes rain.
Cited in Dunwoody, p. 83; Inwards, p. 106.
4149. A cold wind blowing through an open door is said to be a stepmother's breath.

Cited in Wurtele, p. 297.

4150. A frequent change of wind, with agitation in the clouds, denotes a storm.
Cited in Sloane 2, p. 116.

4151. A high wind prevents frost.
Cited in Dunwoody, p. 55; Marvin, p. 212; Humphreys 2, p. 62; Whitman, p. 23.

4152. A storm wind settles in the chimney, but a clear wind coaxes out the smoke.
Cited in Sloane 2, p. 23.

4153. A veering wind indicates fair weather, a backing wind foul weather. Vars.: (a) A veering wind brings fair weather; a backing wind brings foul weather. (b) A veering wind, fair weather; a backing wind, foul weather. (c) A veering wind, fine weather; a backing wind, foul weather.
Cited in Dunwoody, p. 88; Marvin, p. 212; Humphreys 2, p. 60; Whitman, p. 22; Inwards, p. 107; Wilshere, p. 19; Freier, p. 58.

4154. A veering wind will clear the sky; a backing wind says storms are nigh.
Cited in Sloane 2, p. 26; Lee, p. 112.

4155. A west wind always brings wet weather, the east wind cold and wet together; the south wind surely brings us rain, the north wind blows it back again.
Cited in Swainson, p. 220; Dunwoody, p. 88.

4156. A wind generally sets from the sea to the land during the day, and from the land to the sea at the night, especially in hot climates.
Cited in Inwards, p. 104.

4157. An ill wind blows no good.
Cited in Mieder, p. 656.

4158. As old sinners have all points o' the compass in their joints, can by their pangs and aches find all turns and changes of the wind.
Cited in Dunwoody, p. 83; Humphreys 1, p. 443; Humphreys 2, p. 82; Inwards, p. 217; Lee, p. 40.

4159. As the wind blows, so bends the twig. Var.: As the wind blows, the tree lists.
Cited in Mieder, p. 656.

4160. As the wind blows, you must set your sail. Var.: As the wind blows, set your sails.

Cited in Denham, p. 3; Wilson, p. 893; Mieder, p. 656.

4161. Back the wind and front the sun. Comment: This is an old mariners' rule for running from a storm or gale: the wind brings the bad weather and the sun indicates good weather.
Cited in Mieder, p. 656.

4162. Before the rising of a wind, the lesser stars are not visible, even on a clear night.
Cited in Alstad, p. 168.

4163. Blow the wind never so fast, it will lower at last. Var.: Blow the wind never so fast, it will fall at last.
Cited in Denham, p. 11; Cheales, p. 19; Wilson, p. 70; Lee, p. 113.

4164. Drop a penny in the sea and it will bring a wind.
Cited in Smith, p. 12.

4165. Every wind has its weather. Var.: Come wind, come weather.
Cited in Swainson, p. 218; Dunwoody, p. 88; Humphreys 1, p. 441; Humphreys 2, p. 59; Whitman, p. 21; Wilson, p. 135; Mieder, p. 656.

4166. Every wind is ill to a crazy ship. Comment: Crazy means broken.
Cited in Wilson, p. 231.

4167. Great winds blow upon high hills.
Cited in Wilson, p. 335.

4168. Greater winds' chance in the day than in the night.
Cited in Inwards, p. 104; Sloane 2, p. 128.

4169. High altitude winds soon descend to earth.
Cited in Sloane 2, p. 62.

4170. Hoist your sail when the wind is fair.
Cited in Wilson, p. 376.

4171. If cold wind reach you through a hole, say your prayers and mind your soul. Vars.: (a) If wind blows on you through a hole, say your prayers and mind your soul. (b) Take heed of wind that comes in at a hole, and a reconciled enemy.
Cited in Denham, p. 16; Wilson, p. 894; Mieder, p. 656.

4172. If it blows (wind) in the day, it generally hushes toward evening.
Cited in Dunwoody, p. 43.

4173. If sailors can catch a louse, and put him on the leech of the mainsail, wind is promised, provided the louse crawls upward.
Cited in Hand, p. 458.

4174. If the boom of a sailboat creaks while the boat is in motion, the wind will soon die out.
Cited in Hand, p. 458.

4175. If the wind backs against the sun, trust it not, for back it will run. Vars.: (a) If the wind backs the sun, trust her not for back she'll come. (b) The wind's backing, more's coming.
Cited in Dunwoody, p. 83; Whitman, p. 22; Sloane 2, p. 26.
Recorded in Bryant (Can. [Ont.]).

4176. If the wind be hushed with sudden heat, expect heavy rain.
Cited in Marvin, p. 212.

4177. If the wind follows the sun's course, expect fine weather. Var.: If the wind follows the sun's course, fair weather.
Cited in Swainson, p. 228; Inwards, p. 107; Sloane 2, p. 128; Wilson, p. 893.

4178. If the wind goes down with the sun, it will be a clear day tomorrow.
Recorded in Bryant (Ohio).

4179. If the wind increases during a rain, fair weather may be expected soon.
Cited in Dunwoody, p. 85.

4180. If the wind is in the southwest at Martinmas, it remains there till after Candlemas. Comment: Martinmas is November 11th, Candlemas is February 2nd.
Cited in Swainson, p. 144; Dunwoody, p. 103; Inwards, p. 66.

4181. If we go with the wind, we shall soon be gone with the wind.
Cited in Mieder, p. 656.

4182. If you sail with a bad wind, you must understand tacking.
Cited in Mieder, p. 656.

4183. In by day and out by night. Comment: Refers to wind coming and going.
Cited in Inwards, p. 104.

4184. It is a sign of continued fine weather when the wind changes during the day so as to follow the sun.
Cited in Dunwoody, p. 83; Inwards, p. 107.

4185. It's a bad wind that never changes. Vars.: (a) It's a bad wind that never shifted. (b) It's an ill wind that has no turning.
Cited in Whiting 1, p. 485; Whiting 2, p. 685; Mieder, p. 656.
Recorded in Bryant (Fla.).

4186. It's a mighty good wind that blows nobody ill.
Cited in Whiting 2, p. 685.

4187. It's an ill wind that blows nobody good. Vars.: (a) It's a bad wind that blows nobody good. (b) It's an ill wind that blows no good. (c) It is an ill wind that blows no one any good. (d) It's an ill wind that blows nobody any good. (e) It's an ill wind that blows nobody good luck. (f) It's an ill wind that doesn't blow apples into someone's garden. (g) Never an ill wind blows that it doesn't do someone some good.
Cited in Denham, p. 11; Taylor 2, p. 403; Whiting 1, p. 486; Whiting 2, p. 685; Mieder, p. 656.

4188. It's an ill wind that blows nowhere.
Cited in Taylor 2, p. 403.

4189. Light winds point to pressures low, but gales around the same do blow.
Cited in Whitman, p. 23; Lee, p. 113.

4190. No wind can do him good who steers for no port. Var.: No wind is of service to him who is bound for nowhere.
Cited in Mieder, p. 656.

4191. Puff not against the wind.
Cited in Denham, p. 7; Mieder, p. 656.

4192. Slow wind also brings the ship to harbor.
Cited in Mieder, p. 656.

4193. South or north (wind), sally forth; west or east, travel least.
Cited in Sloane 2, p. 31.

4194. Sow the wind and reap the whirlwind. Var.: If you sow the wind you reap the whirlwind.
Cited in Mieder, p. 656.

4195. The colder the wind, the warmer the hearth.
Cited in Mieder, p. 656.

4196. The faster the wind blows, the sooner we will have a change in the weather, the gentler the wind, the slower will be the change in the weather.
Cited in Whitman, p. 61.

4197. The harder the wind blows, the deeper the oak grows.
Recorded in Bryant (N.C.).

4198. The hollow winds begin to blow, the clouds look black, the glass (barometer) is low; last night the sun went pale to bed, the moon

in halos hid her head; 'twill surely rain.
Cited in Steinmetz 1, p. 100; Humphreys 2, p. 63; Lee, p. 23.

4199. The sharper the blast (wind), the shorter 'twill last. Vars.: (a) The sharper the blast, the sooner 'tis past. (b) The sharper the blow (wind), the sooner 'tis past.
Cited in Cheales, p. 19; Dunwoody, p. 83; Sloane 2, p. 115; Lee, p. 114; Wilshere, p. 20. Recorded in Bryant (Wis.).

4200. The warst blast (wind) comes in the borrowing days. Comment: The Spanish story about the borrowing days is that a shepherd promised March a lamb if he would temper the winds to suit his flocks; but after gaining his point, the shepherd refused to pay over the lamb. In revenge March borrowed three days from April, in which fiercer winds than ever blew and punished the deceiver.
Cited in Inwards, p. 50.

4201. The west wind, as a father, all goodness doth bring; the east, a forbearer, no manner of things; the south, as unkind, draweth sickness too near; the north, as a friend, maketh all again clear.
Cited in Marvin, p. 214.

4202. The whistling of the wind heard indoors denotes rain.
Cited in Mitchell, p. 230; Dunwoody, p. 18.

4203. The wind blows easily through a leafless tree.
Cited in Mieder, p. 657.

4204. The wind changes its mind with a change in the barometer.
Cited in Alstad, p. 42.

4205. The wind is fresh and free.
Cited in Mieder, p. 657.

4206. The wind is hushed and the moon clears his eye, when frost and dew are drawing nigh.
Cited in Alstad, p. 119.

4207. The wind keeps not always in one quarter.
Cited in Wilson, p. 893.

4208. The winds of the daytime wrestle and fight, longer and stronger than those of the night.
Cited in Swainson, p. 218; Humphreys 2, p. 63; Whitman, p. 23; Inwards, p. 104; Sloane 2, p. 127; Lee, p. 114.

4209. The worst winds are at the end of the storm.
Cited in Sloane 2, p. 63.

4210. To get wind when sailing, stick a knife blade into the mast.
Cited in Hand, p. 458.

4211. Unsteadiness of the wind is an indication of changeable weather.
Var.: Unsteadiness in the wind shows changing weather.
Cited in Dunwoody, p. 88; Sloane 2, p. 116.

4212. Whatever way the wind does blow, some hearts are glad to have it so.
Cited in Mieder, p. 657.

4213. When the wind backs, and the weather glass falls, then be on your guard against gales and squalls.
Cited in Cheales, p. 27; Humphreys 2, p. 60; Whitman, p. 22; Inwards, p. 107; Alstad, p. 41; Sloane 2, p. 281; Freier, p. 25.

4214. When the wind blows hard for three days, it will blow up a rain.
Cited in Reeder, no pp.

4215. When the wind comes before the rain, you may hoist your topsails up again; but when the rain comes before the wind, you may reef when it begins. Vars.: (a) Wind before rain, set your topsails fair again; rain before the wind, keep your topsails snug as sin. (b) Wind before rain, take your topsails down; rain before wind, put them up again.
Cited in Denham, p. 20; Steinmetz 2, p. 41; Northall, p. 465.

4216. When the wind doth feed the clay, England woe and well-a-day; but when the clay doth feed the sand, then it is well for Angle-land. Comment: Feed the clay refers to a wet summer, feed the sand refers to a dry summer.
Cited in Denham, p. 9.

4217. When the wind veers against the sun, trust it not, for back 'twill run.
Cited in Steinmetz 1, p. 90; Swainson, p. 228; Dunwoody, p. 15; Wright, p. 80; Whitman, p. 22; Wilson, p. 894; Lee, p. 112.

4218. When the wind's in the north, hail comes forth; when the wind's in the west, look for a wet blast; when the wind's in the soud, the weather will be fresh and good; when the wind's in the east, cold and snow comes neist. Comment: Soud means south, neist means next.
Cited in Denham, p. 18; Swainson, p. 220; Marvin, p. 214.

4219. When the wind's in the north, you need not go forth; when the wind's in the east, the fish will bite least; when the wind's in the

south, the bait goes into their mouth; when the wind's in the west, the fish will bite the best.
Cited in Cheales, p. 26.

4220. When wind comes before rain, soon you may make sail again.
Cited in Steinmetz 1, p. 80; Dunwoody, p. 86; Lee, p. 113; Mieder, p. 657.

4221. Where the wind is on Martinmas Eve, there it will be for the coming winter. Vars.: (a) Where the wind is on Martinmas Eve there it'll remain till Candlemas. (b) Where the wind is on Martinmas Eve there'll it be for the rest of the winter. Comment: Martinmas Day is November 11th, Candlemas is February 2nd.
Cited in Inwards, p. 67; Wilson, p. 893; Wilshere, p. 19.

4222. Where there's a wind there's a wave.
Cited in Mieder, p. 657.

4223. Who spits against the wind spits in his own face.
Cited in Mieder, p. 657.

4224. Wind and tide wait for no man.
Cited in Mieder, p. 657.

4225. Wind northwest at Martinmas, severe winter to come. Comment: Martinmas is November 11th.
Cited in Inwards, p. 66.

4226. Wind precedes a storm.
Cited in Mieder, p. 657.

4227. Wind right, sun right, fish bite.
Recorded in Bryant (N.Dak.).

4228. Wind roaring in chimney, rain to come.
Cited in Marvin, p. 212; Inwards, p. 104.

4229. Wind storms usually subside about sunset; but if they do not, they will go on for another day.
Cited in Inwards, p. 104.

4230. Winds at night are always bright, but winds in the morning, sailors take warning.
Cited in Dunwoody, p. 86; Inwards, p. 104.

4231. Winds changing from foul to fair during the night are not permanent.
Cited in Dunwoody, p. 83.

4232. Winds that change against the sun are sure to backward run.
Cited in Dunwoody, p. 87; Garriott, p. 44; Inwards, p. 107; Alstad, p. 41; Lee, p. 112.

4233. Winds that swing against the sun and winds that bring the rain are one. Var.: Winds that swing against the sun, and winds that bring the rain are one; winds that swing round with the sun, keep the rain storm on the run. Comment: Winds that swing against the sun are winds that shift from west to east.
Cited in Alstad, p. 69; Sloane 2, p. 26; Lee, p. 113.

4234. You can't hinder the wind from blowing.
Cited in Mieder, p. 657.

wind (east)

4235. A dry east wind raises the spring.
Cited in Inwards, p. 117.

4236. A right easterly wind is very unkind. Var.: The east wind has a bad reputation.
Cited in Northall, p. 468; Inwards, p. 117. Recorded in Bryant (Mich.).

4237. An easterly wind is like a boring guest that hasn't sense enough to leave.
Cited in Sloane 2, p. 62.

4238. An easterly wind's rain makes fools fain. Comment: Fain means happy.
Cited in Marvin, p. 213; Inwards, p. 118.

4239. An eastern wind is followed by rainy weather in summer and by snowy weather in winter; soon say some, within thirty-six hours say others.
Cited in Hyatt, p. 7.

4240. Fish bite least with wind in the east. Vars.: (a) If the wind blows from the east, the fish bite least. (b) When the wind is in the east, the fish bite the least; when the wind is in the west, the fish bite the best. (c) When the wind is in the east, then the fishes bite the least.
Cited in Dunwoody, p. 49; Marvin, p. 214; Inwards, p. 198; Wilshere, p. 19. Recorded in Bryant (Ill., Miss. Can. [Ont.]).

4241. If an east wind blows against a dark, heavy sky from the northwest, the wind decreasing in force as the clouds approach, expect thunder and lightning.
Cited in Dunwoody, p. 80; Inwards, p. 118.

4242. If an east wind veers to the northwest and rain fails to accompany it, there will be no wet weather for a week.

Cited in Hyatt, p. 8.

4243. If the wind becomes fixed in the east for the space of forty- eight hours, expect steady and continuous rain, with driving winds in the southwest during summer.
Cited in Dunwoody, p. 84.

4244. If the wind's in the east on Easter Day, you'll have plenty of grass, but little good hay.
Cited in Wilshere, p. 19.

4245. In early winter or late spring an easterly wind precedes a rain or a snow.
Cited in Hyatt, p. 7.

4246. In summer, if the wind changes to the east, expect cooler weather.
Cited in Dunwoody, p. 83.

4247. The chill of the east wind is conducive to aches and pains. Var.: The east wind brings aches and pains.
Cited in Garriott, p. 19; Hand, p. 19; Sloane 2, p. 62.

4248. There are a hundred days of easterly wind in the first half of the year.
Cited in Inwards, p. 31.

4249. Three days of wind from the east terminate in rain. Var.: If an east wind blows for three days it will rain.
Cited in Hyatt, p. 8; Reeder, no pp.

4250. When the wind is from the east, it is four and twenty hours at least.
Cited in Marvin, p. 213.

4251. When the wind is in the east, the fisher likes it least; when the wind is in the west, the fisher likes it best.
Cited in Swainson, p. 227; Inwards, p. 118.

4252. When the wind is in the east, then the sap will run the least; when the wind is from the west, then the sap will run the best. Var.: North to west, sugar's best; south to east, flow is least. Comment: "Sugar" refers to Maple sap.
Cited in Sloane 2, p. 53.

4253. When the wind's in the east, it's neither good for man nor beast. Vars.: (a) The wind is east and that's not good for man or beast (b) When the wind is in the east, it's good for neither man nor beast. (c) When the wind is in the east, it's neither fit for man

nor beast. (d) Wind from the east is bad for man and beast. (e) Wind from the east is good for neither man nor beast.
Cited in Denham, p. 10; Swainson, p. 226; Cheales, p. 26; Dunwoody, p. 21; Northall, p. 468; Taylor 1, p. 111; Whitman, p. 22; Brunt, p. 72; Alstad, p. 68; Hyatt, p. 8; Smith, p. 12; Wilson, p. 893; Lee, p. 107; Wilshere, p. 18; Freier, p. 60.
Recorded in Bryant (Ill., N.Y., Oreg.).

4254. Wind east or west, is a sign of a blast; wind north or south, is a sign of drought. Vars.: (a) East and west, the sign of a blast; north and south, the sign of a drouth. (b) North and south the sign o' drought; east and west the sign o'blast.
Cited in Denham, p. 17; Swainson, p. 220; Dunwoody, p. 83; Northall, p. 459.

4255. Wind from the east and warm weather are companions.
Cited in Hyatt, p. 7.

4256. Wind in the east, sailors feast.
Cited in Hyatt, p. 8.

wind (equinox)

4257. If the wind is northeast at vernal equinox, it will be a good season for wheat and a poor one for corn; but if south or southwest, it will be good for corn and bad for wheat. Var.: If the wind is in the north when the sun crosses the line on March 21st, there will be a good corn crop. If it is in the south, there will be a poor corn crop.
Cited in Dunwoody, p. 119; Boughton, p. 124.

wind (north)

4258. A nor'wester is not long in debt to a sou'wester.
Cited in Whitman, p. 22; Inwards, p. 116.

4259. A north wind is a broom for the English Channel.
Cited in Inwards, p. 115.

4260. A north wind with the new moon will hold until the full.
Cited in Marvin, p. 213; Inwards, p. 115.

4261. A northern air (wind) brings weather fair.
Cited in Denham, p. 15; Swainson, p. 221; Northall, p. 469; Marvin, p. 212; Inwards, p. 114; Sloane 2, p. 117; Wilson, p. 577; Freier, p. 59.

4262. Except in winter, a north wind over inland New York and New Jersey brings two full days of drear and drizzle.
Cited in Sloane 2, p. 14.

4263. Fisherman in anger froth, when the wind is in the north; for fish bite best, when the wind is in west.
Cited in Dunwoody, p. 50.

4264. North winds send hail, south winds bring rain, east winds we bewail, west winds blow amain; northeast is too cold, southeast not too warm, northwest is too bold, southwest doth no harm.
Cited in Swainson, p. 220; Marvin, p. 214; Lee, p. 107.

4265. Northerly wind and blubber, brings home the Greenland lubber.
Comment: Lubber means a clumsy or unskilled seaman.
Cited in Denham, p. 20; Northall, p. 469; Wilson, p. 577.

4266. Rain does not fall during a north wind.
Cited in Hyatt, p. 8.

4267. The cold of the north wind is bracing. Var.: Northern winds are always cold.
Cited in Garriott, p. 19; Hand, p. 19; Hyatt, p. 8.

4268. The north is a noyer to grass of all suits; the east a destroyer to herbs and all fruits. Comment: Noyer is an annoyance or nuisance.
Cited in Marvin, p. 214.

4269. The north wind doth blow, and we shall have snow.
Cited in Denham, p. 18; Northall, p. 469; Whitman, p. 21; Inwards, p. 115; Wilson, p. 577; Wilshere, p. 18.

4270. The north wind makes men more cheerful, and begets a better appetite towards meat.
Cited in Inwards, p. 114; Sloane 2, p. 117.

4271. The wind at north and east, is neither good for man nor beast; so never think to cast a clout, until the end of May is out.
Cited in Northall, p. 469.

4272. There's little use in praying for rain if the wind is in the north.
Cited in Freier, p. 59.

4273. When the wind is in the north, the skillful fisher goes not forth.
Vars.: (a) If the wind blows from the north, the fisherman never comes forth. (b) If the wind's in the north, the skillful fisherman goes not forth. (c) When the wind is in the north, the skillful fisher goes not forth; when the wind is in the east, 'tis good for neither man nor beast; when the wind is in the south, it blows the

flies in the fish's mouth; but when the wind is in the west, there it is the very best.

Cited in Denham, p. 17; Swainson, p. 221; Dunwoody, p. 85; Whitman, p. 21; Sloane 2, p. 37; Hyatt, p. 8; Smith, p. 12; Wilson, p. 893; Wilshere, p. 19; Freier, p. 60. Recorded in Bryant (Ala., Ill., Iowa. Can. [Ont.]).

4274. When the wind is in the north, then the fishes do come forth.
Cited in Marvin, p. 214. Recorded in Bryant (Ala., Ill., Iowa, Can. [Ont.]).

4275. When the wind's in the north, you mustn't go forth.
Cited in Inwards, p. 114.

4276. Whenever the wind first blows from the north, after having been for some days in another direction, a fine day or two will be almost sure to follow.
Cited in Inwards, p. 115.

4277. Wind from the north scares the fishes off.
Recorded in Bryant (Ala., Ill., Iowa. Can. [Ont.]).

4278. With a north wind it seldom thunders.
Cited in Dunwoody, p. 81; Inwards, p. 115.

wind (northeast)

4279. A northeast wind in winter is the forerunner of a big snowstorm.
Cited in Hyatt, p. 8.

4280. If the wind changes to the northeast or north, expect cold weather.
Cited in Dunwoody, p. 85.

4281. If the wind is from the northeast, there will be a bad storm.
Cited in Smith, p. 12.

4282. If the wind is northeast three days without rain, eight days will pass before south wind again.
Cited in Dunwoody, p. 21; Wilson, p. 893. Recorded in Bryant (Mich.).

4283. If there be northeast or east winds in the spring, after a strong increase of heat, and small clouds appear in the different parts of the sky, or if the wind changes from east to south at the appearance of clouds preceded by heat, expect heavy rains.
Cited in Dunwoody, p. 84.

4284. Northeast is bad for man and beast; northwest is much the best.
Var.: Northwest is far the best.

Cited in Cheales, p. 26; Marvin, p. 212; Humphreys 2, p. 61; Brunt, p. 72; Inwards, p. 117.

wind (northwest)

4285. An honest man and a northwest wind generally go to sleep together. Comment: The northwest wind generally abates about sunset.
Cited in Mitchell, p. 228; Swainson, p. 223; Inwards, p. 116; Sloane 2, p. 117.

4286. Do business with men when the wind is from the northwest.
Cited in Dunwoody, p. 15; Marvin, p. 212; Humphreys 2, p. 62; Whitman, p. 22; Sloane 1, p. 150; Alstad, p. 175; Wilson, p. 893; Wurtele, p. 297.

4287. If the northwest or north winds blow with rain or snow during three or four days in the winter and then the wind passes to the south through the west, expect continued rain.
Cited in Dunwoody, p. 86.

4288. If the wind is from the northwest or southwest, the storm will be short; if from the northeast, it will be a hard one. Var.: Northwest wind brings a short storm; a northeast wind brings a long storm.
Cited in Dunwoody, p. 85; Inwards, p. 117.

4289. If there be a change of wind from the northwest or west to the southwest or south, or else from the northeast or east to the southeast or south, expect wet weather.
Cited in Dunwoody, p. 86; Inwards, p. 116.

4290. In summer if the wind changes to the northwest, expect cooler weather.
Cited in Dunwoody, p. 86; Inwards, p. 116.

4291. Northwest wind brings only rain showers.
Cited in Dunwoody, p. 86; Inwards, p. 116.

4292. When the wind is in the northwest, the weather is at its best; but if the rain comes out of the east, 'twill rain twenty-four hours at least. Vars.: (a) When the wind is northwest, the weather is at the best; if the rain comes out of east, 'twill rain twice twenty-four hours at the least. (b) When the wind is in the northwest, the weather is best.
Cited in Dunwoody, p. 85; Northall, p. 469; Whitman, p. 22; Inwards, p. 116; Alstad, p. 174.

wind (south)

4293. A southerly wind and a cloudy sky proclaim a hunting morning.
Cited in Denham, p. 8; Marvin, p. 204; Inwards, p. 120; Wilshere, p. 19.

4294. A southerly wind and a fog, bring an east wind home snog.
Comment: Snog means with certainty.
Cited in Northall, p. 470; Inwards, p. 119.

4295. A southerly wind with showers of rain, will bring the wind from west again.
Cited in Swainson, p. 224; Northall, p. 469; Whitman, p. 22; Inwards, p. 121; Alstad, p. 69; Lee, p. 106.

4296. Brisk winds from the south for several days in Texas are generally followed by a "norther."
Cited in Dunwoody, p. 87; Inwards, p. 119.

4297. If a wind blows three days in the south, it will afterward blow three days in the north.
Cited in Hyatt, p. 8.

4298. If the wind continue any considerable time in the south, it is an infallible sign of rain.
Cited in Inwards, p. 119.

4299. If the wind is in the south when the sun crosses the line in March, there will be mild weather.
Cited in Boughton, p. 124.

4300. If the wind shifts around to the south and southwest, expect warm weather. Var.: A south wind is accompanied by warm weather.
Cited in Dunwoody, p. 87; Hyatt, p. 8.

4301. If there be dry weather with a light south wind for five or six days, it having previously blown strongly from the same direction, expect fine weather.
Cited in Dunwoody, p. 87.

4302. It never rains while the wind is southerly and the sky cloudy.
Cited in Hyatt, p. 8.

4303. South winds bring rain. Var.: The south wind surely brings us rain, the north wind blows it back again.
Cited in Dunwoody, p. 69; Marvin, p. 214; Whitman, p. 22.

4304. The south wind brings wet weather, the north wind wet and cold together; the west wind always brings us rain, the east wind

blows it back again.
Cited in Denham, p. 10; Hyatt, p. 8.

4305. The south wind veering to the northwest portends bad weather.
Cited in Hyatt, p. 8.

4306. The south, with his showers, refreshes the corn; the west to all
flowers may not be forlorn.
Cited in Marvin, p. 214.

4307. The warmth of the south wind is enervating.
Cited in Garriott, p. 19; Hand, p. 19; Freier, p. 58.

4308. The wind of the south will be productive of heat and fertility; the
wind of the west, of milk and fish; the wind from the north, of
cold and storm; the wind from the east, of fruit on the trees.
Cited in Wright, p. 9.

4309. When the wind is in the south, it is in the rain's mouth. Var.: A
wind in the south is in the rain's mouth.
Cited in Swainson, p. 224; Dunwoody, p. 15; Wilson, p. 893.

4310. When the wind is south, it blows the bait to the fish's mouth.
Vars.: (a) If the wind blows from the south, it blows the hook in
the fish's mouth. (b) If the wind is in the south, it blows the fly
in the fish's mouth. (c) When the south wind blows, it blows the
fly in the fish's mouth. (d) When the wind is in the south, it
blows the bait in the fish's mouth.
Cited in Denham, p. 10; Swainson, p. 224; Dunwoody, p. 51;
Marvin, p. 214; Whitman, p. 22; Inwards, p. 121; Sloane 1, p.
150; Sloane 2, p. 37; Hyatt, p. 8; Smith, p. 12; Wilson, p. 893;
Freier, p. 60. Recorded in Bryant (Ala., Ark., Ill., Kans., Miss.,
Ohio, Oreg. Can. [Ont.]).

4311. When the wind shifts around to the south and southwest expect
warm weather.
Cited in Inwards, p. 122.

4312. When the wind's in the south, the rain's in its mouth. Vars.: (a)
A wind in the south, has rain in her mouth. (b) When the wind's
in the south, it has rain clouds in its mouth.
Cited in Cheales, p. 25; Humphreys 1, p. 441; Wright, p. 84;
Whitman, p. 22; Inwards, p. 120; Lee, p. 106; Wilshere, p. 19;
Freier, p. 60. Recorded in Bryant (Mich.).

wind (southeast)

4313. If a southeast wind brings rain, the latter is expected to last some time.
Cited in Swainson, p. 225.
4314. When cliffs and promontories on distant shores appear higher, then the southeast wind is blowing.
Cited in Sloane 2, p. 118.

wind (southwest)

4315. If in unsettled weather the wind veers from southwest to west or northwest at sunset, expect fine weather for a day or two.
Cited in Inwards, p. 108.
4316. If the wind is southwest at Martinmas, it keeps there till after Christmas. Comment: Martinmas Day is November 11th.
Cited in Inwards, p. 122.
4317. In fall and winter if the wind holds a day or more in the southwest, a severe storm is coming; in summer, same of northeast wind.
Cited in Dunwoody, p. 87; Inwards, p. 121.
4318. In fall and winter if the wind holds a day or more in the southwest there will be a gale, and wind will veer to northwest between 1:00 and 2:00 a.m. (in winter) with increasing force.
Cited in Dunwoody, p. 87.
4319. In Southern Indiana a southwest wind brings rain in thirty-six hours.
Cited in Inwards, p. 121.
4320. Three southwesters, then one heavy rain.
Cited in Dunwoody, p. 87; Inwards, p. 121.

wind (west)

4321. A west wind and an honest man go to bed together. Vars.: (a) A west wind like an honest man, goes to bed at sundown. (b) The west wind is a gentleman, and goes to bed.
Cited in Denham, p. 2; Swainson, p. 228; Inwards, p. 122; Sloane 2, p. 210.
4322. A west wind is a favorable wind.
Cited in Sloane 2, p. 62.

4323. A west wind north about never long holds out. Var.: A west wind north about never hangs long out. Comment: A wind which goes round (north) from east to west rarely continues.
Cited in Swainson, p. 228; Dunwoody, p. 88; Marvin, p. 213; Inwards, p. 122.

4324. A western wind carries water in his hand.
Cited in Marvin, p. 213; Whitman, p. 21; Inwards, p. 122.

4325. Cold weather attends a west wind.
Cited in Hyatt, p. 8.

4326. Do business with men when the wind is in the west. Var.: Do business with men when the wind is from the westerly; for then the barometer is high.
Cited in Swainson, p. 222; Garriott, p. 19; Hand, p. 19.

4327. It will not rain during a west wind and a cloudy sky. Var.: There is never any rain during a west wind.
Cited in Hyatt, p. 8.

4328. When the wind is west, the fish bite best.
Cited in Wilson, p. 893.

4329. When the wind's in the west, the weather's always best. Vars.: (a) The wind in the west suits everyone best. (b) When the wind is in the west, the weather is at the best. (c) When the wind is in the west, there it is the very best. (d) When the wind is in the west, things are at their best. (e) Wind in the west, weather at the best. (d) When wind is west, health is best.
Cited in Denham, p. 12; Humphreys 1, p. 441; Marvin, p. 212; Humphreys 2, p. 61; Whitman, p. 61; Inwards, p. 122; Hyatt, p. 8; Wilson, p. 893; Freier, p. 59. Recorded in Bryant (Ill., Miss., Can. [Ont.]).

4330. Wind west, rain's nest.
Cited in Swainson, p. 228; Marvin, p. 213; Inwards, p. 122.

wind gall

4331. The wind gall or prismatic coloring of the clouds is considered by sailors a sign of rain.
Cited in Inwards, p. 127.

windflower

4332. The windflower closes its petals and droops before rain.
Cited in Lee, p. 73.

window

4333. Do not wash windows on moving day; it will surely rain.
Cited in Hyatt, p. 32.
4334. If windows stick to the frames, rain may be forecasted.
Cited in Hyatt, p. 30.
4335. It always rains after windows have been washed.
Cited in Hyatt, p. 32.
4336. Moisture on the windows at dawn indicates cold weather in winter and fair weather in summer.
Cited in Hyatt, p. 30.
4337. Rattling windows are a token of a change in the weather.
Cited in Hyatt, p. 30.

winter

4338. A cold winter, a hot summer, an open winter, a cool and rainy summer.
Cited in Hyatt, p. 14.
4339. A dry winter brings a poor harvest.
Recorded in Bryant (Can. [Ont.]).
4340. A fair day in winter is the mother of a storm.
Cited in Wright, p. 10; Inwards, p. 34; Wilson, p. 239. Recorded in Bryant (Wis.).
4341. A frosty winter, and a dusty March, and a rain about April, and another about Lammas time when the corn begins to fill, is worth a plough of gold and all her pins theretill. Comment: Lammas time refers to a harvest festival held on or about August 1st.
Cited in Wright, p. 26; Inwards, p. 48.
4342. A good winter brings a good summer.
Cited in Wilson, p. 326; Wilshere, p. 17.
4343. A green winter makes a fat churchyard. Var.: A green winter, a full graveyard.
Cited in Denham, p. 2; Swainson, p. 9; Garriott, p. 45; Inwards, p. 33; Hand, p. 45; Wilson, p. 337; Freier, p. 80; Mieder, p. 659.
4344. A hard winter, an early spring.
Cited in Hyatt, p. 14.
4345. A hard winter makes a good crop year.
Cited in Mieder, p. 659.

4346. A hard winter when bear eats bear.
Cited in Whiting 1, p. 489.

4347. A lengthy winter and tardy spring are both good for hay and grain, but bad for corn and garden.
Cited in Dunwoody, p. 92; Inwards, p. 34.

4348. A misty winter brings a pleasant spring; a pleasant winter, a misty spring.
Cited in Mieder, p. 659.

4349. A rainy winter is followed by a dry autumn in the following year; and a dry spring by a rainy autumn.
Cited in Steinmetz 1, p. 136.

4350. A warm and open winter portends a hot and dry summer.
Cited in Whitman, p. 45; Hand, p. 46.

4351. A warm winter and cold summer never brought a good harvest.
Cited in Garriott, p. 46.

4352. A winter beginning early will be long and cold, beginning late it will soon end.
Cited in Hyatt, p. 14.

4353. A winter's calm is as bad as a summer's storm.
Cited in Whiting 1, p. 489.

4354. After a rainy winter, a plentiful summer.
Cited in Wilson, p. 897.

4355. After a rainy winter follows a fruitful spring.
Cited in Dunwoody, p. 90; Garriott, p. 45; Marvin, p. 204; Whitman, p. 45; Inwards, p. 34; Hand, p. 45; Mieder, p. 659. Recorded in Bryant (N.Y., Utah).

4356. An early winter, a surly winter.
Cited in Denham, p. 61; Swainson, p. 8; Dunwoody, p. 91; Northall, p. 477; Wright, p. 118; Inwards, p. 34; Freier, p. 78.

4357. Every mile is two in winter.
Cited in Wilson, p. 531.

4358. In winter, if the sun rises with a red sky, expect rain that day; in summer, expect showers and wind.
Cited in Dunwoody, p. 78.

4359. In winter when the sky about midday has a greenish appearance to the east or northeast, snow and frost are expected.
Cited in Mitchell, p. 221; Inwards, p. 148.

4360. It is a hard winter when one wolf eats another.
Cited in Wilson, p. 353. Recorded in Bryant (Okla.).

4361. Neither give credit to a clear winter nor a cloudy spring.
Cited in Inwards, p. 34.

4362. One fair day in the winter does not make the birds merry.
Cited in Dunwoody, p. 89; Inwards, p. 34.

4363. Snowy winter, plentiful harvest.
Cited in Whiting 1, p. 489; Mieder, p. 659.

4364. The average winter shows warm periods in February and cold periods in March.
Cited in Hand, p. 39.

4365. The number of snowfalls during any winter is determined by the day of the month on which the first snowfall comes.
Recorded in Bryant (Kans.).

4366. When winter begins early, it ends early.
Cited in Inwards, p. 34.

4367. When winter is not early, it will not be late.
Cited in Dunwoody, p. 92.

4368. Winter fills the ponds before it freezes them.
Cited in Whiting 2, p. 689.

4369. Winter finds out what summer lays up. Vars.: (a) Winter discovers what summer conceals. (b) Winter eats what summer gets. (c) Winter eats what summer lays up.
Cited in Swainson, p. 10; Dunwoody, p. 92; Inwards, p. 33; Wilson, p. 897; Mieder, p. 659.

4370. Winter is summer's heir.
Cited in Swainson, p. 9; Wilson, p. 897.

4371. Winter never died in a ditch.
Cited in Inwards, p. 33.

4372. Winter never rots in the sky.
Cited in Denham, p. 23; Inwards, p. 33; Wilson, p. 897; Whiting 1, p. 489; Whiting 2, p. 690.

4373. Winter spends what summer lends.
Cited in Whiting 1, p. 489.

4374. Winter time for shoeing; peasecod time for wooing.
Cited in Wright, p. 11.

4375. Winter under water, dearth; under snow, bread.
Cited in Garriott, p. 46; Whitman, p. 45; Hand, p. 46.

4376. Winter weather and women's thoughts often change.
Cited in Page, p. 43; Mieder, p. 659.

4377. Winter will not come till the swamps are full.
Cited in Dunwoody, p. 91; Garriott, p. 46; Whitman, p. 45;
Inwards, p. 34; Hand, p. 46.
4378. Winter's back breaks about the middle of February.
Cited in Dunwoody, p. 91; Garriott, p. 46; Whitman, p. 45;
Hand, p. 46.

wintering
4379. Bad wintering will tame both man and beast.
Cited in Denham, p. 6.

wolf
4380. If the wolves howl and foxes bark during the winter, expect cold
weather.
Cited in Dunwoody, p. 33.
4381. If wolves howl in the evening, expect a "norther."
Cited in Dunwoody, p. 34.
4382. Wolves howl more before a storm. Var.: Wolves always howl
more before a storm.
Cited in Dunwoody, p. 33; Lee, p. 66.

woman
4383. If you see many women pushing baby buggies, a rain is predicted.
Cited in Hyatt, p. 32.
4384. If you see many women walking on the street, rain is on its way.
Cited in Hyatt, p. 32.

wood
4385. Burning wood in winter pops more before snow. Var.: Burning
wood pops more before rain and snow.
Cited in Inwards, p. 221; Freier, p. 27.
4386. If any wood be cut off in the last day of December, and on the
first of January, it shall not rot nor wither away, nor be full of
worms, but always wear harder, and in its age get as hard as a
stone.
Cited in Swainson, p. 167.

wood sorrel

4387. Wood sorrel contracts its leaves at the approach of rain.
Cited in Inwards, p. 214.

woodcock

4388. An early appearance of the woodcock indicates the approach of
a severe winter.
Cited in Dunwoody, p. 41; Garriott, p. 40; Inwards, p. 190;
Hand, p. 40.

woodpecker

4389. The ivory-billed woodpecker commencing at the bottom end of
a tree and going to the top, removing all the outer bark, indicates
a hard winter with deep snow.
Cited in Dunwoody, p. 41; Garriott, p. 40; Inwards, p. 192;
Hand, p. 40.

4390. The woodpecker's cry denotes wet.
Cited in Swainson, p. 247.

4391. The yaffle, or green woodpecker, cries at the approach of rain,
and is described as "laughing in the sun, because the rain is
coming." Var.: The yaffle or rainbird laughs at the sun, to say
the rain is coming before the day is done. Comment: The green
woodpecker is also called the "yaffle" or "rainbird."
Cited in Inwards, p. 193; Alstad, p. 78.

4392. When the woodpecker leaves, expect a hard winter.
Cited in Dunwoody, p. 41; Garriott, p. 40; Inwards, p. 192;
Hand, p. 40.

4393. When woodpeckers are much heard, expect rain. (b) The call of
the heigh-ho (woodpecker) forbodes rain. (c) When woodpeckers
are unusually noisy, rain will come.
Cited in Inwards, p. 192; Sloane 2, p. 130; Lee, p. 64; Wilshere, p. 22.

4394. When woodpeckers peck low on the trees, expect warm weather.
Cited in Dunwoody, p. 41; Garriott, p. 40; Inwards, p. 192;
Hand, p. 40.

woodtick

4395. If you find a woodtick, stick a pin through it and stick it on the
side of a wall or tree and it will rain in twenty-four hours.

Cited in Hyatt, p. 19.

wooly worm

4396. If the wooly worm has mostly brown on it, the coming winter is supposed to be a cold one.
Cited in Smith, p. 12.

4397. If the wooly worm is mostly white, the winter will be warm.
Cited in Smith, p. 12.

4398. When wooly worms turn black, a hard severe winter is ahead.
Cited in Smith, p. 12.

worm

4399. If, after some days of dry weather, fresh earth is seen which has been thrown up by worms, expect dry weather.
Cited in Dunwoody, p. 72.

4400. If worms live near the top of the ground in late autumn or early winter, good weather is indicated for winter; if deep in the ground, bad weather.
Cited in Hyatt, p. 19.

4401. Those who live along the banks of the Mississippi say worms coming to the surface of the ground in early spring denote high water for the river.
Cited in Hyatt, p. 19.

4402. When worms creep out of the ground in great numbers, expect wet weather.
Cited in Dunwoody, p. 72.

4403. Worms creep out of the ground before rain. Vars.: (a) Worms come forth more abundantly before rain, as do snails, slugs, and almost all our limaceous reptiles. (b) Worms rise in greater numbers before rain.
Cited in Steinmetz 1, p. 112; Steinmetz 2, p. 290; Dunwoody, p. 58.

worm hole

4404. If you see worm holes in the ground or a large number of worms crawling about, rain will fall within twenty-four hours.
Cited in Hyatt, p. 19.

4405. One of the things showing the arrival of spring is worm holes in the ground.

Cited in Hyatt, p. 19.

wren
4406. When wrens are seen in winter, expect snow.
Cited in Dunwoody, p. 41; Garriott, p. 40; Inwards, p. 195;
Hand, p. 40.

Y

year

4407. A bad year comes in swimming.
Cited in Garriott, p. 46; Whitman, p. 46.

4408. A cherry year, a merry year; a plum year, a dumb year. Vars.:
(a) A plum year, a dumb year. (b) In the year that plums flourish
all else fails. Comment: Dumb refers to the silence of death.
Cited in Denham, p. 52; Swainson, p. 2; Cheales, p. 22;
Dunwoody, p. 88; Northall, p. 487; Garriott, p. 46; Hand, p.
46; Wilson, p. 117; Simpson, p. 35; Mieder, p. 685.

4409. A cow year, a sad year; a bull year, a glad year.
Cited in Whitman, p. 46.

4410. A dry year never beggars the master.
Cited in Marvin, p. 207.

4411. A dry year never made a dear peck.
Cited in Swainson, p. 15.

4412. A dry year never starves itself.
Cited in Dunwoody, p. 92; Marvin, p. 207.

4413. A fertile year is foretold by violent north winds in February.
Cited in Hyatt, p. 34.

4414. A good hay year, a bad fog year.
Cited in Denham, p. 49.

4415. A good nut year, a good corn year. Var.: A great store of nuts,
a good corn year.
Cited in Denham, p. 55; Swainson, p. 4; Dunwoody, p. 89;
Freier, p. 80.

4416. A good year is always welcome.
Cited in Humphreys 1, p. 429.

4417. A haw year, a braw year. Comment: Haw is a blossom on the
hedgerow, braw means pleasant.
Cited in Swainson, p. 4; Dunwoody, p. 89.

4418. A haw year is a snaw year. Comment: Snaw is snow.
Cited in Mitchell, p. 226; Swainson, p. 4; Cheales, p. 22;
Dunwoody, p. 89.

4419. A pear year, a dear year.
Cited in Swainson, p. 4; Dunwoody, p. 90; Northall, p. 491.

4420. A snow year, a rich year.
Cited in Denham, p. 2; Swainson, p. 5; Cheales, p. 20; Dun-
woody, p. 76; Wright, p. 10; Wilson, p. 748; Wilshere, p. 16;
Freier, p. 80; Mieder, p. 685.

4421. A wet year will make a full barn, but not of corn.
    Cited in Marvin, p. 207.
4422. A year of grass, good for nothing else.
    Cited in Whitman, p. 46.
4423. A year of snow, a year of plenty. Var.: Year of snow, year of plenty.
    Cited in Humphreys 1, p. 430; Wright, p. 11; Humphreys 2, p. 9; Whitman, p. 45; Lee, p. 134; Wilshere, p. 16.
4424. After a wet year, a cold one.
    Cited in Garriott, p. 46; Marvin, p. 207; Hand, p. 46; Mieder, p. 685.
4425. Frost year, fruit year.
    Cited in Humphreys 1, p. 430; Humphreys 2, p. 9; Whitman, p. 45.
4426. Frost year, good year. Var.: Snow year, good year.
    Cited in Garriott, p. 46; Whitman, p. 45; Hand, p. 46.
4427. If the old year goes out like a lion, the new year will come in like a lamb.
    Cited in Dunwoody, p. 90.
4428. It's a mighty dry year when the crabgrass fails.
    Cited in Mieder, p. 124.
4429. Leap year was ne'er a good sheep year.
    Cited in Garriott, p. 46.
4430. Rainy year, fruit dear.
    Cited in Garriott, p. 46; Whitman, p. 45; Hand, p. 46.
4431. Say no ill of the year till it be past.
    Cited in Denham, p. 12.
4432. Wet and dry years come in triads.
    Cited in Dunwoody, p. 92; Garriott, p. 46; Whitman, p. 46; Hand, p. 46.
4433. Year of radishes, year of health.
    Cited in Garriott, p. 46; Whitman, p. 46; Hand, p. 46.
4434. Year of snow, crops will grow.
    Cited in Lee, p. 30.
4435. Year of snow, fruit will grow.
    Cited in Humphreys 1, p. 430; Humphreys 2, p. 9; Smith, p. 10; Lee, p. 134.

List of Sources

Alstad   Ken Alstad. *Weather Proverbs That Work and Why They Do*. Lake Forest, Illinois: Walden Woods Press, 1955.

Arora   Shirley L. Arora. "Weather Proverbs: Some Folk Views." *Proverbium* 8 (1991): 1-17.

Boughton   Audrey Boughton. "Predictions of Rain." *New York Folklore Quarterly* 1 (1945): 123-125.

Brunt   D. Brunt. "Meteorology and Weather Lore." *Folklore* (London) 57 (1946): 66-74.

Bryant   Margaret M. Bryant, comp. "Collection of the Committee on Proverbial Sayings of the American Dialect Society (c. 1945-1980)." Archives of the University of Missouri (Columbia). Unpublished texts.

Cheales   Alan B. Cheales. "Weather Wisdom." *Proverbial Folk-Lore*, 2nd ed. London: Simpkin, Marshall & Co., 1875; rpt. Folcroft, Pennsylvania: Folcroft Library Edition, 1976. 17-27.

Denham   Michael A. Denham. "A Collection of Proverbs and Popular Sayings Relating to the Weather, and Agricultural Pursuits; Gathered Chiefly from Oral Tradition." *Percy Society's Publication* 20 (1846): 1-73.

Dundes   Alan Dundes. "On Whether 'Weather Proverbs' Are Proverbs." *Folklore Matters*. Knoxville, Tennessee: The University of Tennessee Press, 1989. 92-97.

Dunwoody   H. H. C. Dunwoody. *Weather Proverbs*. Washington, D.C.: Government Printing Office, 1883.

Freier   George D. Freier. *Weather Proverbs*. Tucson, Arizona: Fisher Books, 1989.

Garriott   Edward B. Garriott. *Weather Folk-Lore and Local Weather Signs*. Washington, D.C.: Government Printing Office, 1903; rpt. Detroit: Grand River Books, 1971.

Halpert   Herbert Halpert. "The Devil is Beating His Wife." *Kentucky Folklore Record* 1 (1955): 105-106.

Hand   Wayland D. Hand. *Popular Beliefs and Superstitions from North Carolina*, vol. VI of the Frank C. Brown Collection of North Carolina Folklore. Durham, North Carolina: Duke University Press, 1961.

Humphreys 1   W. J. Humphreys. "Some Weather Proverbs and Their Justifications." *Popular Science Monthly* 28 (1911): 428-444.

Humphreys 2   W. J. Humphreys. *Weather Proverbs and Paradoxes*. Baltimore, Maryland: Williams and Wilkins, 1923.

Hyatt   Harry M. Hyatt. *Folk-Lore from Adams County, Illinois*. Hannibal, Missouri: Alma Egan Hyatt Foundation, 1965.

Inwards   Richard Inwards. *Weather Lore. A Collection of Proverbs, Sayings, and Rules Concerning the Weather*, 3rd ed. London: Elliot Stock, 1898. The page references in the citations are from the reprint entitled *Weather Lore: The Unique Bedside Book*. New York: Rider, 1972.

Lee   Albert Lee. *Weather Wisdom*. Garden City, New York: Doubleday & Company, Inc., 1976.

Marvin   Dwight E. Marvin. "Weather Proverbs." *Curiosities in Proverbs*. New York: G. P. Putnam's Sons, 1916; rpt. Folcroft, Pennsylvania: Folcroft Library Editions, 1980. II, 203-218.

Mieder   Wolfgang Mieder, Stewart A. Kingsbury, and Kelsie B. Harder. *A Dictionary of American Proverbs*. Oxford: Oxford University Press, 1992.

Mitchell   Arthur Mitchell. *On the Popular Weather Prognostics of Scotland*. Edinburgh: Blackwood, 1863.

Northall   G. F. Northall. *English Folk-Rhymes*. London: Kegan Paul, Trench, Trubner and Company, 1892.

Page   Robin Page. *Weather Forecasting the Country Way*. New York: Summit Books, 1977.

Reeder   William Reeder. "Rain Signs." Fredericksburg, Texas (1992): no pp. Unpublished texts.

Simpson   John Simpson, ed. *The Concise Oxford Dictionary of Proverbs*. Oxford: Oxford University Press, 1982.

Sloane 1   Eric Sloane. *Almanac and Weather Forecaster*. New York: Duell, Sloane and Pearce, 1955.

Sloane 2   Eric Sloane. *Folklore of American Weather*. New York: Duell, Sloane and Pearce, 1963.

Smith   Elmer L. Smith. "Virginia Folklore, Weather Lore." Unpublished manuscript. Harrisonburg, Virginia (1966).

Steinmetz 1   Andrew Steinmetz. *Manual of Weathercasts: Storm Prognostics on Land and Sea*. London: George Routledge and Sons, 1866.

Steinmetz 2   Andrew Steinmetz. *Sunshine and Showers: Their Influences Throughout Creation*. London: Reeve and Company, 1867.

Swainson   C. Swainson. *A Handbook of Weather Folk-Lore*. London: William Blackwood and Sons, 1873; rpt. Detroit: Gale Research Company, 1974.

Taylor 1   Archer Taylor. "Weather Proverbs." *The Proverb*. Cambridge, Massachusetts: Harvard University Press, 1931; rpt. Hatboro, Pennsylvania: Folklore Associates,

1962; rpt. with an introduction and bibliography by Wolfgang Mieder. Bern: Peter Lang, 1985.

Taylor 2   Archer Taylor and Bartlett Jere Whiting. *A Dictionary of American Proverbs and Proverbial Phrases (1820-80).* Cambridge, Massachusetts: The Belknap Press of Harvard University, 1958.

Walton   Ivan Walton. "Weather Lore." *Motor Boating* (December, 1939): no pp.

Whiting 1   Bartlett Jere Whiting. *Early American Proverbs and Proverbial Phrases.* Cambridge, Massachusetts: The Belknap Press of Harvard University Press, 1977.

Whiting 2   Bartlett Jere Whiting. *Modern Proverbs and Proverbial Sayings.* Cambridge, Massachusetts: Harvard University Press, 1989.

Whitman   Frances E. Whitman. "A Determination of the Weather Proverbs which are True or False in Terms of Scientific Principles." Master's Thesis. Boston University School of Education, 1934.

Wilshere   Jonathan Wilshere. *Leicestershire Weather Sayings.* London: Leicester Research Section of Chamberlain Music and Books, 1980.

Wilson   F. P. Wilson. *The Oxford Dictionary of English Proverbs*, 3rd ed. London: Oxford University Press, 1970.

Wright   M. E. S. Wright. *Medley of Weather Lore.* Bournemouth: Horace G. Commin, 1913.

Wurtele   M. G. Wurtele. "Some Thoughts on Weather Lore." *Folklore* (London) 82 (1971): 292-303.

Index of Indications

714, 721, 723, 736, 756,
757, 759, 761, 763, 765,
767, 771, 776, 787, 788,
791, 794, 811, 813, 819,
833, 836, 837, 838, 845,
870, 874, 875, 879, 880,
882, 883, 885, 886, 890,
894, 895, 896, 898, 899,
904, 906, 910, 914, 918,
919, 926, 929, 933, 934,
938, 939, 940, 943, 944,
945, 946, 948, 952, 953,
954, 956, 965, 975, 977,
981, 983, 985, 988, 993,
1034, 1037, 1045, 1046,
1047, 1049, 1050, 1053,
1057, 1059, 1068, 1069,
1074, 1075, 1076, 1077,
1081, 1083, 1085, 1088,
1089, 1092, 1095, 1096,
1099, 1100, 1101, 1102,
1104, 1118, 1119, 1120,
1121, 1123, 1125, 1126,
1128, 1130, 1132, 1139,
1140, 1141, 1147, 1148,
1155, 1156, 1159, 1190,
1253, 1259, 1267, 1275,
1276, 1283, 1288, 1289,
1291, 1292, 1299, 1300,
1304, 1305, 1308, 1314,
1316, 1317, 1318, 1319,
1320, 1321, 1323, 1326,
1329, 1332, 1335, 1336,
1339, 1342, 1343, 1345,
1354, 1356, 1370, 1385,
1401, 1410, 1412, 1414,
1416, 1418, 1419, 1420,
1422, 1425, 1431, 1434,
1435, 1443, 1456, 1460,

1461, 1462, 1463, 1469,
1472, 1473, 1475, 1476,
1477, 1480, 1482, 1487,
1492, 1497, 1510, 1512,
1513, 1515, 1518, 1532,
1541, 1543, 1554, 1555,
1556, 1557, 1561, 1562,
1563, 1571, 1573, 1574,
1575, 1576, 1579, 1581,
1582, 1584, 1585, 1587,
1589, 1591, 1605, 1609,
1610, 1612, 1619, 1620,
1621, 1624, 1625, 1640,
1649, 1655, 1667, 1669,
1671, 1672, 1673, 1674,
1687, 1689, 1695, 1697,
1704, 1707, 1711, 1713,
1714, 1717, 1718, 1719,
1721, 1722, 1727, 1730,
1732, 1761, 1763, 1765,
1766, 1767, 1768, 1769,
1773, 1774, 1775, 1776,
1778, 1779, 1783, 1799,
1801, 1802, 1808, 1809,
1811, 1812, 1911, 1912,
1918, 1922, 1923, 1924,
1931, 1939, 1941, 1944,
1954, 1955, 1959, 1961,
1964, 1965, 1966, 1973,
1979, 1993, 2001, 2003,
2010, 2021, 2022, 2031,
2032, 2035, 2038, 2133,
2134, 2225, 2231, 2232,
2239, 2243, 2247, 2258,
2260, 2273, 2284, 2286,
2289, 2294, 2303, 2304,
2309, 2312, 2314, 2320,
2322, 2327, 2328, 2329,
2331, 2356, 2382, 2383,

2394, 2398, 2401, 2405,
2406, 2408, 2410, 2412,
2440, 2452, 2455, 2459,
2466, 2473, 2475, 2480,
2481, 2494, 2523, 2580,
2592, 2594, 2595, 2599,
2601, 2602, 2603, 2614,
2616, 2621, 2625, 2627,
2631, 2637, 2639, 2640,
2643, 2646, 2661, 2662,
2667, 2668, 2676, 2677,
2678, 2680, 2681, 2682,
2683, 2684, 2688, 2689,
2690, 2691, 2693, 2695,
2696, 2701, 2702, 2717,
2718, 2721, 2722, 2729,
2920, 2929, 2931, 2932,
2939, 2943, 2949, 2952,
2964, 2968, 2970, 2972,
2976, 2977, 2981, 2982,
2984, 2986, 2987, 2988,
2993, 2998, 2999, 3000,
3001, 3004, 3007, 3009,
3015, 3016, 3018, 3020,
3022, 3025, 3026, 3027,
3028, 3029, 3031, 3032,
3035, 3052, 3058, 3060,
3083, 3096, 3122, 3126,
3139, 3145, 3148, 3149,
3156, 3163, 3164, 3165,
3168, 3169, 3178, 3181,
3196, 3197, 3199, 3200,
3201, 3202, 3203, 3207,
3208, 3209, 3210, 3217,
3218, 3221, 3228, 3229,
3231, 3233, 3238, 3239,
3242, 3243, 3244, 3245,
3247, 3248, 3249, 3251,
3252, 3253, 3255, 3256,

3258, 3308, 3324, 3326,
3327, 3329, 3330, 3334,
3336, 3337, 3339, 3341,
3344, 3346, 3347, 3348,
3349, 3353, 3355, 3356,
3357, 3358, 3359, 3360,
3361, 3370, 3371, 3372,
3376, 3377, 3378, 3380,
3381, 3392, 3394, 3396,
3415, 3416, 3443, 3448,
3452, 3454, 3471, 3508,
3571, 3587, 3588, 3593,
3594, 3595, 3596, 3600,
3601, 3619, 3625, 3689,
3690, 3692, 3704, 3718,
3731, 3732, 3733, 3736,
3737, 3738, 3750, 3757,
3762, 3765, 3767, 3777,
3778, 3795, 3796, 3797,
3800, 3801, 3802, 3806,
3812, 3823, 3824, 3825,
3826, 3827, 3828, 3829,
3832, 3834, 3835, 3848,
3851, 3852, 3853, 3860,
3873, 3896, 3925, 3926,
3933, 3964, 3971, 3976,
3978, 3979, 3980, 3986,
3991, 3993, 3994, 3996,
3999, 4000, 4001, 4002,
4004, 4008, 4014, 4015,
4016, 4019, 4021, 4032,
4033, 4034, 4035, 4036,
4038, 4046, 4047, 4055,
4056, 4058, 4059, 4062,
4066, 4067, 4068, 4099,
4111, 4112, 4131, 4132,
4133, 4134, 4136, 4137,
4139, 4148, 4155, 4198,
4202, 4214, 4228, 4298,